Mathematics Textbooks for Science and Engineering

Volume 3

For further volumes:
http://www.springer.com/series/10785

Shapoor Vali

Principles of Mathematical Economics

Shapoor Vali
Department of Economics
Fordham University
New York, NY
USA

ISBN 978-94-6239-035-5   ISBN 978-94-6239-036-2   (eBook)
DOI 10.2991/978-94-6239-036-2

Library of Congress Control Number: 2013951796
Published by Atlantis Press, Paris, France www.atlantis-press.com

Printed on acid-free paper

Series Information

Textbooks in the series 'Mathematics Textbooks for Science and Engineering' will be aimed at the broad mathematics, science and engineering undergraduate and graduate levels, covering all areas of applied and applicable mathematics, interpreted in the broadest sense.

Series editor

Charles K. Chui
Stanford University, Stanford, CA, USA

Atlantis Press
8, square des Bouleaux
75019 Paris, France

For more information on this series and our other book series, please visit our website www.atlantis-press.com

Editorial

Recent years have witnessed an extraordinarily rapid advance in the direction of information technology within the scientific, engineering, and other disciplines, in which mathematics play a crucial role. To meet such urgent demands, effective mathematical models as well as innovative mathematical theory, methods, and algorithms have to be developed for data information manipulation, understanding, visualization, communication, and other applications.

The Atlantis book series, *"Mathematics Textbooks for Science and Engineering (MTSE)"*, is founded to meet the need of textbooks on the fundamental and core of such mathematical development, as well as their applications, that can be used for classroom teaching at the college level and beyond. For the benefit of students and readers from the interdisciplinary areas of mathematics, economics, computer science, physical and biological sciences, and various engineering specialties, contributing authors are requested to keep in mind that the writings for the MTSE book series should be elementary and relatively easy to read, with sufficient examples and exercises. We welcome submission of such book manuscripts from all who agree with us on this point of view.

This third volume, authored by Prof. Shapoor Vali, on the principles of mathematical economics, with emphasis on non-linear mathematical models, is a very valuable contribution to the MTSE series. Written for students in economics, business, management, and related fields, this textbook is self sufficient and self contained for the reader with only a basic knowledge of pre-college algebra as well as introductory micro- and macro-economics. A distinct and important feature of this nicely written textbook is the connection of the mathematical models, developed and formulated in each chapter, to specific real-world problems. We welcome this textbook to the MTSE book series.

Menlo Park, CA   Charles K. Chui

To Firoozeh, Behrang, and Seena

Preface

This book has evolved from over many years of teaching Mathematical Economics (Math-Econ, for short) and other quantitative courses at Fordham University. In my Math-Econ classes, I have used different books as text or reference. However, I always felt compelled to complement the texts by a large volume of my own notes and handouts covering areas and topics that I considered important but either not covered or not fully developed in the existing text books. So in 2009, I decided to collect my notes and develop them into an organized text book. The result was a manuscript which I have used as a text in my Math-Econ courses for the last 4 years and annually revised it based on students' feedbacks.

This textbook is written for students in economics, business, management, and related fields, for both undergraduate and introductory level graduate courses. Chaps. 1 through 7 and parts of Chaps. 8 and 9 could be covered in a one-semester undergraduate course. The entire book can be covered in a one-semester graduate course, with emphasis on Chaps. 6 through 13.

Mathematical Economics is generally a required course for students majoring in economics. It is also a widely recommended course for business schools students. This book is designed to be useful in both areas.

Economics is a quantitatively oriented discipline where mathematics plays a fundamental role in all of its related fields. Working with and manipulating numbers is an indispensable and inescapable part of economics and business studies, even in the field of economic history. A 1993 Nobel Prize in Economics was awarded to Robert Fogel of University of Chicago and Douglass North of Washington University, both economic historians, for their contribution to the application of quantitative methods to the field of economic history. Fogel and North advanced "Cliometrics",[1] a new quantitative technique for analyzing historical data and solving historical puzzles.

Lack of proficiency in mathematics is a great obstacle to learning economics. Lack of proficiency should not be interpreted only as lack of skill in manipulating algebraic or arithmetic expressions, but rather a lack of power of abstraction and

---

[1] Combination of 'Clio', muse and daughter of Zeus in Greek mythology, and 'metrics'. There is a Cliometric society which publishes *Cliometrica—Journal of Historical Economics and Econometric History*.

generalization. In economics, and in many other fields for that matter, power of abstraction is much more important than mechanical skills.

Mathematical Economics is not a distinct area in economics, like *International Economics, Money and Banking*, or *Urban and Regional Economics*. It is simply an application of mathematical tools in economics, all areas of economics, in the process of formulation and test of theories (or their purely logical aspects called *models*), and stating and providing solutions to economic "problems."

I consider Math-Econ as not simply a specific area of applied mathematics. I am sure many economists will agree with me that the main purpose of Math-Econ is not to learn more mathematics, but rather use mathematics as an aid for learning more economics. This principle has been the guiding star in writing this book. Naturally, the text contains a lot of *economics*.

A glance at various topics covered in different chapters shows that some of these topics are classic Math-Econ repertoire. But in addition to covering traditional topics, I used the same mathematical tools for formulating and solving some important Economics models that are either completely ignored or not fully explored by other textbooks, in some cases for the simple reason that they lead to nonlinearity. Almost all of mathematical text books published in the last 74 years, following R. G. D. Allen's classic *Mathematical Analysis for Economists* published in 1938, have avoided nonlinearity by framing the models such that solutions do not go beyond solving quadratic equations.

A very simple example arises in the context of standard market equilibrium model. Here we have a supply function, a demand function, and an equilibrium condition. As long as the supply and demand functions are linear or quadratic, we have no problem determining the equilibrium price and quantity, but if the supply and demand functions are nonlinear we may run into computational problem. Consider the following simple model

$$Q_d = \frac{100P + 1500}{(P^2 + 10)}$$
$$Q_s = 10P^{0.7}$$

Here demand is a rational function and supply is a polynomial of degree 0.7.

To find the equilibrium price algebraically, we follow the familiar routine

$$Q_d = Q_s \ \rightarrow \ \frac{100P + 1500}{(P^2 + 10)} = 10P^{0.7}$$

leading to

$$P^{2.7} + 10P^{0.7} - 10P - 150 = 0$$

This equation is not a run of the mill familiar equation, like linear or quadratic, and does not yield an analytical solution.

Another typical problem arises in the context of a firm production decision. Given a firm's total cost function (*TC*) and the demand function for its product, we typically want to determine the firm's profit maximizing output/price combination, its break-even level(s) of output/price and the normal profit level of output and price. We can safely do all three if the total cost function is linear or quadratic and the demand function is linear. But as soon as we introduce a more realistic cubic total cost function, a function that produces all the nice average total, average variable, and marginal cost curves, which are used in all the existing economics textbooks, then the determination of break-even and normal profit output/price leads to solving cubic equations. If we assume

$$TC = aQ^3 + bQ^2 + cQ + d$$

$$P = e - fQ$$

then for break-even(s) we need to solve $TC - TR = 0$ (where TR is the total revenue), that is the following cubic equation

$$aQ^3 + (b+f)Q^2 + (c - e)Q + d = 0$$

And for the long run normal profit scenario, not only we need $TC - TR = 0$, but we also need $MC - MR = 0$ (where MC and MR are the marginal cost and marginal revenue). This leads to solving the following cubic equation

$$2aQ^3 + (b+f)Q^2 - d = 0$$

Determining a firm's normal profit output/price when the market is competitive leads to a similar situation. In this case, we must solve

$$\frac{dATC}{dQ} = 2aQ^3 + bQ^2 - d = 0$$

So far we simply assumed that a noncompetitive firm faces a linear demand function in the market. What if the demand function is not linear? What if it is a quadratic function? A semi-log function? Or a log-linear function (a model so widely used in practice for estimating demand functions due to its desirable property of having constant price and nonprice elasticity)? In the log-linear case,

$$\ln P = \ln d - \beta \ln Q$$

a simple profit optimization routine leads to solving an equation of the form

$$3aQ^{(2+\beta)} + 2bQ^{(1+\beta)} + cQ^\beta - (1 - \beta)d = 0$$

if the total cost function is cubic, or

$$2aQ^{(1+\beta)} + bQ^\beta - (1 - \beta)d = 0$$

if the cost function is quadratic.

Modeling and solving numerous real world problems invariably leads to a nonlinear scenario, both in economics and in business. I believe in the validity of the view expressed by Naseem Talib, which is quoted in the opening page of Chap. 7 "Nonlinear Models." The last sentence of the quote reads:

> Yesterday afternoon I tried to take a fresh look around me to catalog what I could see during my day that was linear. I could not find anything.

The introduction of nonlinearity may also lead to interestingly unexpected results; an issue ignored in current textbooks and economics literature.

We have reached a level of computational sophistication that many of these nonlinear models can be handled fairly easily. The currently available tools, some like WolframAlpha, **R**, and SAGE freely available online, or Microsoft Mathematics that can be downloaded, remove the barrier to the introduction of this type of models in economics and Math-Econ textbooks and free us from exclusively graphic presentation of many important economic models. Even in the absence of access to Internet-based mathematical software, an inherently nonlinear system can be transformed into a linear or quadratic model and subsequently solved, in many cases with small or tolerable errors. In an appendix to Chap. 6, some of these tools are introduced and in Chap. 7, and the subsequent chapters, a large number of nonlinear models are presented and solved.

My own experience is that students prefer the use of computer-based learning to paper/pencil. They only use paper/pencil when we insist that they must show their work on their weekly homework assignments. They come to class with laptops, tablets, and smart phones which are all capable of accessing the Internet and thus the full scope power of the web.

Based on my observations, it takes students not more than 30 min to learn how WolframAlpha works, along with the basics of graphing and solving equations.[2] In my undergraduate Math-Econ classes, I cover Chaps. 1 through 7 plus logarithms and exponential functions (Chap. 9) along with some sections from Chap. 8 (Additional Topics in Perfect and Imperfect Competition). After a short review of optimization models in Chap. 6, I cover the remaining chapters in a graduate Math-Econ course. I introduced both classes to WolframAlpha and Microsoft Mathematics, which both sets of students easily learn. In my graduate class, I introduced some of the tools of **R**, especially for matrix operations and applications of matrices covered in Chap. 13, along with basics of Maple. Nowadays, almost all 4-year universities with a Mathematics department have license for some mathematical software including Maple. I share with some of my grad students, who are eager to learn Maple, a series of Maple snippets (a short collection of Maple commands for solving specific problems). I have included a sample of Maple snippets at the end of Chap. 8.

---

[2] Besides computational and graphic power, WolframAlpha provides students with up to date economics and business information and data.

All of these tools will undoubtedly enhance the text. No other text currently available covers this important area and uses any technological tools to enhance learning. I am confident that the economics profession welcomes a new generation Math-Econ textbook with both theoretical and computational orientation, utilizing some of the readily available computer tools for formalizing and solving many economic models that currently limited to graphic presentation. We should also note that a text of this nature will be even more attractive to new generation of students who are computationally very savvy. Each instructor will, of course, emphasize different topics covered in the book. It has been my intention to put together a book with rich ingredients to allow Math-Econ teachers to select the menu that will be closest to their students level and need.

The book is self-sufficient and self-contained. Chaps. 3 and 5 in the book cover all the mathematical tools that students need in order to understand topics covered in other chapters. It is only assumed that students have a good knowledge of algebra and have taken introductory micro- and macro-economics. Basic macro- and micro-economics are universally taken by students majoring in economics or business and in many cases by nonmajors as a part of social science core requirement.

I have attempted to connect mathematical models developed, formulated, and discussed in each chapter to real world problems by using, as much as possible, actual data in the process of operationalizing and solving the models. I consider this approach to be a distinct feature of this text.

In addition to a large number of economic examples specified and solved in each chapter, exercise sections of chapters contain a large number of problems covering application of chapter's material in various real world situations and fields of economics. As an example, exercise section of Chap. 6 that deals with various optimization problems and short- or long-run equilibrium is nine pages long. This is another desirable feature of this book.

I owe thanks to many people, including some of my students and colleagues at Fordham University, who, in one form or another, have been helpful in the process of writing this book over several years. I would like to thank all of them, especially Keith Tipple, Shanu Bajaj, Mohammed Khalil, and Yuichi Yokoyama. I would like to thank Joseph Bertino and Michael Malenbaum for proofreading, editing, and offering many helpful comments and technical assistance. I also wish to express my gratitude to Dr. Gregory Bard of the Department of Applied Mathematics and Computer Science at the University of Wisconsin-Stout, for his encouragement and valuable suggestions.

Above all, I am indebted to my wife Firoozeh and my sons Behrang and Seena for their unwavering support and care.

Spring 2013   Shapoor Vali

Contents

Figures

Tables

Chapter 1
Household Expenditure

1.1 Consumer's Expenditure and Budget Constraint

A household allocates its income to variety of goods and services. A collection
of goods and services that a household purchases and consumes over a specific
time period or horizon (a week, a month, or a year) is called a *consumption bundle*
(bundle for short), or a *basket*. Assume that a household's annual bundle consists of
n different goods and services. If we denote the quantity of the ith item purchased
by this household by Q_i $i = 1, 2, ..., n$, and the corresponding price of each unit
of Q_i by P_i, then we can write the household's expenditure E as

$$E = P_1 Q_1 + P_2 Q_2 + P_3 Q_3 + \cdots + P_n Q_n$$

or more compactly, using *summation* or *sigma notation*,[1] as

$$E = \sum_{i=1}^{n} P_i Q_i$$

If we denote the household's annual income by I, then

$$E = P_1 Q_1 + P_2 Q_2 + P_3 Q_3 + \cdots + P_n Q_n \leq I$$

or, using sigma notation

$$E = \sum_{i=1}^{n} P_i Q_i \leq I \tag{1.1}$$

is the household *budget constraint*. Strict equality in (1.1) implies that this household
spends all its income on its bundle and finishes the year without any savings. In

---

[1] If you are not familiar with sigma notation you should read the Appendix to this chapter first.

S. Vali, *Principles of Mathematical Economics*,
Mathematics Textbooks for Science and Engineering 3,
DOI: 10.2991/978-94-6239-036-2_1, © Atlantis Press and the authors 2014

this case future improvement in the standard of living of this household, as partly measured by purchase of additional quantity of current goods and services and/or consumption of new items not currently in the bundle, must be supported by future additional income.

Strict inequality, on the other hand, implies that this household saves part of its income. In that case we can include an item S, dollar amount or quantity of savings with the associated price of 1, to the household's bundle and rewrite (1.1) as

$$\sum_{i=1}^{n} P_i Q_i + S = I \tag{1.2}$$

If household's unspent income is invested in interest bearing, income generating, and wealth enhancing asset(s), then future improvement of household's standard of living could be partially financed by savings. For example, if rate of interest is r, then the next period income of this household would increase from I to $I + rS$.

The third (and not uncommon) possibility for household's budget equation is

$$\sum_{i=1}^{n} P_i Q_i > I \tag{1.3}$$

This implies that this household runs a "budget deficit" and its current consumption must be partially supported by "borrowing". By denoting the dollar amount of this borrowing by B with the associated price of 1, we can rewrite (1.3) as

$$\sum_{i=1}^{n} P_i Q_i = I + B \quad \rightarrow \quad \sum_{i=1}^{n} P_i Q_i - B = I \tag{1.4}$$

If there is no prospect for future higher income, this situation is, of course, not sustainable. B in (1.4) does not necessarily represent only the amount of borrowing, but rather the interest payment and partial repayment of borrowed amount. A typical example is home mortgage loan. When a household purchase a house it typically borrows a substantial portion of value of the house from a bank or a mortgage company at certain interest rate, called mortgage rate. In this case household monthly expenditures will include a new item "mortgage payment", which is a combination of interest and partial repayment of initial amount borrowed, called the *principal*. In a 25-year fixed mortgage loan, the fixed monthly mortgage payment is structured such that over 300 payments the borrower pays off the loan, principal plus interest (learn more about mortgage loans in Chap. 12 *Mathematics of Interest Rate and Finance*).

exp. budget   household income

$$E = \sum_{i=1}^{n} P_i Q \le I$$

Savings

$$E + S = I \qquad\qquad E \ge I \rightarrow E - B = I$$

Tables 1.1 and 1.2 give the results from the Consumer Expenditure Survey released by the Bureau of Labor Statistics (BLS) of the U.S. Department of Labor (USDL), for years 2008, 2009, and 2010.[2]

Bureau of Labor Statistics, Consumer Expenditure

Table 1.1 Consumer expenditure survey 2008–2010

	2008	2009	2010	% change 2008–2009	% change 2009–2010
Average annual expenditures	$50,486	$49,067	$48,109	−2.8	−1.95
Food	6,443	6,372	6,129	−1.1	−3.8
At home	3,744	3,753	3,624	0.2	−3.4
Away from home	2,698	2,619	2,505	−2.9	−4.35
Housing	17,109	16,895	16,557	−1.3	−2.0
Apparel and services	1,801	1,725	1,700	−4.2	−1.4
Transportation	8,604	7,658	7,677	−11.0	0.25
Health-care	2,976	3,126	3,157	5.0	1.0
Entertainment	2835	2,693	2,504	−5.0	−7.0
Personal insurance and pensions	5,605	5,471	5,373	−2.4	−1.8
All other expenditures	5,060	5,113	5,127	0.01	0.27
Number of consumer units (000's)	120,770	120,847	121,107		
Income before taxes	$63,563	$62,857	$62,481	−1.1	−0.6
Average number in consumer unit					
Persons	2.5	2.5	2.5		
Earners	1.3	1.3	1.3		
Vehicles	2.0	2.0	1.9		
Percent homeowner	67	66	66		

Consumer Expenditure Survey data records how consumers allocate their spending to different consumer goods and services. According to Table 1.1 the number of households (in BLS's terminology number of consumer units) in the US in 2010 was 121,107,000 with average size of 2.5 persons. Average income before tax (average gross income) of the households is reported as $62,481 and the average expenditure as $48,109. As a clear evidence of the "great recession", for the first time since 1984 consumer spending registered a drop of 2.8 % in 2009 from 2008 and 1.95 % in 2010 from 2009. Loss of income have forced consumers to further cut back in their consumption expenditure in 2010, specially for entertainment and food. With the average saving rate of the American households negligibly small, it is safe to say that a major portion of the difference between average gross income and expenditure is the average annual amount of taxes households paid to various levels of government— local, state, and federal.

---

[2] The most recent data available are for 2011, released by BLS on February 2013. I am using 2008, 2009, and 2010 to highlight the impact of the great recession on consumers.

Table 1.2 Consumer expenditure survey 2010

	2010	Percentage
Average annual expenditures	$48,109	100.0
Food	6,129	12.7
At home	3,624	7.5
Away from home	2,505	5.2
Housing	16,557	34.4
Apparel and services	1,700	3.5
Transportation	7,677	15.96
Vehicle purchase	2,588	5.4
Other vehicle expenses	2,464	5.1
Gasoline and motor oil	2,132	4.4
Public transportation	493	1.0
Health care	3,157	6.6
Entertainment	2,504	5.2
Personal insurance and pensions	5,373	11.2
Other expenditures	5,012	10.4

BLS classifies household expenditure on consumer goods and services in eight broad categories, like Food, Housing, and Transportation. From Table 1.2 it is clear that the largest household expenditure item is Housing; 34.4 % of the total annual expenditure in 2010. Housing expenditure generally consists of mortgage payment for homeowners, rental payment for households that do not own their homes, maintenance and repair costs, fuel and utility costs, and homeowner insurance cost.

Transportation and food are the next two big items. While the share of Transportation from the household's total expenditure is 15.96 % (an 11 % decline in 2009 from 2008), gasoline and public transportation costs constitute only 4.4 and 1.0 % of expenses, respectively. A combination of deep recession and historically high gas price has lead, for the first time, to a decline in gasoline consumption in 2009 compared to 2008. The data, however, indicates that while other industrially advanced nations have cut their oil consumption since 1980 (Sweden and Denmark by as much as 33%), U.S. oil consumption has increased by more that 21 %. The United States still has the lowest gasoline price in the industrial world.

6.7 % of household expenditure is related to Health Care and 11.2 % to personal insurance and pension. As Table 1.1 indicates health care is still, and almost the only, growing expenditure item for the household. An increase of 5 % from 2008 to 2009 comes on the heels of 4.3 % increase from 2007 to 2008, 7.9 % increase from 2004 to 2005, and 18.9 % increase from 2003 to 2004.

1.1.1 A Simple Two-Commodity Model

Assume a household consumes two goods. Let X denote the units of good 1 and Y the units of good 2 consumed by this household in a year. Assume P_1 and P_2 are the market prices of good 1 and 2, respectively. Also assume that this household earns $ I as annual income. With these assumptions we can write this household's budget equation as

$$P_1 X + P_2 Y = I$$

Solving for Y, we have

$$Y = \frac{1}{P_2} I - \frac{P_1}{P_2} X$$

If price of good 1 is $100, price of good 2 is $200, and household annual income is $50,000, then the household budget equation is

$$100X + 200Y = 50000$$

leading to

$$Y = \frac{50000}{200} - \frac{100}{200} X = 250 - 0.5X \tag{1.5}$$

Figure 1.1 represents the graph of this linear budget equation. Points on the budget line signify combinations of units of good 1 and 2 that the household can buy, spending its entire annual income. Points A (250 units of Y and 0 unit of X) and B (500 units of X and 0 unit of Y) illustrate the cases when the household spends its income entirely on good 2 or on good 1.

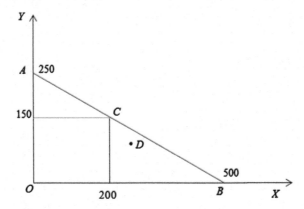

Fig. 1.1 Budget line

Assume this household picks the combination of 200 units of good 1 and 150 units of good 2. This is the point C on the line. This household can choose a combination of X and Y at a point like D below the line. In this case the household is not spending its entire income and has some savings. All the points on the sides and inside of the triangle AOB constitute the *feasible consumption* set, i.e. all combinations of X and Y that the household can buy. Given the current income of this household, it cannot buy any combinations of X and Y above the budget line unless it borrows additional fund. Points above the budget line are *non-feasible consumption* set for this household.

P_1, P_2, and I are parameters of the budget constraint. Changes in any of these parameters lead to a new budget equation. The slope of the budget line is $-\dfrac{P_1}{P_2}$. Changes in P_1 and P_2, individually or simultaneously, will impact the household budget and change its budget equation. For example, if the price of good X increases from \$100 to \$110 *ceteris paribus*, the new budget equation will be

$$110X + 200Y = 50000 \quad \longrightarrow \quad Y = 250 - 0.55X$$

Here the slope of the budget line changes from 0.5 to -0.55, rotating inward and reducing the size of the feasible consumption set. This situation is depicted in Fig. 1.2.

To examine the impact of change in income, assume the household income increases from \$50,000 to \$60,000. In this case $100X + 200Y = 60000$ and the new budget equation is

$$Y = 300 - 0.5X$$

The X-and Y-intercepts of the new budget line are now 600 and 300 while the slope of the line remains the same at -0.5. An increase in income generates an outward parallel shift in the budget line and leads to a new and expanded feasible consumption set. The reverse is also true; a loss of income shrinks the feasible set. If the household income declines by 20 % to \$40,000 a year, then the new budget

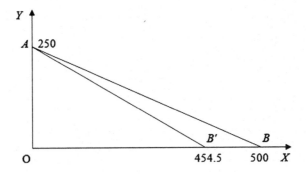

Fig. 1.2 Change in budget line due to increase in price of X

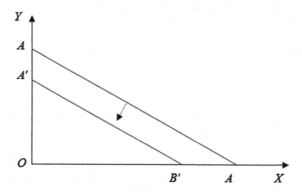

Fig. 1.3 Shift in budget line due to loss of income

equation is

$$Y = 200 - 0.5X \tag{1.6}$$

with the new budget line $A'B'$ in Fig. 1.3

While a change in income leads to a parallel shift in the budget line, a proportional change in prices has the same effect. Assume that the inflation rate (rate of increase in the general level of prices) is 20 %. Because of 20 % inflation prices of good 1 and 2 rise to \$120 and \$240, respectively. The new budget equation is

$$Y = \frac{50000}{240} - \frac{120}{240}X = 208.33 - 0.5X \tag{1.7}$$

Notice that due to the proportional changes in price, the slope of this line (-0.5) is the same as line in (1.5) and (1.6), i.e. the new budget line is parallel to $A'B'$ and AB. This creates an impression that the impact of inflation must be similar to the loss of income—a parallel downward shift in the budget line. But does inflation have the same impacts on household's consumption as the loss of income? In order to examine this question we must make certain assumptions about the household preference. Here we assume that the household consumes X and Y in the same proportions as the combination of 150 units of Y and 200 units of X, i.e. in the ratio of $150/200 = 0.75$.[3] This means that the number of units of Y consumed is always equal to 0.75 units of X. This can be expressed as:

$$Y = 150/200X \quad \text{or} \quad Y = 0.75X \tag{1.8}$$

The graph of this equation is a ray from the origin to point C on the budget line and is called the *income-consumption path* (see Fig. 1.4). Any change in the household income or in the prices of X and Y would lead to parallel shifts or rotations of the

---

[3] This is a rather restrictive assumption. We must consider this problem in more detail in the context of consumers' welfare maximization strategy.

Fig. 1.4 Linear income-consumption path

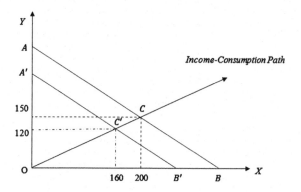

budget line. In response to the new budgetary realities, the household must adjust its bundle and pick a new combination of X and Y. The assumption of an equal proportion of consumption of X and Y dictates that these new combinations must be points on this line. This assumption adds a new condition or constraint; not only must a bundle be on the budget line, it must also be on this line. So the new combination must be at the intersection of these two lines.

The new bundle that the household purchases after a 20 % loss of income, would consist of 160 units of X and 120 units of Y (point C' on $A'B'$ and OC.) We obtain this result by solving (1.6) and (1.8) as a set of simultaneous equations.

$$\begin{cases} Y = 200 - 0.5X \\ Y = 0.75X \end{cases}$$

Solving this system leads to

$$200 - 0.5X = 0.75X \longrightarrow 1.25X = 200 \text{ and } X = \frac{200}{1.25} = 160 \text{ units}$$

$$Y = 200 - 0.5(160) = 120 \text{ units}$$

The income-consumption path and the result of the loss of income are depicted in Fig. 1.4

The new bundle resulting from 20 % inflation would consist of 166.67 units of X and 125 units of Y, obtained by solving (1.7) and (1.8) simultaneously in the same manner.

$$\begin{cases} Y = 208.33 - 0.5X \\ Y = 0.75X \end{cases}$$

It is clear that inflation and loss of income both adversely affect the standard of living of this household, measured in real terms in the number of units of good 1 and good 2 purchased. With the assumption of a linear income-consumption path, it is also clear that the impact of a loss of income is greater than that of inflation. Unemployment and underemployment may have a much larger negative impact on welfare and standard of living of households than a comparable decline in purchasing power due to inflation.

Next let's examine the effect of a non-proportional change in the prices. Let assume the price of X increases from $100 to $110 and the price of Y increases from $200 to $210 per unit, while the household income remains at $50,000. The new budget equation is

$$110X + 210Y = 50000 \longrightarrow Y = 238.1 - 0.524X \qquad (1.9)$$

Compare the new budget equation (1.9) with the original equation (1.5). There are changes in both the intercept and the slope. To determine the new consumer's equilibrium, we must find the intersection of this budget line with the income-consumption path by solving (1.9) and (1.8) simultaneously.

$$\begin{cases} Y = 238.1 - 0.524X \\ Y = 0.75X \end{cases}$$

We obtain $X = 186.9$ units and $Y = 140.2$ units

1.2 Exercises

1. A household allocates its $2,000 monthly income to the purchase of three goods. Prices of these goods are $30, $40, and $20 per unit.

    (a) Write the household monthly budget constraint.
    (b) If this household purchases 40 units of good three each month, write its budget equation and graph it. What is the slope of the budget line?

2. Assume a household with a monthly income of $5,000. This household allocates its income to the purchasing of food and nonfood products. If the average price of food products is $20 per unit and nonfood items costs $150 per unit

    (a) write the households budget equation.
    (b) If this household consumes 100 units of food products, how many units of nonfood items it can buy?
    (c) Assume that the price of nonfood products increases to $160. Write the new budget equation.
    (d) If this household wants to purchase food and nonfood items in the same proportion as in part (b), what is the household's new bundle in part (c)?

3. Assume the household's income in Problem 2 increases by 5 %. Repeat parts (a) through (d) of problem (2).

4. Assume that due to competition prices of food and nonfood products in Problem 2 decline by 5% while the household's income remains the same. Repeat parts (a) and (b) of problem (2). Compare your result with problem (3).

5. A household splits its $4,000 monthly income between necessity and luxury goods. The average price of necessities is $30 per unit and that of luxuries is $100 per unit.

   (a) Write the household budget constraint.
   (b) Determine the household equilibrium bundle if its proportion of necessity and luxury goods purchases is 10 to 1.
   (c) What is the equation of household income-consumption path?
   (d) Assume the household income declines by 10%. What is the household's new bundle?
   (e) Assume no loss of income but an inflation rate of 10%. What is the household's new bundle?
   (f) Compare your answers in part (b) and (c).
   (g) Assume the original household income increases by 10% to $4,400. What is the household's new bundle?
   (h) Assume no change in income but the price level declines by 10%. What is the household's new bundle?
   (i) Compare your answers in parts (g) and (h)
   (j) Compare your answers to parts (f) and (i). Are you surprised?

Appendix A: A Short Note on Sigma Notation

An economical and convenient method for writing long sums is the use of *summation* or *sigma* notation. The upper case Greek letter sigma $\sum$ (counterpart of S, the first letter of Sum) is employed to denote sums, hence the 'sigma notation'.

Assume $x_1, x_2, \ldots, x_n$ (or alternatively $x_i i = 1, 2, \ldots, n$) are n real numbers. The sum of this n numbers $x_1 + x_2 + \cdots + x_n$ can be compactly expressed[4] as $\sum_{i=1}^{n} x_i$. The symbol $\sum_{i=1}^{n} x_i$ (or its *in-line* or *in-text* version $\sum_{i=1}^{n} x_i$) is the instruction for conducting the summing of numbers over the whole range of values of i, called the *index of summation*. The index i takes consecutive integer numbers beginning with, in this example, 1 and ending with n. 1 is referred to as the *initial value* or *lower bound* and n as the *terminal value* or *upper bound* of the index. Besides i other letters such as $j, k,$ and l are used to denote the summation index. Note that conversely the expansion of $\sum_{i=1}^{n} x_i$ is $x_1 + x_2 + \cdots + x_n$.

---

[4] Note that in $x_1 + x_2 + \cdots + x_n$, the use of ellipses '$\cdots$', representing the omitted terms, in itself is one step in the direction of more economical and time- and space-saving expression of long sums.

Examples

Expand and Evaluate

1.

$$\sum_{j=5}^{10} a_i = a_5 + a_6 + a_7 + a_8 + a_9 + a_{10}$$

In the above example, the initial value of the summation index is 5 and its terminal value is 10.

2.

$$\sum_{k=3}^{8} k = 3 + 4 + 5 + 6 + 7 + 8 = 33$$

3.

$$\sum_{i=1}^{4} (a_i + 2)^2 = (a_1 + 2)^2 + (a_2 + 2)^2 + (a_3 + 2)^2 + (a_4 + 2)^2$$

4.

$$\sum_{i=1}^{4} 5 = 5 + 5 + 5 + 5 = 20$$

5.

$$\sum_{k=0}^{4} (2k + 2^k) = 1 + (2 + 2) + (4 + 4) + (6 + 8) + (8 + 16) = 51$$

6.

$$\sum_{j=n}^{n+2} j^2 = n^2 + (n + 1)^2 + (n + 2)^2 = 3n^2 + 6n + 5$$

7.

$$\sum_{k=0}^{5} (-1)^k = (-1)^0 + (-1)^1 + (-1)^2 + (-1)^3 + (-1)^4 + (-1)^5$$
$$= 1 - 1 + 1 - 1 + 1 - 1 = 0$$

Write the sum in sigma notation

8.

$$1 + 1/2 + 1/3 + 1/4 + 1/5 + \cdots + 1/100 = \sum_{n=1}^{100} \frac{1}{n}$$

9.

$$1 + 1/4 + 1/9 + 1/16 + 1/25 + 1/36 + \cdots + 1/100 = \sum_{n=1}^{10} \frac{1}{n^2}$$

10.

$$1 + 2^3 + 3^3 + 4^3 + \cdots + n^3 = \sum_{i=1}^{n} i^3$$

11.

$$2^4 + 3^4 + 4^4 + \cdots + n^4 = \sum_{i=2}^{n} i^4 \quad \text{or alternatively} \quad \sum_{i=1}^{n-1} (i+1)^4$$

$$\text{or even} \quad \sum_{i=0}^{n-2} (i+2)^4$$

12.

$$1 - x + x^2 - x^3 + x^4 - x^5 + \cdots + (-1)^n x^n = \sum_{j=0}^{n} (-1)^j x^j$$

13. Write the sum of odd numbers from 1 to 99 in sigma notation

$$1 + 3 + 5 + 7 + \cdots + 99 = \sum_{i=1}^{50} (2i - 1) \quad \text{or} \quad \sum_{i=0}^{49} (2i + 1)$$

Three Rules of Sigma Notation

(1) If c is a constant then

$$\sum_{i=1}^{n} c = nc$$

We obtain this result by simply expanding $\sum_{i=1}^{n} c$

$$\sum_{i=1}^{n} c = \underbrace{c + c + \cdots + c}_{n} = nc$$

For example $\sum_{i=1}^{4} 5 = 4 * 5 = 20$

A more generalized form of this rule is that if c is a constant and $n > m$, then

$$\sum_{i=m}^{n} c = (n - m + 1)c$$

for example

$$\sum_{i=4}^{10} 5 = (10 - 4 + 1)5 = 35$$

(2) If c is a constant, we can expand and show that

$$\sum_{i=1}^{n} cx_i = c \sum_{i=1}^{n} x_i$$

$$\sum_{i=1}^{n} cx_i = cx_1 + cx_2 + \cdots + cx_n = c(x_1 + x_2 + \cdots + x_n) = c \sum_{i=1}^{n} x_i$$

For example if $\sum_{i=1}^{n} x_i = 45$, then $\sum_{i=1}^{n} 2x_i = 2 \sum_{i=1}^{n} x_i = 2(45) = 90$
The generalized version of this rule is that if c is a constant and $n > m$, then

$$\sum_{i=m}^{n} cx_i = c \sum_{i=m}^{n} x_i$$

(3) Sum and difference rule

$$\sum_{i=1}^{n} (a_i \pm b_i) = \sum_{i=1}^{n} a_i \pm \sum_{i=1}^{n} b_i$$

We show the sum rule by expanding $\sum_{i=1}^{n} (a_i + b_i)$

$$\sum_{i=1}^{n} (a_i + b_i) = (a_1 + b_1) + (a_2 + b_2) + \cdots + (a_n + b_n)$$

$$= a_1 + b_1 + a_2 + b_2 + \cdots + a_n + b_n$$

$$= a_1 + a_2 + \cdots + a_n + b_1 + b_2 + \cdots + b_n = \sum_{i=1}^{n} a_i + \sum_{i=1}^{n} b_i$$

Similarly, the difference rule is shown by expanding $\sum_{i=1}^{n} (a_i - b_i)$

$$\sum_{i=1}^{n} (a_i - b_i) = (a_1 - b_1) + (a_2 - b_2) + \cdots + (a_n - b_n)$$

$$= a_1 - b_1 + a_2 - b_2 + \cdots + a_n - b_n$$

$$= a_1 + a_2 + \cdots + a_n - b_1 - b_2 - \cdots - b_n = \sum_{i=1}^{n} a_i - \sum_{i=1}^{n} b_i$$

If, for example, $\sum_{i=1}^{10} x_i = 45$ then

$$\sum_{i=1}^{10}(3x_i - 2) = \sum_{i=1}^{10} 3x_i - \sum_{i=1}^{10} 2, \quad \text{using rule 3}$$

$$= 3\sum_{i=1}^{10} x_i - 10(2), \quad \text{using rules 2 and 1}$$

$$= 3(45) - 20 = 115$$

Most sigma notation applications can be handled by a combination of the above three rules. For example $\bar{x}$, the *mean* or average of a set of n numbers x_i $i = 1, 2, \ldots, n$, is computed by dividing the sum of the n numbers by n. By using sigma notation we can express $\bar{x}$ as

$$\bar{x} = \frac{\sum_{i=1}^{n} x_i}{n}$$

Also s^2, the *variance* of x_i, is defined as

$$s^2 = \frac{\sum_{i=1}^{n}(x_i - \bar{x})^2}{n - 1}$$

The so-called short cut formula for variance

$$s^2 = \frac{\sum_{i=1}^{n} x_i^2 - n\bar{x}^2}{n - 1}$$

is derived by using rules of sigma notation and manipulating the numerator of the variance formula

$$\sum_{i=1}^{n}(x_i - \bar{x})^2 = \sum_{i=1}^{n}(x_i^2 + \bar{x}^2 - 2x_i\bar{x}) = \sum_{i=1}^{n} x_i^2 + \sum_{i=1}^{n} \bar{x}^2 - \sum_{i=1}^{n} 2x_i\bar{x}$$

$$= \sum_{i=1}^{n} x_i^2 + n\bar{x}^2 - 2\bar{x}\sum_{i=1}^{n} x_i$$

From the formula for the mean $\bar{x} = \frac{\sum_{i=1}^{n} x_i}{n}$ we have $\sum_{i=1}^{n} x_i = n\bar{x}$. Substituting $n\bar{x}$ for $\sum_{i=1}^{n} x_i$ in the above formula, we have

$$\sum_{i=1}^{n}(x_i - \bar{x})^2 = \sum_{i=1}^{n}x_i^2 + n\bar{x}^2 - 2\bar{x}\sum_{i=1}^{n}x_i = \sum_{i=1}^{n}x_i^2 + n\bar{x}^2 - 2\bar{x}(n\bar{x})$$

$$= \sum_{i=1}^{n}x_i^2 + n\bar{x}^2 - 2n\bar{x}^2 = \sum_{i=1}^{n}x_i^2 - n\bar{x}^2$$

As an example, assume the sum of the grades of 20 students $\sum_{i=1}^{20}x_i$ is 1200 and the sum of the square of the grades $\sum_{i=1}^{20}x_i^2$ is 72475. What is the average and variance of the grades?

$$\bar{x} = \frac{1200}{20} = 60 \quad \text{and} \quad s^2 = \frac{72475 - 20(60)^2}{19} = 25$$

Since the class average is low, the instructor decides to give every student an additional 5 points. What is the new average and variance?
Here each x_i is changed to $x_i + 5$, therefore the new mean is

$$\bar{x}_{new} = \frac{\sum_{i=1}^{20}(x_i + 5)}{20} = \frac{\sum_{i=1}^{20}x_i + 20(5)}{20} = \frac{\sum_{i=1}^{20}x_i}{20} + 5 = \bar{x} + 5 = 60 + 5 = 65$$

Giving each student an extra 5 points increases the mean by the same 5 points. As for the variance

$$s^2_{new} = \frac{\sum_{i=1}^{20}[(x_i + 5) - \bar{x}_{new}]^2}{19} = \frac{\sum_{i=1}^{20}[(x_i + 5) - (\bar{x} + 5)]^2}{19} = s^2$$

indicating no change in the variance.

Exercises

1. Write the following sums in sigma notation
   (a) $1 + 2 + 3 + 4 + \cdots + 20$
   (b) $a + a^2 + a^3 + \cdots + a^m$
   (c) $1/2 + 1/4 + 1/8 + \cdots + 1/64$
   (d) $\sqrt{5} + \sqrt{6} + \sqrt{7} + \cdots + \sqrt{25}$
   (e) $1 + 1/3 + 1/5 + 1/7 + 1/9 + \cdots + 1/31$
   (f) $2 + 4 + 6 + 8 + \cdots + 100$ (the sum of even numbers from 2 to 100)
   (g) $1 + 4 + 9 + 16 + 25 + \cdots + 100$

2. Expand and find the sum

$$(a) \ \sum_{j=2}^{5} (j+1)^2 \qquad (b) \ \sum_{i=1}^{5} i^2$$

$$(c) \ \sum_{j=-2}^{3} 2^{j-1} \qquad (d) \ \sum_{k=0}^{3} \frac{3k-2}{3k+2}$$

3. Show that

$$(a) \ \sum_{i=1}^{n} (a_i + b_i)^2 \neq \sum_{i=1}^{n} a_i^2 + \sum_{i=1}^{n} b_i^2$$

$$(b) \ \sum_{i=1}^{n} (x_i - y_i)^2 \neq \sum_{i=1}^{n} x_i^2 - \sum_{i=1}^{n} y_i^2$$

4. Assume n numbers $x_1, x_2, \ldots, x_n$ have mean $\bar{x}$ and variance s^2. Show that if we subtract a constant c from each number the mean changes to $\bar{x} - c$ but variance stays the same.

5. Assume that in problem 4 instead of adding a constant c to each number, we multiply the numbers by c. What would be the new mean and variance?

6. Show that the sum of the first n positive integers $\sum_{i=1}^{n} i$ is $\dfrac{n(n+1)}{2}$.
   [Hint: use the trick that Karl Gauss, a nineteenth century German mathematician, used to solve this problem when he was about 9 years old. Write the sum twice, first in the usual order and second in the reverse order, and then add both sides of the sums.] Use the result and find

$$(a) \ \sum_{j=1}^{100} j \qquad (b) \ \sum_{j=1}^{100} (j+2)$$

7. Find the number n such that $\sum_{i=1}^{n} i = 20100$

8. True or false

$$\sum_{j=1}^{n} a_j + a_{n+1} + a_{n+2} = \sum_{j=1}^{n+2} a_j$$

9. Evaluate

$$(a) \ \sum_{k=1}^{5} (-1)^k k \qquad (b) \ \sum_{j=1}^{4} (j+2)^j$$

Chapter 2
Variables, a Short Taxonomy

Variables

A phenomenon that changes over time, space, objects, or living things is called a variable. If a variable's changes could be measured, that is, expressed by numbers or quantities, then we have a *quantitative variable*; otherwise, we have a *qualitative variable*. Blood pressure is an example of a quantitative variable—because it takes different values from individual (living thing) to individual. It can be measured, and has its own scale of measurement. Tree Hight is another example of a quantitative variable. It varies from tree to tree and can be measured (in centimeters or meters in the metric system and in inches and feet in the English system)[1] and expressed by a numerical value.

Changes in qualitative variables can only be *observed*. Gender is an example of a qualitative variable. Color of automobiles is another example of a qualitative variable. It varies from automobile to automobile. Qualitative variables taking only two different values are known as *dichotomous* or *binary* variables. Those taking more than two values are called *multinomial*. Gender is an example of a dichotomous variable while color is an example of a multinomial variable. In this book we do not deal with any qualitative variables.

The U.S. Gross Domestic Product (GDP) varies from year to year. The Bureau of Economic Analysis (BEA), an agency of the Department of Commerce, produces some of the most closely watched economic statistics, including annual GDP. Besides annual estimates, BEA also produces semi-annual and quarterly estimate of GDP and its growth rate. The Bureau of Labor Statistics (BLS) of the US Department of Labor measures the size of the labor force and number of people unemployed every month and reports the monthly Unemployment rate (U-rate). The GDP and U-rate are examples of quantitative variables measured and reported for a certain time. In each case the measurement of the variable is conducted over an equally spaced time interval, a year for GDP and a month for the unemployment rate.

---

[1] There are many other measurement systems in the world. Metric and English systems are by far the most widely used.

S. Vali, *Principles of Mathematical Economics*,
Mathematics Textbooks for Science and Engineering 3,
DOI: 10.2991/978-94-6239-036-2_2, © Atlantis Press and the authors 2014

Data generated by measuring values of a quantitative variable that changes over time are called *time series* data. Frequency of measurement of a variable or the time interval over which measurement is made is called the *periodicity* of the variable. There are variables with different periodicities, as the following examples demonstrate. The size of a country's population is measured every 10 years, through a census of the population. It is a decennial variable and its values generate decennial data. Publicly traded corporations are required to report, among other things, their revenue for each quarter. The Labor Department reports jobless claims (the number of newly laid-off workers filing claims for unemployment benefits) each week. Department stores also report their sales volume weekly. At the end of a business day the Dow Jones Industrial Average (DJIA) for closing prices of the 30 major US corporations that comprise the index, is widely reported. During a trading day, the DJIA is measured, and posted on the New York Stock Exchange (NYSE) big board almost every 15 seconds. Meteorological instruments of various weather stations continuously measure and record changes in temperature and atmospheric pressure in various parts of the country. Maintaining heat inside a blast furnace is an important factor in the production of high quality steel. In modern steel making processes, infra red instruments continuously measure the heat inside the furnace and display it on a screen in the control room of the factory.

As the above examples should indicate, there are two types of time series variables; *discrete* and *continuous*. In the continuous case values of the variable are measured or observed at every moment of time, i.e. over infinitesimally small time intervals to the extent that periodicity of the variable approaches zero; however, most time series variables, particularly in economics, are discrete. Values of a discrete time series variable are recorded or measured at predetermined equal time interval. The most common periodicities of economic variables are annual, semi-annual, quarterly, monthly, and weekly. Changes in a time series variable may happen with an irregular time interval or periodicity. A good example is the change in price of shares of a company traded on a stock exchange. These changes may occur over different unequal time intervals, but the price of the company's stock is still considered a time series variable. There are numerous examples of time series variables, with or without regular periodicity. What they have in common is that they all evolve over time and are all indexed by time, irrespective of whether the index is equally spaced or not.[2]

Most macroeconomic variables are time series variables. A very good, updated and comprehensive source of macroeconomics time series is the Annual Report of the President, prepared by the Council of Economic Advisors (CEA). Appendix B of the latest report issued (at the time of writing this passage) February 2011 contains a collection of about 130 pages of statistical tables produced by various departments and agencies of the US government. The statistical tables of the report cover historical and the latest available data related to income, population, employment, wages and productivity, production, prices, money stock, corporate profit and finance,

---

[2] For a formal presentation of time series variables see Chap. 11 *"Economics Dynamics and Difference Equations"*.

agriculture, and international trade and finance. The entire report can be accessed at the US Government Printing Office site http://www.gpo.gov.

BLS not only measures and reports the national unemployment rate each month, but also provides the same information for each state and a large number of major metropolitan areas like New York City, Philadelphia, and Los Angeles. In this case the variable unemployment rate takes different values over space or geographic areas. This is an example of a *cross section* variable. Another example would be the second quarter earnings reported by companies in *Standard and Poor's* (S&P) 500 index. Here variable 'earning' takes 500 different values, one for each of the 500 companies in the index at the same time (quarter). Data generated by changes in a cross section variable are called "cross section" data.

The third class of variables, called *mixed variables*, are *mixed* of time series and cross section or mixed of cross sections. Typical examples are monthly unemployment or inflation rates of 50 states of the United States, annual growth rates of the G7 (group of 7 most advanced industrialized countries) in the last 10 years, and average income of households of different size over different geographic areas. Mixed variables are sometimes referred to as *panel variables* and their values called *panel data*. This designation is due to the fact that presentation of the values of a mixed variable is best achieved in a rectangular array (panel) format, i.e. a table. For example, to present the growth rate of the G7 countries for the last 10 years we must construct a table with 7 rows, one row for each country, and 10 columns, one column for each year. This table contains $7*10 = 70$ cells. Each cell that is at the intersection of a row (country) and a column (year) will contain the growth rate of that country for that year.

Before we go any further in our discussion of variables we must distinguish between *flow* and *stock* variables. This distinction is very important in economics. Values of a flow variable can only be measured between two points in time, that is, over a specific time interval, while values of stock variables can be measured at every moment in time. Examples of flow variables are GDP, income, revenue, output, and rainfall. Monthly household income is the amount a household earns over a month, and the GDP is the market value of all finished goods and services produced by labor and property located in a country in a given period of time, generally one year.

Examples of stock variables are capital stock of a firm, balance of your saving account, wealth of an individual, and market value of assets of a corporation. All these variables must be measured at one point in time. Measuring these variables at different time may lead to a different value for the variable. Unemployment rate is a stock and inflation rate is a flow variable. All of the items appearing on the *balance sheet* of a company are stock variables, while items appearing on a company's *income statement* are flow variables.

Since variables can take a range of values, they must be represented by symbols. In fact Bertrand Russell—a giant of philosophy, mathematics and literature of the twentieth century—defined mathematics as "a game played with symbols". Use of the first letter, or an abbreviation, of a variable's name for denoting that variable is very common. In economics, for example, it is a well established convention that price, quantity of output, cost, profit, and revenue (just name a few) are represented by P, Q, C, Π, and R. In general and abstracting from any specific context, the uppercase letters of the English (or Greek) alphabet like X, Y, and Z are used to denote or name a variable. The corresponding lowercase letters x, y, and z are used to denote specific values that they may take.

Values that a quantitative variable assumes are *numbers*, so we need to say a few words about the number system.

The Decimal (base 10) number system starts with 10 digits 0, 1, 2, ..., 9.[3] The *Natural* or "counting" numbers start from 1 and extend beyond the last digit 9 to 10, 11, and so on with the help of an algorithm for making these new numbers. The symbol commonly used to denote the structure of the natural numbers is N. Addition and multiplication are fundamental Arithmetic operations in N. Subtraction, as long as the first number is greater than the second, and division with remainder are considered implied arithmetic operations in N, because they are actually inverse of addition and multiplication ($10 - 2 = 8$ because $2 + 8 = 10$ and $20/5 = 4$ because $5 * 4 = 20$).

By adding zero and negative integers (or nonzero natural numbers with a "negative sign") to N, we arrive at the collection or set [4] of *integer numbers*. This set is generally denoted by Z. The rules and operations in Z are a little more complicated. First, we have a new order. In N, 50 is greater than 40. In Z, -50 is smaller than -40, and -20 is less than 2. Second, we have *algebraic* operations, which mean that we must consider the signs of the numbers.

While 2 and 3 are in Z, 2.5 is not. How can we divide 3 apples between two children? If we conduct this task (division in Z) we'll be left with an undivided (remainder) apple. It is most prudent to cut the third apple in half so each kid can have 1.5 apples. We need a new set containing numbers like 1.5. This set is the set of *rational numbers*. Rational here is *ratio-nal*, which means all the numbers in this set can be expressed as a ratio of two integer numbers. 1.5 is 3/2 or 6/4. Integer numbers are in this set too. 5 is 10/2. Commonly Q is used as a symbol to denote the set of rational numbers. Z is included in or a subset of Q, as N is a subset of Z.

What if a number cannot be expressed as a ratio of two integers? An example is a number like $\sqrt{2} = 1.4142...$, which is non-terminating and non-repeating decimal. Another classic example is the number π, the ratio of the circumference of a circle to its diameter. To four decimal places π is 3.1415, but it has a non-terminating and non-repeating decimal part. Numbers like $\sqrt{2}$ and π, that cannot be expressed as ratio of two integers, comprise the set of *irrational numbers*. Irrational numbers fill in the gaps between rational numbers.

There is no commonly used symbol for the set of irrationals. The set of irrationals does not have a structure similar to N or Z. An important property of N and Z is that they are *closed* under addition and multiplication. *Closure* under addition means that sum of any two numbers in Z is also a number in $Z(10 + 5 = 15)$. The product of any two numbers in Z is another number in $Z(10 * 5 = 50)$, so it is also closed under multiplication. The set of irrationals does not have these properties. For example, if

---

[3] The Octal (base 8) system has 8 digits 0, 1, ..., 7. The Binary (base 2) system has only two digits, 0 and 1.

[4] See Chap. 3 for a formal discussion of set and related topics.

we add two irrational numbers $2 + \pi$ and $3 - \pi$ we get 5 which is not an irrational number. Similarly if we multiply $2\sqrt{2}$ and $5\sqrt{2}$, the resultant number 20 is an integer. For this reason no specific symbol is devoted to the irrationals.

A union of the set of rational numbers and the set of irrational numbers creates the set of *real numbers*, universally denoted by R. R has all the nice properties of a well-structured set. It is not only closed under operations like addition, subtraction, multiplication, and division (except by zero), it is also closed under operations like integer exponentiation or raising to the power of a whole number ($2.53^3 = 15.625$).

Is R closed under fractional exponentiation or 'taking roots'? It is not. $\sqrt{-5} = (-5)^{1/2}$ does not exist in R, nor does the even root of any negative number. Note that we can express $\sqrt{-5}$ as $\sqrt{5} * \sqrt{-1}$. Denoting $\sqrt{-1}$ by i, so $i^2 = -1$, then we can write $\sqrt{-5} = \sqrt{5}\,i$, which is an example of an *imaginary number*. In general a number of the form $a + bi$, where a and b are real numbers, is called a complex number. In $a + bi$, a is the real part and bi is the imaginary part of the number. For all values of a and b, $a + bi$ generates the set of *complex* numbers, denoted by C. R is a subset of C. In this book we only deal with variables that take on real numbers.

Following our short discussion of the number system, we can introduce another classification of quantitative variables; *discrete* and *continuous*. A discrete variable can take only whole numbers or integer values, that is, the set of possible values that this variable can assume is a subset of Z. Examples of discrete variables are the number of children in a family, the number of passengers on a bus, the number of tourists visiting Grand Canyon during summer, and the number of employees of a company.

Unlike a discrete variable, values that a continuous variable can take is not restricted to integer numbers, it can also take rational and irrational numbers. The set of all possible values of a continuous variable is a subset of the set of all real numbers R (see Fig. 2.1).

By a time-honored convention, all economic variables are treated as continuous. It may be a little counter intuitive, but there are reasonably good justifications for this treatment. If a variable is measured in dollars then the smallest unit must be a cent. How could it be continuous, you might ask? Let's consider an example: the exchange rate between Dollar and Euro. Assume that the exchange rate between Euro and dollar is 0.67120, this is how much a dollar is worth in Euro. We can express the same relationship by specifying one Euro in terms of units of US currency, that is as $1.48987. It must be clear that if you want to exchange one Euro for dollar, you can only get $1.48 (don't think of rounding it to $1.49!) How about if you want to exchange 1000 Euros? Obviously you expect to receive $1489.87, $9.87 more than trading your Euro for $1.48 each! Exchanging 10,000,000 Euro generates $14,898,700. This is $98,700 more than exchange on the basis of one $Euro \simeq \$1.48$ (here the symbol $\simeq$ is used to indicate approximately "equivalent").

Given the astronomical volume of dollars and Euros traded every day in the international foreign exchange market, it is clear that there is a great deal of financial motivation to treat the exchange rate as a continuous variable and to express it to 6 or 7 significant digits (an alternative way of saying to 5 or 6 decimal places). But if this works for variables measured in monetary terms, does it work for variables

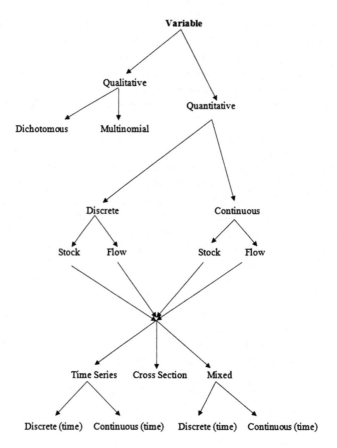

Fig. 2.1 Variable chart

expressed in physical terms? How can we treat the volume of output of an automobile manufacturer as continuous variable?

A typical automobile has 1000s of components that must be assembled on an assembly line. An assembly line is an arrangement of machinery and equipments used for the continuous flow of work pieces in mass production operations. Given that at different stages on an assembly line for automobile production, a group of components are added, then an automobile at various stages of its assembly must be considered a different fraction of a complete car, and by extension, output of an automobile assembly plant can be considered a continuous variable.

Another justification for treating economic variables as continuous is purely "mathematical". Mathematical economics employs tools and techniques of calculus of variation, a part of differential calculus, for solving a series of optimization (finding maximum or minimum) or equilibrium problems. A continuous variable lends itself much easier to these techniques.

2.1 Exercises

1. Classify the following as quantitative or qualitative variables

   (a) Height   (b) Ethnicity   (c) Number of passengers on a bus

   (d) Weight   (e) Language   (f) Religion

   (g) Party affiliation   (h) US government budget deficit

   (i) US trade deficit with China   (j) Time takes to run 100 meter

   (k) Record of the top 100 runners in New York Marathon

   (l) Consumer credit outstanding   (m) Corporate profit

   (n) Closing price of stocks listed on NYSE on January 28, 2008

   (o) Population of major cities in the US based on 2000 census.

   (p) Unemployment rate of 27 members of European Union in the last 10 years.

   (q) Unemployment rate of 27 members of European Union in 2007

   (r) Closing price of stocks listed on NYSE from Jan. 2, 1990 to Jan. 2, 2008.

   (s) Record of the top 100 runners in New York Marathon in the last 5 years.

2. Classify the qualitative variables in Exercise 1 as dichotomous or multinomial.

3. Classify the quantitative variables in Exercise 1 as discrete or continuous.

4. Classify the economic variables in Exercise 1 as time-series, cross-section, or mixed.

5. Classify k and s in Problem 1 as time-series, cross-section, or mixed.

6. Classify the economic variables in Exercise 1 as stock or flow.

7. Determine the size of the table—number of rows, columns, and cells—you need in order to present the variables that you identified as mixed in Exercise 1 (for (r) assume 3500 stocks listed on NYSE and 252 trading days).

8. How would you present the unemployment rate of white and non-white population in the last 20 years for 50 states of the United States?

Chapter 3
Sets and Functions

3.1 Sets and Set Presentation

A *set* is a collection of distinct objects with a precise description or mechanism for determining whether a given object belongs to the collection. If an object belongs to the collection it is called an *element* or a *member* of the set. Upper case letters of the English or Greek alphabet are commonly used to denote a set, while lower case letters are used to denote elements or members of the set. When x is an element of A, we write $x \in A$ and say "*x belongs* to A." We can also express this concept of *belonging* by saying "x is an *element* of A," or "x is a *member* of A," or "x is *contained* in A," or "x is in A." The symbol $\in$ (a slightly redesigned version of the Greek letter ϵ, counterpart of "e," the first letter of *element*) represents a visual shorthand for all these phrases. When x is not in A, we write $x \notin A$.

The implication in writing $x \in A$ or $x \notin A$ is to express that A is a set without indicating the nature of x. Indeed, x itself may be a set. Consider the following example. The Fordham University graduating class of 2012 is a set of graduating seniors. This set itself is an element of the set of graduating classes of 2012 within all universities and colleges in New York City, which in turn is an element of the set of graduating classes of 2012 colleges and universities in the United States.

A set can be expressed in three different ways: by *enumeration*, by *description*, and by *specification*

A set is expressed by *enumeration* when we list all of its elements individually. Ordinarily the elements are separated by commas and enclosed in braces {}. For example, the set of decimal digits can be written as

$$D = \{0, 1, 2, 3, 4, 5, 6, 7, 8, 9\}$$

If the context of discussion is clear, the presentation of D can be even abbreviated by use of ellipsis (…), to

$$D = \{0, 1, 2, 3, \ldots, 9\}$$

S. Vali, *Principles of Mathematical Economics*,
Mathematics Textbooks for Science and Engineering 3,
DOI: 10.2991/978-94-6239-036-2_3, © Atlantis Press and the authors 2014

Other examples of enumerations are

$$\Phi = \{a, b, c, \ldots, z\}, \qquad \Omega = \{A, B, C, \ldots, Z\}, \qquad A = \{\alpha, \beta, \gamma, \ldots, \xi\}$$

where Φ and Ω are the sets of the lower- and uppercase English alphabet, and A is the set of the lowercase Greek alphabet. But how one does enumerate the set of natural numbers N? There are infinitely many elements in N. One could use an open-ended ellipsis and write $N = \{1, 2, 3, \ldots\}$. However, there are cases when the use of an ellipsis may create a great deal of ambiguity, violating the spirit of mathematical clarity.

We use *description* when we say "the English alphabet," "A is the set of the lowercase Greek alphabet," "the natural numbers," or "students in Fordham University's mathematical economics class." These are phrases that describe or define membership of these sets. Sometimes a symbol is a sufficient description. Consider the symbol N: the whole set of natural numbers is described by this symbol.

In general, set enumeration and description are well-suited for presenting sets with nonnumerical elements. By comparison the "information content" of an enumerated set is much more than the set described. By describing A as "the set of companies selected by the editorial board of the Wall Street Journal to be included in the Dow Jones Industrial Average index," it is not clear whether Apple Inc. belongs to this set. But when A is enumerated—i.e. the names of all 30 companies in the index are listed—it is easy to establish membership status of Apple Inc. Enumeration has, of course, a big disadvantage. If the number of elements of the set, which is called its *cardinality*, is large, its enumeration will be time-consuming or even impossible.

The most widely used set presentation is *specification*. Set specification is sometimes referred to as *set-builder notation*. This method is well-suited for presenting sets with quantitative elements. It specifies elements of a set in terms of another known or defined set or property. If A is a known set and P is a certain property, then the interpretation of specifying a set X as

$$X = \{x \in A \mid x \text{ has property } P\}$$

is that X consists of elements x belonging to A "given" or "such that" they all satisfy property P. The character "$\mid$" is used in place of "given" or "such that." We will illustrate this method with some examples:

(1) $S = \{s \in N \mid s < 5\}$ means that the set S consists of elements s all natural numbers (the known set A) such that s is less than 5 (property P). This set can be expressed by description as "the set of natural numbers less than 5," and enumerated as $S = \{0, 1, 2, 3, 4\}$.

(2) $A = \{a \in R \mid -5 \le a \le 10\}$. The set A consists of elements a that are the real numbers between -5 and 10, inclusive. Although it is possible to describe this set as "real numbers between -5 and 10, inclusive," one would be hard-press to come up with a short enumerative presentation for this set.

(3) $Y = \{y \in N \mid y < 10$ and y is divisible by 2$\}$. This set consists of all natural numbers less that 10 and divisible by 2. Using enumeration, $Y = \{2, 4, 6, 8\}$.

(4) $C = \{c \in Z \mid \mid c \mid < 4\}$. This set can be described as "integers with absolute values less than 4," and enumerated as $C = \{-3, -2, -1, 0, 1, 2, 3\}$.

(5) $X = \{x \in Z \mid x$ is odd$\}$ means that the set X contains element like x which are odd integers.

Sets S in (1), Y in (3), and C in (4) are examples of *finite* sets. These sets have a finite number of elements. Sets A in (2) and X in (5) are examples of *infinite* sets. These sets have an infinite number of elements. Obviously an infinite set cannot be enumerated.

In the definition of a set repetition of elements and/or the order in which elements are listed are irrelevant. Based on this definition $\{5, 6, 7\}$, $\{6, 5, 7\}$, and $\{5, 5, 7, 6\}$ are all the same sets. All three sets contain the same numbers 5, 6, and 7.

Two sets *containing the same elements* are equal. In more formal terms if for all x in A, $x \in B$; and for all x in B, $x \in A$ then $A = B$. Mathematician abbreviate "all" or "for all" symbolically by an inverted upper case letter "A" as $\forall$, so the equality of two sets can be expressed formally as: if $\forall x$ in A, $x \in B$; and $\forall x$ in B, $x \in A$; then $A = B$.

The set with no elements is called the *empty set* or the *null set* and is denoted by $\emptyset$ or $\{\}$. A good example of the null set is, at the time of writing this passage, the set of female presidents of the United States.

3.2 Set Inclusion and Set Relationships

Suppose A and B are two sets. If every elements of A is also an element of B, the set A is called a *subset* of B or is *included* in B. The symbol for "inclusion" is $\subset$. More formally,

$$\text{If } \forall x \in A, \ x \in B \quad \text{then } A \subset B$$

If in set inclusion we want to include the possibility of equality between A and B, then the symbol used is $\subseteq$. Otherwise, we write $A \subset B$. Here A is called a *proper subset* of B. As an example, suppose $A = \{2, 3, 4\}$ and $B = \{0, 1, 2, 3, 4, 5, 6\}$, then A is a proper subset of B. But if $B = \{2, 3, 4\}$ then $A \subseteq B$. If A is not a subset of B this is symbolically expressed as $A \not\subset B$.

The null set $\emptyset$ is considered to be a subset of every set. If the null set is not a subset of a set A, then it must contain an element that is not in A. And this contradicts the definition of $\emptyset$, which is a set with no elements.

The set whose elements are all possible subsets of a set A, including the null set and the set itself, is called the *power set* of A. Suppose $A = \{a, b\}$ then the power set of A, denoted by P, is $P = \{\emptyset, \{a\}, \{b\}, \{a, b\}\}$. Here the number of elements in A, or A's *cardinality*, is 2. The cardinality of P is $4 = 2^2$. It can be shown that if a set has n elements, its power set contains 2^n elements.

If two sets have no elements in common, they are said to be *disjoint*. If, for example, $A = \{0, 1, 2, 3\}$ and $B = \{6, 10, 20\}$, then they are disjoint.

3.3 Set Operations

Before discussing various set operations we must define the concept of a *universal set*. The common notation for "universal set" is U. The universal set cannot be defined in abstract. The definition of U must be connected to a given context. Lets assume that elements of all sets that are subject of our discussion are integer numbers. In this context the universal set is Z, the set of all integer numbers. In other words, within the given context, we are restricting our considerations to subsets of the given universal set Z. Here the frame of reference is Z, that is, U is Z. If we want to study mergers among US corporations, our universal set or frame of reference must be companies listed on organized exchanges, national or regional (like the New York Stock Exchange or Philadelphia Stock Exchange), or Over-The-Counter (OTC) exchanges like NASDAQ. The only sets that the German mathematician George Cantor , who did pioneering work in set theory in the late nineteenth century, considered in his studies were sets of real numbers. Therefore the universal set or the frame of discourse for Cantor was R.

There are three basic set operations: *union, intersection,* and *complement.* Given sets A and B, the union of A and B, denoted $A \cup B$, is a new set C defined as

$$C = A \cup B = \{x \mid x \in A \quad \text{or} \quad x \in B\} \tag{3.1}$$

Let $A = \{u, v, w, x\}$ and $B = \{x, y, z\}$, then $C = A \cup B = \{u, v, w, x, y, z\}$. It is important to understand that the set C formed by the union of A and B does not contain sets A and B as elements. It only contains elements of A and B. Union is a binary operation on sets, the same way that addition is on numbers. "$\cup$" and "$+$" require two arguments. We add *two* numbers, and likewise we find the union of *two* sets. $A \cup B$ is the counterpart of $a + b$.

In spite of their similarities, there is a major distinction between the operations of addition and union. In addition both elements must belong to the same "universe" or universal set. We cannot add oranges and houses, for example. But there is theoretically no restriction in finding the union of two sets belonging to two different universes. As an example consider the set of students in an introductory economics class. Let's enumerate this set as

$$A = \{Jack, Jill, Bill, Max, Brian, Gale\}$$

Consider also $D = \{0, 1, 2, 3, 4\}$, the set of digits less than 5.

The universal set of A could be the set of all introductory economics students of the college, or alternatively the set of all students of the college. The universal set of D is R. The union of these two set is

$$C = A \cup D = \{Jack, Jill, Bill, Max, Brian, Gale, 0, 1, 2, 3, 4\}$$

But what is the universal set of C? Here elements of C belong to two different universal sets. To avoid situations like this we should modify the definition of union of two sets in (3.1) to

$$C = A \cup B = \{x \in U \mid x \in A \quad \text{or} \quad x \in B\} \tag{3.2}$$

The modification here is that A and B must belong to the same universal set.

If we desire union of more than two sets, like $A \cup B \cup C$, we must first find $D = A \cup B$ and then find $D \cup C$. (or alternatively $D = B \cup C$ and then $D \cup A$ or $D = A \cup C$ then $D \cup B$)

The second basic operation on sets is *intersection* of two sets. Here instead of combining elements of sets, we take only their common elements. The intersection of two sets A and B is denoted by $A \cap B$ and is defined by a new set C as

$$C = A \cap B = \{x \in U \mid x \in A \quad \text{and} \quad x \in B\} \tag{3.3}$$

The intersection of A and B contains elements that are common in both A and B. As an example, let $A = \{0, 1, 6, 10, 12, 15\}$ and $B = \{2, 3, 6, 10\}$ then their intersection $C = A \cap B = \{6, 10\}$.

If A and B have no element in common—that is if they are disjoint—their intersection is an empty set.

The third basic set operation is a set's *complement*. Given universal set U and a set A, a subset of U, the complement of A, denoted by A' is defined as $A' = \{x \in U \mid x \notin A\}$.

For example, if U is the set of digits, then the complement of $A = \{2, 3, 4\}$ is $A' = \{0, 1, 5, 6, 7, 8, 9\}$.

If the universe U is R, and $A = \{x \in R \mid x < 0\}$ is the set of negative real numbers, then its complement $A' = \{x \in R \mid x \geq 0\}$ is the set of non-negative real numbers.

If $A = \{x \in R \mid x > 0\}$ is the set of positive real numbers, then $A' = \{x \in R \mid x \leq 0\}$ is the set of non-positive real numbers.

It should be clear that $(A')' = A$—that is, the complement of a complement is the original set. Also, it is easy to see that $U' = \emptyset$ and $\emptyset' = U$. Because of $\emptyset' = U$ we were forced to include the restriction in (3.3) that in the intersection of two sets both sets must belong to a common universal set. If A and B belong to two different universal sets, their intersection is invariably a null set. But inconsistency arises when one considers the complement of this null set: it must be the universal set, but which one? This inconsistency is avoided when both sets are subsets of the same U.

The English mathematician John Venn (1834–1923) introduced the systematic use of certain graphs, which are now called Venn diagrams, as visual aids in the study of sets and set operations. Figure 3.1 below shows some examples of Venn Diagrams :

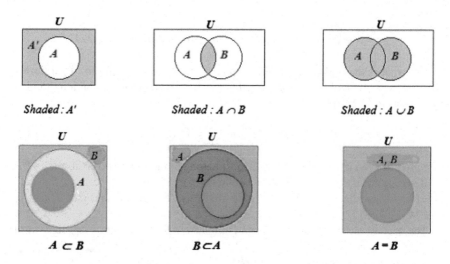

Fig. 3.1 A sample of Venn diagrams

An *n-way partition* of a set A is a collection of n nonempty and mutually disjoint subsets of A whose union is A. To illustrate this concept let $A = \{3, 6, 7, 10, 12\}$. We form three subsets of A as $A_1 = \{3, 6\}$, $A_2 = \{7, 10\}$, and $A_3 = \{12\}$. These sets are pairwise disjoint and their union is the set A. This is a *3-way partition* of A.

In an *n-way* partition of a set A we must generate n subsets of A satisfying the two stated conditions. Let's denote these subsets by $A_1, A_2, \ldots, A_n$ or more compactly by A_i for $i = 1, 2, \ldots, n$. For these sets to be a partition they must satisfy the conditions (1) of being mutually disjoint and (2) their union must be the set A. We can write these conditions as

(1) $A_i \cap A_j = \emptyset \quad \forall\, i, j;\; i \neq j, \quad i = 1, 2, \ldots, n\,;\; j = 1, 2, \ldots, n$

(2) $A_1 \cup A_2 \cup A_3 \cup \ldots \cup A_n = A$, or using the shorthand $\displaystyle\bigcup_{i=1}^{n} A_i = A$

In the shorthand expression in condition (2) $\bigcup$ represents "union notation" in the same way $\sum$ represents "summation notation".

Examples

1. Assume the following sets are subsets of Z:

$$A = \{1, 5, 7, 9, 11\} \quad B = \{3, 4, 5, 7, 12\} \quad C = \{-4, 1, 11, 12\}$$

Then

(a) $A \cup B = \{1, 3, 4, 5, 7, 9, 11, 12\}$

(b) $A \cap B = \{5, 7\}$

(c) $A \cup C = \{-4, 1, 5, 7, 9, 11, 12\}$

(d) $A \cup B \cup C = (A \cup B) \cup C = \{1, 3, 4, 5, 7, 9, 11, 12\} \cup \{-4, 1, 11, 12\}$
$= \{-4, 1, 3, 4, 5, 7, 9, 11, 12\}$

(e) $A \cap B \cap C = \{5, 7\} \cap \{-4, 1, 11, 12\} = \emptyset$

(f) $(A \cup B) \cap C = \{1, 3, 4, 5, 7, 9, 11, 12\} \cap \{-4, 1, 11, 12\} = \{1, 11, 12\}$

2. What is A's complement? It is impossible to enumerate A', but we can specify it as

$$A' = \{a \in Z \mid a \notin A\}$$

3. If $A = \{x \in Z \mid x \le 5\}$ and $B = \{x \in Z \mid x > 0\}$ we can enumerate the elements of $A \cap B$ as

$$A \cap B = \{1, 2, 3, 4, 5\}$$

Note that $A \cup B = Z$

4. If $S = \{\{1\}, \{2, 3\}, \{0\}\}$ and $T = \{\{2, 3, 8\}, \{1\}, \{6\}\}$ then

$$S \cap T = \{\{1\}\} \quad \text{and} \quad S \cup T = \{\{0\}, \{1\}, \{6\}, \{2, 3\}, \{2, 3, 8\}\}$$

Note that the elements of sets S and T are themselves sets.

3.4 Cartesian Product

The *Cartesian product* or *direct product* of two sets A and B, denoted $A \times B$ (or sometimes with the more exotic notation $A \otimes B$), is defined as

$$A \times B = \{(a, b) \mid a \in A \text{ and } b \in B\} \qquad (3.4)$$

If $A = \{x, y\}$ and $B = \{5, 10, 12\}$ then

$$A \times B = \{x, y\} \times \{5, 10, 12\}$$
$$= \{(x, 5), (x, 10), (x, 12), (y, 5), (y, 10), (y, 12)\}$$

The Cartesian product of A and B is formed by paring elements of A with elements of B.[1] The elements of the set $A \times B$ are these pairs enclosed in parentheses, with

---

[1] Notice there is no restriction that A and B belong to the same U. They could come from two totally different universes.

elements of A in the first position and elements of B in the second position. In other words, they are *ordered pairs* where the order of elements in the pairs is determined by the way the Cartesian product is formed. If instead of the Cartesian product A by B in (3.4) we form the Cartesian product of B by A, then

$$B \times A = \{(b, a) \mid b \in B \text{ and } a \in A\}$$

which generates a completely different set of ordered pairs. From the above example,

$$\begin{aligned} B \times A &= \{5, 10, 12\} \times \{x, y\} \\ &= \{(5, x), (5, y), (10, x), (10, y), (12, x), (12, y)\} \end{aligned}$$

Given that in general $(a, b) \neq (b, a)$ unless $a = b$, $A \times B \neq B \times A$ unless $A = B$. That is, the Cartesian products don't commute.

The classic example of a Cartesian product is the standard deck of playing cards. The Cartesian product of the 13-element set of standard playing card ranks

$$\{Ace, King, Queen, Jack, 10, 9, 8, 7, 6, 5, 4, 3, 2\}$$

and the 4-element set of card suits $\{\spadesuit, \heartsuit, \diamondsuit, \clubsuit\}$ generates the 52-element set of playing cards

$$\{(Ace, \spadesuit), (King, \spadesuit), \ldots, (2, \spadesuit), (Ace, \heartsuit), \ldots, (3, \clubsuit), (2, \clubsuit)\}$$

This Cartesian product has 52 elements which is the product of 13 and 4.

The notion of direct product of sets can be easily extended to more than two sets. The Cartesian product of three sets A, B, and C is

$$A \times B \times C = \{(a, b, c) \mid a \in A \text{ and } b \in B \text{ and } c \in C\}$$

where elements of $A \times B \times C$ are *ordered triples*. If the number of elements (cardinality) of A, B, and C are l, m, and n respectively, then the number of elements in their direct product set is lmn.

The Cartesian product of set A by itself, $A \times A$, is called the *2-fold Cartesian product* of A and is simply expressed as A^2. In the same manner the *k-fold* Cartesian product of A is A^k. If D is the set of decimal-digits, then D^2 is the set of 2-digit natural numbers

$$\begin{aligned} D^2 &= \{0, 1, \ldots, 9\} \times \{0, 1, \ldots, 9\} \\ &= \{(00), (01), \ldots, (10), (11), \ldots, (20), \ldots, (99)\} \end{aligned}$$

(Here we removed "," between elements in a pair.) By extension, D^3 is the set of 3-digit numbers, D^4 the set of 4-digit numbers and D^n the set of n-digit numbers in N. Notice that the cardinality of D is 10, so the cardinality of D^2 is $10 * 10 = 100$

(there are 100 numbers from 0 to 99). the cardinality of D^3 is $10 * 10 * 10 = 1,000$ (there are 1,000 numbers from 0 to 999), and so on. Incidentally if we replace D with B, the set of *binary digits* $\{0, 1\}$, then B^2 would be $\{0, 1\} \times \{0, 1\} = \{00, 01, 10, 11\}$ where "10" and "11" are the binary representations of 2 and 3, respectively.

Another application of a k-fold Cartesian product is in statistics and probability. An experiment is generally defined as the process of conducting a test whose outcomes cannot be determined in advance. The set of all possible outcomes of an experiment is called the *sample space* of that experiment and is commonly denoted by S. Flipping a coin is an example of an experiment. This experiment has two possible outcomes, heads (H) or tails (T). The sample space of flipping a coin once is $S = \{H, T\}$.

What is the sample space of tossing a coin twice? Here the experiment consists of flipping a coin twice, or conducting two trials. The outcomes of this experiment are the elements of the 2-fold direct product of S:

$$S^2 = \{H, T\} \times \{H, T\} = \{(H, H), (H, T), (T, H), (T, T)\}$$

Similarly, the sample space of flipping a coin three times is S^3.

$$S^3 = S^2 \times S = \{(H, H), (H, T), (T, H), (T, T)\} \times \{H, T\}$$
$$= \{(H, H, H), (H, H, T), (H, T, H), (H, T, T), (T, H, H), (T, H, T), (T, T, H), (T, T, T)\}$$

Notice that S contains 2, S^2 contains $2^2 = 4$, and S^3 contains $2^3 = 8$ elements. In general, if we flip a coin n times the sample space will have 2^n elements.

The most interesting and important application of the k-fold Cartesian product of a set arises when the set is R, the set of real numbers. When $k = 1$ then $R^k = R$. A geometric representation of R signifies a *one-dimensional space*. Here we associate R with a *line*.

The real number system can be visualized as a horizontal line. A point on the line is designated as the origin. The line extends to the left and to the right of the origin forever (to infinity). Associated with the line is a unit of length for measuring the distance of points on the line from the origin. Since the distance of the origin from itself is zero, the origin corresponds to the number 0. A point x units away from the origin to the right, corresponds to the positive real number x. A negative real number $-x$ corresponds to point x units to the left of the origin. All of this is illustrated in Fig. 3.2.

Now let's consider the case of the k-fold Cartesian product when $k = 2$. Here we have R^2, the set of all possible ordered pairs of real numbers.

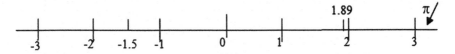

Fig. 3.2 The real number line

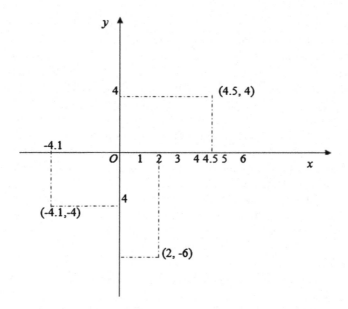

Fig. 3.3 A sample of ordered pairs in R^2

$$R^2 = R \times R = \{(x, y) \mid x, y \in R\}$$

The geometric representation of R^2 is a *two-dimensional space* or a *plane*. Each ordered pair in R^2 corresponds to a point on this plane. We next introduce the Cartesian or rectangular coordinate system on the plane. This consists of a horizontal real line, called the X-axis, and a vertical real line perpendicular to the X-axis, called the Y-axis. The intersection of X and Y is designated as the origin of the coordinate system. Each ordered pair (x, y) in R^2 is represented by a point on the plane x units away from the origin on the X axis (the first dimension) and y units away from the origin on the Y axis (the second dimension). Since the origin is 0 units away from itself in both cases, it represents the ordered pair (0,0). In Fig. 3.3 a number of points with their coordinates are graphed.

When $k = 3$ the 3-fold Cartesian product of R is R^3. Elements of R^3 are ordered triples. The geometric presentation of R^3 is the set of all points in a *three-dimensional space*, with each triple giving the coordinates of the point in a 3-dimensional Cartesian coordinate system or cubic coordinate system. This idea could be extended beyond $k = 3$ to $k = 4$, $k = 5$, and in general $k = n$. The n-fold Cartesian product of R is R^n. Elements of R^n are ordered n-tuples with a geometric presentation in n-*dimensional space*.

3.5 Relation, Correspondence, and Function

Consider two sets A and B. A *relation* between A and B is any subset of $(A \times B) \cup (B \times A)$. Of course $A \times B$ and $B \times A$ are sets of ordered pairs $\{(a, b) \mid a \in A \text{ and } b \in B\}$ and $\{(b, a) \mid a \in A \text{ and } b \in B\}$. As an example of relations, consider a family of four. Let A denote this family as $A = \{father, mother, son, daughter\}$. $A \times A$ or A^2 has 16 elements. Let F be a subset of A^2 such that

$$F = \{(father, mother), (father, son), (father, daughter), (mother, son),$$
$$(mother, daughter), (son, daughter)\}$$

F is a family relation.

Now consider the telephone directory of your area. It is indeed a set of ordered pairs; a subset of the Cartesian product of a set of names, call it set A, and set of 7-digit numbers, denote it by B. Select the name of a person (an element) from A, say Mr. Smith. We want to find Mr. Smith's telephone number from the phone directory. If Mr. Smith does not have a telephone, or has an unlisted phone, his name will not appear in the directory. Otherwise we can find his phone number in the directory. Mr. Smith may have more than one telephone number, so his name and his phone numbers may appear more than once in the directory. This is an example of a *correspondence*. In general, given two sets A and B, a correspondence g from A to B, denoted $g : A \to B$, is defined as a subset of $A \times B$. Set A is the *domain* and set B is the *co-domain* of the correspondence g.

Consider a very simple example:

$$A = \{\text{Smith, Doe, Ray}\} \quad \text{and}$$
$$B = \{721 - 9018, 734 - 8941, 775 - 0124, 768 - 4321\}$$

The telephone directory,

$$TD = \{(\text{Smith}, 734 - 8941), (\text{Smith}, 768 - 4321), (\text{Ray}, 775 - 0124)\}$$

which is a subset of $A \times B$, is a correspondence from A to B. In real world TD is presented as a table, like

Name	Number
Smith	734–8941
Smith	768–4321
Ray	775–0124

Notice that the element Doe $\in A$ does not appear anywhere in the pairs of TD, nor does 721–9018 $\in B$. "Doe" does not have a corresponding element in B and 721–9018 does not correspond to any element in A. The set of elements from co-domain

Y which *participate* in g is called the *range* of g. In our telephone directory example the range is $\{734 - 8941, 768 - 4321, 775 - 0124\}$.

If every x in X has only one corresponding element in Y, then the correspondence g is a *function* or *mapping*. In general, given two sets X and Y, *a function or mapping form X to Y* is a rule that associates each element of X to *one and only one* element of Y. The functional relation between X and Y is expressed as

$$f : X \to Y \qquad \text{or alternatively} \qquad y = f(x)$$

The terminology is f *maps x to y*, or *x maps to y under f*.[2] We also say that y is an *image* of x, or *y corresponds* to x. The function $f : X \to Y$ leads to a set of ordered pairs (x, y).

A "rule" in a function or a correspondence which describes how elements in X are mapped to elements in Y could be a formula or mathematical expression, like $y = f(x) = x^2 + 2$. But readers should be aware that a "formula" is only one way of expressing a function or correspondence. There are functions and correspondences that are impossible to be expressed by formulas. A mapping from a set of names to a set of Social Security numbers is one such example. So is a mapping from Bank of America 16-digit checking account numbers to the Bank's customers' names. I am sure that the Social Security Administration and management of Bank of America are not worried that somebody could pass people's name through a formula and come up with their Social Security or checking account numbers.

A function can be represented in words, in a table, by a diagram or a graph, and by an equation. A common method for visualizing a function is its graph. The graph of a function f with domain D is a set of ordered pairs $\{(x, f(x)) \mid x \in D\}$ plotted on a coordinate system, like in Fig. 3.4.

Note that in Fig. 3.4 each value of x in the domain of the function is mapped to only one value of y in the range. As the graph shows, x_1 in the domain is mapped only to y_1 in the range. If x is mapped to two or more values of y in the range, then we no longer have a function but a correspondence between x and y. In the graph in Fig. 3.5 x_1 is mapped to two values of y, y_1 and y_2, so there is not a functional but a correspondence relation between x and y.

In many areas of the sciences and social sciences, including Economics, a function $y = f(x)$ arises when values that the variable y takes are determined by values that the variable x may take. Generally, this functional relationship is captured by a formula or mathematical expression. In this context x is called the *independent variable* and y is called the *dependent variable*. x is also referred to as the *argument* and y the *value* of the function.

In our discussion of functions so far we have expressed y as a function of only one variable, x. We will adopt the terminology of calling these kinds of functions *univariate functions*. If f maps x_1 and x_2 to y, then y is a function of two variables, i.e. $y = f(x_1, x_2)$. We will refer to this function as a *bivariate function*. If f maps

---

[2] The same terminology is used for a correspondence.

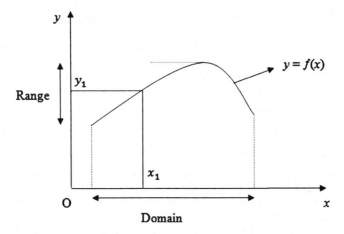

Fig. 3.4 Plot of the function $y = f(x)$

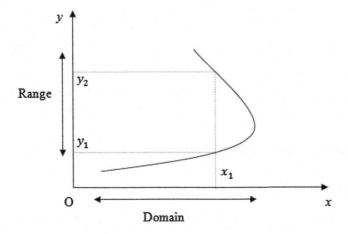

Fig. 3.5 Plot of a correspondence between x and y

three or more variables to y we will call it a *multivariate function*. Alternatively, a univariate function is a function with only one independent variable, $y = f(x)$. A bivariate function is a function with two independent variables, $y = f(x_1, x_2)$. A function with three or more independent variables, $y = f(x_1, x_2, x_3, \ldots, x_n)$, is a multivariate function.

Examples

1. If $f = \{(-4, 5), (-2, 8), (0, 5), (5, 12), (7, 2)\}$ what is the function's domain and range?
   The domain is $D = \{-4, -2, 0, 5, 7\}$ and the range is $R = \{2, 5, 8, 12\}$.

2. Let $X = \{2, 5, 6\}$ and $Y = \{4, 25, 36\}$ be two sets of real numbers. The Cartesian product of X and Y is

$X \times Y = \{(2, 4), (2, 25), (2, 36), (5, 4), (5, 25), (5, 36), (6, 4), (6, 25), (6, 36)\}$

The Cartesian product of Y and X is

$Y \times X = \{(4, 2), (4, 5), (4, 6), (25, 2), (25, 5), (25, 6), (36, 2), (36, 5), (36, 6)\}$

A subset of $X \times Y$ is $f = \{(2, 4), (5, 25), (6, 36)\}$, which is a mapping from X to Y. This mapping takes the functional form of $y = f(x) = x^2$.
A subset of $Y \times X$ is $g = \{(4, 2), (25, 5), (36, 6)\}$, which is a mapping from Y to X. This mapping takes the functional form of $x = g(y) = |\sqrt{y}|$ (each element in the range is the positive square root of an element in the domain).

3. Let $M = \{Jill, Jim, Jack\}$ be a set of names and $W = \{145, 185, 173\}$ be the set of weight of individuals in M. The Cartesian product of M and W is

$$M \times W = \{(Jill, 145), (Jill, 185), (Jill, 173), (Jim, 145), (Jim, 185),$$
$$(Jim, 173), (Jack, 145), (Jack, 185), (Jack, 173)\}$$

A subset of $M \times W$ is $g = \{(Jill, 145), (Jim, 185), (Jack, 173)\}$, which is a function from M to W associating each individual with his or her unique weight.

3.6 Forms of Univariate and Bivariate Functions

The following univariate function

$$y = a_0 + a_1 x + a_2 x^2 + a_3 x^3 + \cdots + a_n x^n \tag{3.5}$$

is called an *nth-degree polynomial function*. It is a multiple-term function with terms consisting of nonnegative integer exponents of x each multiplied by a parameter (coefficient). If n in (3.5) is zero then the polynomial would be reduced to $y = a_0 x^0 = a_0$ which is a *constant* function. The graph of this function is a line parallel to the X-axis, intersecting the Y-axis at a point a_0 units distance from the origin. Similarly, if $n = 1$, then

$$y = a_0 + a_1 x$$

is a *first-degree polynomial*, which is a *linear function*. It is called linear because the graph of the function is a line with intercept a_0 and slope a_1. If $n = 2$ then

$$y = a_0 + a_1 x + a_2 x^2$$

is a *second-degree polynomial* and is called a *quadratic function*. The graph of this function is a *parabola*. If $n = 3$ then

$$y = a_0 + a_1 x + a_2 x^2 + a_3 x^3$$

is a *third-degree polynomial* and is called a *cubic function*. If n is 4, 5, or higher the polynomial does not have a special name and is simply referred to as a 4th-degree, 5th-degree, etc polynomial.

A function of the form

$$y = \frac{g(x)}{f(x)}$$

which is the ratio of two polynomial functions of x, is called a *rational (ratio-*nal) function. A special case of a rational function is $y = \dfrac{a}{x}$ or $xy = a$. The graph of this function is a *rectangular hyperbola* with the X and Y axes as its *asymptotes*. For example, Average Fixed Cost (AFC) curve is a rectangular hyperbola because

$$AFC = \frac{FC}{Q} \quad \text{and} \quad AFC * Q = FC \qquad Q > 0$$

Functions of the form

$$y = [f(x)]^{\frac{1}{n}} = \sqrt[n]{f(x)}$$

are *root* functions. If $n = 2$ then y is a square root function. For $n = 3$, y is a cubic root function, and so on for higher roots.[3]

A bivariate function of the form

$$y = a_0 + a_1 x_1 + a_2 x_2$$

is a linear function. The graph of this function is a *plane* intersecting the vertical Y axis of the three-dimensional coordinate system at a point a_0 units distance from the origin. The intersection of this plane with the YX_1 plane of the three-dimensional coordinate system is a line with equation $y = a_0 + a_1 x_1$ and slope a_1. Similarly, the intersection of this plane with YX_2 plane of the three-dimensional coordinate system is another line with equation $y = a_0 + a_2 x_2$ and a_2.

The graph of nonlinear bivariate functions generate variety of *surfaces*.

The functions discussed so far all belong to the set of *algebraic functions*. The set of functions classified as *non-algebraic* or *transcendental* includes *logarithmic functions* such as

$$y = \log x$$

---

[3] Even roots of numbers, like the square root or forth root, have two answers ($\sqrt{4} = \pm 2$). To stick with the definition of "function", we must consider only one of the possible answers. In the text, if it is not stated otherwise, we will only consider the positive root of an even-root of a number.

exponential functions such as

$$y = c\,a^x$$

and *trigonometric functions* such as

$$y = cos3x + 2sinx^2$$

Finally, if $y = f(x)$ and $z = g(y)$ then $z = g[f(x)]$ is a *composite function* or a *composite mapping*. For example, if $y = f(x) = x^2 - 2$ and $z = g(y) = 3y - 5$ then

$$z = g[f(x)] = g(x^2 - 2) = 3(x^2 - 2) - 5 = 3x^2 - 11$$

Examples

1. If $f = \{(2, 5), (3, 6), (4, 5), (5, 6)\}$

    (a) what is the function's domain and range?
        The function's domain is $D = \{2, 3, 4, 5\}$ and its range is $R = \{5, 6\}$.
    (b) What is $f[f(2)]$?
        $f(2) = 5$ so $f[f(2)] = f(5) = 6$

2. If $f(x) = 3x^3 - 5x + 4$ what is $f(-2x)$?

$$f(-2x) = 3(-2x)^3 - 5(-2x) + 4 = -24x^3 + 10x + 4$$

3. If x and y are real numbers, determine the domain of the function

$$y = \sqrt{x^2 - 81}$$

    For y to be a real number we must have $x^2 - 81 \geq 0$. Solving $x^2 - 81 = 0$, we get

$$x^2 = 81 \quad \longrightarrow \quad x = \sqrt{81} \quad \text{or} \quad x = \pm 9$$

    Thus the domain of the function is

$$D = \{x \in R \mid x \leq -9 \quad \text{or} \quad x \geq 9\}$$

4. Find the domain of the function

$$y = \frac{1}{\sqrt{x^2 - 81}}$$

Here $x^2 - 81$ must be strictly positive, since for $x^2 - 81 \leq 0$ the function is not defined. Therefore the domain of the function must be specified as

$$D = \{x \in R \mid x < -9 \quad \text{or} \quad x > 9\}$$

5. Find the domain of the function

$$y = \sqrt{x^2 - 5x + 6}$$

For y to be a real number $x^2 - 5x + 6$ must be non-negative, that is $x^2 - 5x + 6 \geq 0$. We first solve the quadratic equation $x^2 - 5x + 6 = 0$ using the quadratic formula

$$x = \frac{-(-5) \pm \sqrt{(-5)^2 - 4(1)(6)}}{2(1)} = \frac{5 \pm \sqrt{25 - 24}}{2}$$

leading to the solutions $x = 2$ or $x = 3$. Values of $x^2 - 5x + 6$ over the interval $2 < x < 3$ are negative (see Fig. 3.6, the graph of $x^2 - 5x + 6$), therefore the domain of the function is

$$D = \{x \in R \mid x \leq 2 \quad \text{or} \quad x \geq 3\}$$

Consequently, the range of the function is

$$R = \{y \in R \mid y \geq 0\}$$

6. If the domain of the function

$$f(z) = \frac{z + 5}{z^2 + 2z + 3}$$

Fig. 3.6 Graph of $y = x^2 - 5x + 6$

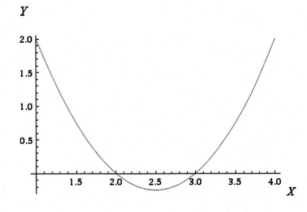

is given as
$$D = \{z \in R \mid 3 \leq z \leq 10\}$$

what is its range?

The values of the function for the lower and upper bounds of the domain, 3 and 10, are

$$f(3) = \frac{3+5}{3^3 + 2*3 + 3} = \frac{4}{9} \quad \text{and} \quad f(10) = \frac{10+5}{10^2 + 2*10 + 3} = \frac{5}{41}$$

then the range must be specified as

$$R = \{z \in R \mid \frac{5}{41} \leq z \leq \frac{4}{9}\}$$

7. Assume x and y are real numbers. Is $G = \{(x, y) \mid y \leq x - 3\}$ a function? $y \leq x - 3$ is not a function. The reason is that for each value of x there are more than one (in this case infinitely many) y values or images. If, for example, $x = 5$ then $y \leq 2$. And there are infinite values of y that satisfy this inequality, like 1, 0.5, 0, or -3. The graph of $y \leq x - 3$ consists of all the points on and below the line $y = x - 3$.

8. If $f(x) = 2x^2 - 10x + 4$ and $g(x) = 3x + 1$,

   (a) what is the composite function $f[g(x)]$?
   $$\begin{aligned} f[g(x)] &= f(3x + 1) = 2(3x + 1)^2 - 10(3x + 1) + 4 \\ &= 2(9x^2 + 6x + 1) - 30x - 10 + 4 \\ &= 18x^2 - 18x - 4 \end{aligned}$$

   (b) What is $g[f(x)]$?
   $$g[f(x)] = g(2x^2 - 10x + 4) = 3(2x^2 - 10x + 4) + 1 = 6x^2 - 30x + 13$$

   (c) What is $g[g(x)]$?
   $$g[g(x)] = 3(3x + 1) + 1 = 9x + 3 + 1 = 9x + 4$$

9. If $f(x) = \dfrac{x-1}{x+1}$

   (a) what is $f(1)$?
   $$f(1) = \frac{1-1}{1+1} = 0$$

   (b) What is $f(-1)$? $f(-1)$ is not defined
   (c) What is $f(x + 2)$?
   $$f(x + 2) = \frac{x+2-1}{x+2+1} = \frac{x+1}{x+3}$$

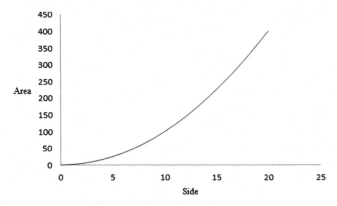

Fig. 3.7 The graph of function $A = s^2$

(d) What is $f[f(x)]$?

$$f\left[f(x)\right] = \frac{\frac{x-1}{x+1} - 1}{\frac{x-1}{x+1} + 1} = \frac{\frac{x-1-x-1}{x+1}}{\frac{x-1+x+1}{x+1}} = \frac{\frac{-2}{x+1}}{\frac{2x}{x+1}} = -\frac{1}{x}$$

10. An example of a simple univariate function is the area of a square, which depends on the length of its side. Denoting area by A and side length by s, the functional relation between A and s can be written as $A = s^2$. We can express this function verbally by "the area of a square is equal to square of its side," or graphically as in Fig. 3.7.

11. As an example of a bivariate function consider the cost C of shipping cargo by air. C depends on the weight w of the cargo and the distance d that the cargo must be shipped. This can be written *implicitly* as $C = f(w, d)$, which expresses that cost depends on weight and distance. An *explicit* functional relationship between C, w, and d may be verbally expressed as "$2 per pound (a minimum of 50 pounds to a maximum of 500 pounds) of the cargo's weight and 5 cents per mile (a minimum of 200 miles to a maximum of 3,000 miles) of distance." The explicit algebraic expression of this cost function is

$$C = f(w, d) = 2w + 0.05d$$

where the set of acceptable weights is $W = \{w \in R \mid 50 \le w \le 500\}$ and the set of acceptable distances is $D = \{d \in R \mid 200 \le d \le 3000\}$. The domain of this function is the Cartesian product $W \times D$, the set of ordered pairs of weights and distances. The range of this function is

$$f(w, d) = \{c \in R \mid \$110 \le c \le \$1150\}$$

3.7 Exercises

1. Consider the sets of real numbers

$$A = \{1, 5, 6, 11\} \quad B = \{-3, 5, 7, 11, 18, 25\} \quad C = \{-4, 1, 15, 19\}$$

   Find the following:

   (a) $A \cup B$ (b) $A \cap B$ (c) $A \cup C$ (d) $C \cup B$

   (e) $A \cup B \cup C$ (f) $A \cap B \cap C$ (g) $(A \cup B) \cap C$ (h) $A \cap C$

2. Write complement of A from Exercise 1 using specification.

3. Write the set of all subsets of C in Exercise 1.

4. Given $A = \{a, b, c, d\}$, $B = \{c, d\}$, and $C = \{a, b, e, f, c, d\}$, verify whether the following statements are valid:

   (a) B is a subset of A.
   (b) A is a subset of C.
   (c) Since B is a subset of A and A is a of subset C, then B must be a subset of C.
   (d) $C \cap A = B$
   (e) $B \cup A = A$
   (f) $B \cap A = B$
   (g) $C \not\subset A$
   (h) $C \cup A = B$

5. Assume x and y are real numbers. Which of the following is not a function?

   $$(a) \; y = x + 2 \qquad (b) \; y = \sqrt{x^2 + 1}$$
   $$(c) \; y = -x^2 + 10 \quad (d) \; y = x + 1$$

6. In general, $A \times B \neq B \times A$. Under what condition they are equal?

7. Find the Cartesian product of $X = \{3, 4, 9\}$ and $Y = \{9, 16, 81\}$. Write the function that maps X to Y.

8. Assume A is the set of names of 100 faculty members at a small college and B is the set of academic ranks a faculty member could have:

   $$B = \{Professor, Associate\ Professor, Assistant\ Professor, Lecturer\}$$

   Is a mapping from A to B a function?

9. Determine the range of the following functions for the given domain:

(a) $y = x + 2$ $D = \{x \in R \mid x = 2\}$

(b) $y = x^2 + 3x + 5$ $D = \{x \in R \mid 5 \leq x \leq 10\}$

(c) $y = x^2 + 5$ $D = \{x \in R \mid -4 \leq x \leq 10\}$

(d) $w = \sqrt{x^2 + 25}$ $D = \{x \in R \mid -4 \leq x \leq 4\}$

(e) $f(z) = \sqrt{z + 5}$ $D = \{z \in R \mid z > -3\}$

(f) $f(w) = \dfrac{w + 2}{\sqrt{w^2 - 16}}$ $D = \{w \in R \mid w \leq -6 \ \text{or} \ w \geq 6\}$

10. Assume x, y, w, and z are real numbers. Determine the domain of the following functions:

(a) $y = \sqrt{x + 5}$ (b) $y = \dfrac{1}{\sqrt{x + 5}}$

(c) $y = \sqrt{x^2 - 25}$ (d) $y = \dfrac{1}{\sqrt{x^2 - 25}}$

(e) $y = (x - 3)^{\frac{1}{2}}$ (f) $f(x) = \sqrt{x^2 - 9x + 20}$

(g) $f(w) = \dfrac{w + 2}{\sqrt{w^2 - 16}}$ (h) $y = (x^2 - 2x + 10)^{\frac{-1}{2}}$

11. Given
$$f(x) = x^2 - 2 \quad \text{and} \quad g(x) = \frac{2x + 1}{3x - 5}$$

find the following:

(a) $f(3)$ (b) $g(3)$ (c) $f(-3)$ (d) $g(-3)$

(e) $f(x + 2)$ (f) $g(x + 2)$ (g) $f(\sqrt{x})$ (h) $g(x^2 + 1)$

(i) $f[f(x)]$ (j) $g[g(x)]$ (k) $f[g(x)]$ (l) $g[f(x)]$

12. Is $g(\frac{5}{3})$ in Exercise 11 defined?

13. Enumerate the elements of the following sets:

(a) $\{x \in Z \mid x < 5\} \cap \{x \in Z \mid x > 0\}$

(b) $\{a \in Z \mid a \geq 10\} \cap \{a \in Z \mid a < 15\}$

(c) $\{y \in N \mid 1 < y < 9\}$

(d) $\{x \in Z \mid |x| < 5\}$

(e) $\{x \in N \mid \sqrt{x} < 10\}$

14. Describe the following sets:

    (a) $\{y \in N \mid 1 < y < 9\}$

    (b) $\{y \in R \mid -4 \leq y < 9\}$

    (c) $\{w \in R \mid w > 0\}$

    (d) $\{x \in R \mid |x| > 10\}$

15. Write the following sets using specification:

    (a) Set of integers less than 10.

    (b) Set of positive integer numbers

    (c) Set of real numbers between -15 and 20

    (d) Set of real numbers square of which is greater than 50.

    (e) Set of real numbers with square roots less than 200.

    (f) Set of natural numbers with square roots less than 10

16. Graph the functions

    (a) $y = 3x + 2$ (b) $y = 10 - 2x$ (c) $y = -5 + 3x$

    Find the coordinate of point of intersection between (a) and (b). Why can't we do the same for (a) and (c)?

Chapter 4
Market Equilibrium Model

4.1 Market Demand and Aggregation Problem

A household's demand for a good or service expresses the quantities of that good or service that the household is willing and able to purchase as a function of a number of variables. These variables include the price of the commodity P, the household disposable income DI, the size of the household S, the prices of complementary and substitute goods P_c and P_s, the expectation of all things relevant to the household's income and the price of the commodity in the future E, and the taste of the household T. Taste is a broad generic name for all cultural and social factors that may influence a household's consumption decision. There are other variables that could possibly influence a household's purchase decision (like color of an automobile or smell of a perfume), but they are mostly household or product specific.

Assuming that there are m households, the ith household's demand for a good Q_i^d can be implicitly expressed as

$$Q_i^d = f_i \ (P, DI, S, Pc, Ps) \qquad\qquad i = 1, 2, \ldots, m \qquad (4.1)$$

A *parsimonious* refinement of (4.1) would be achieved by combining disposable income DI and family size S in a new variable *per capita disposable income $PCDI = DI/S$* and reformulate (2.1) as

$$Q_i^d = f_i \ (P, PCDI, Pc, Ps) \qquad\qquad i = 1, 2, \ldots, m \qquad (4.2)$$

Note that of all the variables listed in the first paragraph, E and T are not included in (4.2). They are not quantitative variables; therefore the impact of their changes on Q_i^d must be measured differently. By concentrating on the remaining quantitative variables and further assuming a linear demand function, we can write (4.2) explicitly as

$$Q_i^d = a_i - b_i P + c_i PCDI_i - d_i P_c + e_i P_s \qquad\qquad i = 1, 2, \ldots, m \qquad (4.3)$$

S. Vali, *Principles of Mathematical Economics*,
Mathematics Textbooks for Science and Engineering 3,
DOI: 10.2991/978-94-6239-036-2_4, © Atlantis Press and the authors 2014

where parameters $a_i, b_i, c_i, d_i,$ and e_i are positive real numbers.[1] As we will see, the impact of changes in a household expectation E or its taste T (or any other variable not included in the model) would be transmitted to Q_i^d through changes in the parameter a_i.

In writing (4.3), it is assumed that this good is a "normal good". If it is considered an "inferior good", then (2.3) must be expressed as

$$Q_i^d = a_i - b_i P - c_i PCDI_i - d_i P_c + e_i P_s \qquad\qquad i = 1, 2, \ldots, m$$

Note that P_c and P_s are entered in (4.3) with negative and positive coefficients. This is in conformity with the standard demand analysis that an increase in price of a complementary good should reduce demand for the good while an increase in price of a substitute good generally leads to higher demand for the good. Also note that c_i, the coefficient of $PCDI$, should be interpreted as the household's per capita *marginal propensity to consume*.

Let us assume some numerical values for the parameters in (4.3) and write it as

$$Q_i^d = 20 - 2P + 0.001 PCDI_i - 0.8P_c + 1.2P_s \qquad\qquad (4.4)$$

The graph of this function, or the household's demand curve for this good, is a hyper-plane in a 5-dimensional space! Lets reduce the complexity of the problem by further assuming that $P_c = \$5$ and $P_s = \$10$. We also assume that the household disposable income is \$40,000 a year[2] and $S = 4$, so $PCDI = \dfrac{DI}{S} = 10000$. With these additional assumptions, (4.4) is reduced to

$$Q_i^d = 20 - 2P + 0.001(10000) - 0.8(5) + 1.2(10)$$
$$Q_i^d = 20 - 2P + 10 - 4 + 12$$
$$Q_i^d = 38 - 2P \qquad\qquad (4.5)$$

which is the more familiar demand equation.

Equation (4.3) is a *behavioral equation*. It explains changes in the ith household demand in response to changes in $PCDI$, P_c or P_s and changes in quantity demanded in response to changes in P. If, for example, this household's income increases to \$44,000 a year *ceteris paribus*—that is, with no change in S, P_c, or P_s and the parameters of the function—the new demand equation for this household would be

$$Q_i^d = 20 - 2P + 0.001(11000) - 0.8(5) + 1.2(10)$$
$$Q_i^d = 39 - 2P \qquad\qquad (4.6)$$

---

[1] In the next section we will discuss restrictions on parameters of both supply and demand function.

[2] Expressing the household's income annually immediately establishes the fact that quantity demanded is for one year period.

Fig. 4.1 Shift of demand curve due to rise in income

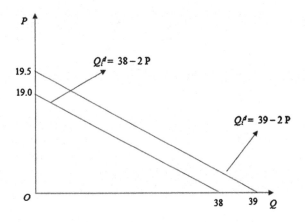

Equation (4.6) has the same slope as Eq. (4.5), so its graph is parallel to (4.5), but its Y-intercept has increased by one unit to 39. Figure 4.1 presents the graph of these functions.

Similarly any change in S, P_c, or P_s ceteris paribus would lead to a new demand equation with the same slope, but a different constant term or intercept. Any Change in a household's expectation E, taste T, or other qualitative variable(s) not included in the model, would similarly leads to a change in the intercept, manifested by parallel shifts in the demand curve.

By summing (4.2) over m households we arrive at the *market aggregate demand*, which is the market or industry demand equation. But this reasonably innocent process may, unfortunately, lead to what is known in economics as the *Aggregation problem*. Lawrence Klein, winner of 1980 Nobel Prize in economics, is generally credited with raising this issue in a formalized manner in late 1940s, while the noted economists Henry Theil further expanded the boundary of the debate.[3]

The aggregation problem generally arises when economists try to arrive at a macro-relation from aggregating a series of micro-relations. A typical example is deriving a market demand for a certain good, a macro-relation, from a series of individual household demands for that good, micro-relations. Consider m households or consumers with the following simple linear demand functions for a good

$$Q_i^d = a_i - b_i P \qquad\qquad i = 1, 2, \ldots, m$$

where a_i and b_i are the parameters of the demand equations that differ for different consumers. Consumers' demand functions are micro-relations. By aggregating over individual consumers we arrive at the market demand function, a macro-relation.

[3] L.R. Klein, "Macroeconomics and the Theory of Rational Behavior", *Econometrica*, 14, 1946a.
L.R. Klein, "Remarks on the Theory of Aggregation" *Econometrica*, 14, 1946b.
H. Theil, *Linear Aggregation of Economic Relations*, North-Holland, Amsterdam, 1954.

$$Q_d = \sum_{i=1}^{m} Q_i^d = \sum_{i=1}^{m}(a_i - b_i P) = \sum_{i=1}^{m} a_i - P \sum_{i=1}^{m} b_i$$

Denoting $\sum_{i=1}^{m} a_i$ by a and $\sum_{i=1}^{m} b_i$ by b we write the market demand equation as

$$Q_d = a - bP$$

The following simple numerical example demonstrates the pitfall of deriving a macro-relation from a set of micro-relations through the process of aggregation. Assume that there are only four consumers, $m = 4$, with the following demand functions

$$Q_1 = 50 - 6P \qquad\qquad\qquad Q_2 = 75 - 5P$$
$$Q_3 = 32 - 10P \qquad\qquad\qquad Q_4 = 110 - 9P$$

The aggregated market demand function is

$$Q_d = 267 - 30P \qquad\qquad\qquad\qquad\qquad\qquad (I)$$

Table 4.1 shows each of the consumers' quantity demanded at various prices, their sum as the market quantity demanded, and the market quantity demanded using the market demand equation (I).

For prices 2 and 3 the sum of the micro-variables quantity demanded by four consumers is equal to the macro-variable market demand. But when the price is 5 there is an 18 unit difference. The difference is even larger when the price is 10.

There is a simple reason for these discrepancies. At $P = 5$ the third household is priced out of the market and subsequently Q_3 is zero. But the result of a purely "mathematical" summation is that the same household "contributes" *negative* 18 units to the aggregated pool of market demand! At the price of 10, the first and third consumers are priced out of the market, but their "algebraic" contribution to the market quantity demanded is -45 units.

In the absence of price discrimination, all consumers face the same price in the market. But when variables such as household income and size, which vary from one household to another, are included in the consumers' demand functions, things

Table 4.1 Aggregation problem

P	Q_1	Q_2	Q_3	Q_4	Q_d as sum of individual demand	Q_d from aggregated demand equation (I)
2	38	65	12	92	207	207
3	32	60	2	83	177	177
5	20	50	0	65	135	117
10	0	25	0	20	45	0

get more complicated. As a simple example, consider m consumers with m linear demand functions relating the ith consumer's demand for a good to its price and the consumer's income

$$Q_i^d = a_i - b_i P + c_i DI_i \qquad\qquad i = 1, 2, \ldots, m$$

Aggregating over all consumers, we have

$$Q_d = \sum_{i=1}^{m} Q_i^d = \sum_{i=1}^{m}(a_i - b_i P + c_i DI_i) = \sum_{i=1}^{m} a_i - P \sum_{i=1}^{m} b_i + \sum_{i=1}^{m} c_i DI_i$$

Here the added problem is how to aggregate m micro-variables for the consumer's income into one macro-variable for total income and arrive at a market demand function of the form

$$Q_d = a - bP + cPCDI$$

A possible way out of the log-jam of aggregation is to assume households have basically the same propensity to consume c_i and denote it by c, that is $c_i = c \ \forall \ i$. c can be treated as an average of the propensity to consume of households in different income groups. With this assumption we can sum over m households in (4.2) and arrive at the market demand function Q_d.

$$Q_d = \sum_{i=1}^{m} Q_i^d = \sum_{i=1}^{m}(a_i - b_i P + cPCDI_i - d_i P_c + e_i P_s)$$

$$Q_d = \sum_{i=1}^{m} a_i - P \sum_{i=1}^{m} b_i + c \sum_{i=1}^{m} PCDI_i - P_c \sum_{i=1}^{m} d_i + P_s \sum_{i=1}^{m} e_i$$

$$Q_d = a - bP + c\,TPCDI - dP_c + eP_s \qquad\qquad (4.7)$$

where

$$a = \sum_{i=1}^{m} a_i \qquad\qquad b = \sum_{i=1}^{m} b_i \qquad TPCDI = \sum_{i=1}^{m} PCDI_i$$

$$d = \sum_{i=1}^{m} d_i \qquad\qquad e = \sum_{i=1}^{m} e_i$$

and *TPCDI* is Total Per Capita Disposable Income.

Equation (4.7) describes the collective behavior of consumers in the market. To get a better picture of the process, let's assume that 1000 households with the following identical demand equation[4] populate the demand side of the market for a commodity

---

[4] This is another nice trick to get rid of the aggregation problem altogether. The downside is, of course, an *I-robot* society!

$$Q_d = 2 - 0.4P + 0.009PCDI_i - 0.05P_c + 0.12P_s \qquad i = 1, 2, \ldots, 1000$$

We arrive at the market demand equation by aggregating the households' demands, i.e.

$$Q_d = \sum_{i=1}^{1000} Q_i^d = \sum_{i=1}^{1000} (2 - 0.4P + 0.009PCDI_i - 0.05P_c + 0.12P_s)$$

$$Q_d = 2000 - 400P + 0.009TPCDI - 50P_c + 120P_s \qquad (4.8)$$

Lets further assume that the households' per capita disposable income *PCDI* is, on the average, \$10,000 and the prices of complement and substitute goods are \$10 and \$20, respectively. With these assumptions, the market demand equation for the commodity is

$$Q_d = 2000 - 400P + 0.009 * (1000 * 10000) - 50 * 10 + 120 * 20$$
$$Q_d = 93900 - 400P$$

As another example assume the monthly demand for hamburger sold by a fast food chain in a mid-size city is given by

$$Q_d = 22000 - 5500P + 0.06POP + 0.05DI + 0.4Adv$$

where P is the price of hamburger, POP is the population of the city, DI is the household residents average annual disposable income, and Adv is the fast food chain's monthly expenditure on advertising. If, for example, the population of a city is 300000, the household DI is \$50,000, and the chain spends \$20,000 monthly on advertising, the demand equation will be

$$Q_d = 22000 - 5500P + 0.06(300000) + 0.05(50000)$$
$$+ 0.4(20000) = 50500 - 5500P$$

It is easy to verify that if the chain charges \$3 for a hamburger, the monthly quantity demanded would be 34000 hamburgers.

To see a practical application of demand analysis consider the following empirical study of demand for cable TV conducted in 1979–1980, when cable TV was fairly new.[5] After a study of 40 top television market, the following demand equation was estimated

$$Sub = -6.808 + 0.406 \, Home - 0.526 \, Inst + 2.039 \, Svc + 0.757 \, TV$$
$$+ 1.194 \, Age - 5.111 \, Air + 0.0017 \, Y$$

where the variables are

[5] Taken from Ramu Ramanathan *Introductory Econometrics with Application*, 1992, second edition, the Dryden Press.

- Sub = Number of subscribers
- Home = Number of homes (in thousand)
- Inst = Installation fee
- Svc = Monthly service charge
- TV = Number of stations carried by each cable system
- Age = Age of the system in years
- Air = Number of TV signals received without cable
- Y = Per capita income in the area.

4.1.1 Market Supply

On the supply side, a supply function relates the quantity of output supplied by the ith firm (assuming there are n suppliers $i = 1, 2, 3, \ldots, n$) Q_i^s over a specific period of time to the price of the product in the market P and a number of other factors. Notable among these factors are the cost of production, which in turn depends on prices of factor inputs (raw materials, energy, capital, labor, ...) and technology, prices of related products, rate of return of alternative activities, and number of competitors. To make it manageable lets express the firm's supply equation as a linear function of price of the product P and prices of labor P_L and energy P_E

$$Q_i^s = -h_i + j_i P - k_i P_L - l_i P_E \qquad\qquad i = 1, 2, \ldots, n \qquad (4.9)$$

where parameters h_i, j_i, k_i, and l_i are positive real numbers. Equation (4.9) is also a _behavioral equation_. It describes the behavior of the firm in the market; that is how the firm changes its output in response to changes in the price of the good and/or the prices of inputs. Assuming some numerical values for parameters of the function, we express (4.9) as

$$Q_i^s = -50 + 10P - 0.5P_L - 0.2P_E$$

By further assuming that the _wage rate_ (price of labor) is \$10 an hour and the price of energy, measured by the price of a barrel of oil, is \$70, we can simplify the supply equation to

$$Q_i^s = -50 + 10P - 0.5(10) - 0.2(70) = -69 + 10P \qquad (4.10)$$

At the price of \$10 per unit, this producer is willing to offer 31 units of output to the market. If, due to an increase in demand for oil by India and China, the price of a barrel of oil jumps to \$140, the supply equation of this firm will change to

$$Q_i^s = -50 + 10P - 0.5(10) - 0.2(140) = -83 + 10P$$

Under the new circumstance, at the market price of \$10 per unit this firm will only produce 17 units of output. Figure 4.2 presents both supply curves.

The parallel leftward shift in the supply curve of the firm in Fig. 4.2 is due to an increase in production cost. Because of higher energy costs this firm, with its current *working capital*, would not be able to hire the same amount of labor and buy the same amount of energy.

By summing (4.9) over n firms we get the market supply function.

$$Q_s = \sum_{i=1}^{n} Q_i^s = \sum_{i=1}^{n} (-h_i + j_i P - k_i P_L - l_i P_E)$$

$$Q_s = -h + jP - kP_L - lP_E \qquad (4.11)$$

where $h = \sum_{i=1}^{n} h_i \quad j = \sum_{i=1}^{n} j_i \quad k = \sum_{i=1}^{n} k_i \quad l = \sum_{i=1}^{n} l_i$.

Equation (4.11) describes the collective behavior of suppliers in the market. This equation explains how the quantity supplied changes in response to changes in the price of the good and/or the prices of inputs. A numerical example would be

$$Q_s = -5000 + 800P - 50P_L - 8P_E \qquad (4.12)$$

As another example of supply function consider supply of an agricultural commodity (in bushel per year)as function of its price P, hourly wage or price of farm workers P_L, price or hourly rental cost of capital (farm machinery) P_K, and the amount of rain fall (in inches) in the growing season RF

$$Q_s = -100000 + 200000P - 3500P_L - 2500P_K + 3000RF$$

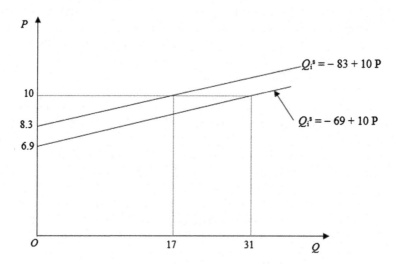

Fig. 4.2 Shift of supply curve due to increase in input price

If we assume that wage of farm workers is \$7 an hour, rental cost of farm machinery is \$50 an hour, and the amount of rainfall in the growing season is 18 inches, then the supply function will be

$$Q_s = -195500 + 200000P$$

(Note that RF must be within certain range. Too much rain may damage the crop and a flood totally destroys it).

4.2 Market Equilibrium

To complete the model we need to describe the requirement(s) for *market equilibrium* or the *equilibrium condition(s)*. The standard condition for reaching equilibrium in a single and isolated market is that the *excess demand* and *excess supply* in the market are eliminated; that is, the market *clears*. Market equilibrium is established at the price where the quantity of goods supplied equals the quantity of goods demanded. Thus the equilibrium condition can be expressed as

$$Q_d = Q_s = Q^* \quad \text{or} \quad |Q_d - Q_s| = 0 \tag{4.13}$$

Taken together, Eqs. (4.7), (4.11) and (4.13) are called a *structural model*; they describe the structure of the market. Notice that (4.13) describes a condition for a *competitive* market equilibrium. If there are sufficiently large numbers of suppliers and consumers in this market, no barriers to entry and exit, and all market participants have access to full information, then the market would reach a market clearing equilibrium.[6] Otherwise, the market could reach an equilibrium in which excess supply and demand are not fully eliminated.

Our supply and demand model for a single market would consist of

$$Q_d = a - bP + cTPCDI - dP_c + eP_s$$
$$Q_s = -h + jP - kP_L - lP_E$$
$$Q_d = Q_s = Q^* \tag{4.14}$$

Models are simplified representations of complex systems which we hope to understand. Sometimes models consist of computer programs, like a weather model that simulates changes in weather patterns and provides a weather forecast. The National Hurricane Center of the National Weather Service has several computer models for tracking and projecting the paths of hurricanes. Sometimes models are small scale versions of an underlying process or a real product that are more accessible to observation and experiment, like the model airplanes that designers test in wind

---

[6] In addition, for market equilibrium to be *socially optimal*, there should be no *externalities*.

tunnels.[7] Economic models mostly consists of systems of equations and identities; simplified anatomy designed to study the real world complex economic processes.

The above system (4.14) is an example of a simple economic model. The main objective of constructing this model is to find the market equilibrium price and quantity for a commodity. These two variables, whose solution we seek from the model, are called *choice* or *endogenous variables* (originating from within). The values of the endogenous variables are determined by the model. There are five additional variables and nine parameters in this model. In order to be able to solve for the two endogenous variables we must have the values of these variables and parameters. Variables whose magnitudes are not determined by the model, but supplied to the model as data, are called the *exogenous variables* (originating from without). The values of exogenous variables are "input" in our model. They could be, of course, "output" of another model in which they are entered as endogenous variables.

If a model seeks an empirical or numerical solution for its endogenous variables, the numerical values of its parameters, referred to as *coefficients*, must be known. A *parameter* is a hybrid of a "constant" and a "variable". A constant never changes. A variable is in a permanent state of flux. A parameter is a variable that changes very slowly and gradually over a long period of time, to the extent that within a certain time frame it could be considered a constant. Researchers who construct the model must, of course, qualify "slow and gradual" change and "long period of time" and specify the "time frame". Several examples may help to understand the delicate nature of a parameter.

In Eq. (4.7) parameter c, attached to the total per capita disposable income *TPCDI* measures the average impact of a dollar change in *TPCDI* on Q_d. It is the marginal propensity to consume in the aggregate. In the numerical example (4.8), c is given as 0.009. We refer to this numerical value of the parameter as the *coefficient*; that is, the coefficient of *TPCDI* is 0.009. The value 0.009 signifies the fact that an increase of \$1000 in *TPCDI* would, on the average, lead to a 9 unit increase in quantity demanded. Coefficients, as numerical estimates of a model's parameters, are generally products of extensive statistical analysis and studies conducted by *Statisticians* or *Econometricians*. As long as the underlying mechanism relating variables of a model remains the same, there is no reason to expect significant change in the estimated values of the parameters. In other words, only structural changes in a system will lead to changes in the system's parameters. In our example, 0.009 embodies the budget allocation and consumption "culture" of households in this market. These things can change, but gradually and over a relatively long period of time. So long as there is no new statistical evidence that c is different from 0.009, we can safely continue using 0.009 as the coefficient of *TPCDI*.

A similar example from Macroeconomics is the estimate of the *average aggregate marginal propensity to consume MPC*. A standard consumption function in a Keynesian macroeconomic model relates the level of consumption C linearly to disposable income Y_D, $C = C_0 + bY_D$. Ignoring technical complications related to estimating b—which is the *MPC*—it suffices to say that in the last 70 years there

---

[7] Paul Krugman, *The Return of Depression Economics and the Crisis of 2008*, p. 18.

have been various estimates of this parameter. Using late 1920s and early 1930s data economist Robert Rosa estimated b to be 0.615. Arthur Smithies, a noted economist of the 1950s and 1960s, used data from 1923 to 1940 and estimated b to be 0.76. Estimates of b in the 1980s were close to 0.92. James Doti and Esmael Adibi used quarterly data from 1965 to 1983 and estimated b to be 0.93.[8] Given the continued decline in the saving rate in the United States, b is now believed to be more than 0.95. Changes in estimates of average national propensity to consume over time clearly indicate structural shifts in the US economy and changes in the pattern of consumption and savings of US households over time periods of 15–25 years.

Utilizing the equilibrium condition in a supply and demand model represented by (4.14), we can write

$$a - bP + cTPCDI - dP_c + eP_s = -h + jP - kP_L - lP_E$$

Solving for the market clearing level of price P (denoted by P^*), we have

$$(b + j)P = a + h + cTPCDI - dP_c + eP_s + kP_L + lP_E$$

and dividing through by $(b + j)$

$$P = \frac{(a+h)}{(b+j)} + \frac{c}{(b+j)}TPCDI - \frac{d}{(b+j)}P_c + \frac{e}{(b+j)}P_s + \frac{k}{(b+j)}P_L + \frac{l}{(b+j)}P_E$$
$$P^* = \alpha_1 + \alpha_2 TPCDI - \alpha_3 P_c + \alpha_4 P_s + \alpha_5 P_L + \alpha_6 P_E \qquad (4.15)$$

$$\text{where} \quad \alpha_1 = (a+h)/(b+j) \qquad \alpha_2 = c/(b+j) \qquad \alpha_3 = d/(b+j)$$
$$\alpha_4 = e/(b+j) \qquad \alpha_5 = k/(b+j) \qquad \alpha_6 = l/(b+j)$$

a, b, c, d, e, h, j, k, and l are *structural parameters* of the model and $\alpha_1, \alpha_2, \alpha_3$, α_4, α_5, and α_6 are the *reduced form parameters*. It should be noted that P^* is expressed entirely in terms of parameters and exogenous variables. By substituting for P from (4.15) in either (4.7) or (4.11), the equilibrium quantity Q^* would be determined.

Using the numerical example for the demand and supply equations in (4.8) and (4.12), our model is

$$Q_d = 2000 - 400P + 0.009\ TPCDI - 50P_c + 120P_s$$
$$Q_s = -5000 + 800P - 50P_L - 8P_E$$
$$Q_d = Q_s = Q^*$$

Using the equilibrium condition, we have

---

[8] James Doti and Esmael Adibi, *Econometric Analysis: An Applications Approach*, Prentice-Hall 1988.

$$2000 - 400P + 0.009 \ TPCDI - 50P_c + 120P_s = -5000 + 800P - 50P_L - 8P_E$$

$$1200P = 7000 + 0.009 \ TPCDI - 50P_c + 120P_s + 50P_L + 8P_E$$

Assuming the following values for the exogenous variables
$TPCDI = \$1,200,000$, $P_c = \$5$, $P_s = \$10$, $P_L = \$10$, and $P_E = \$70$ we have

$$1200P = 7000 + 0.009(1200000) - 50(5) + 120(10) + 50(10) + 8(70) = 19810$$

$$P^* = \frac{19810}{1200} = \$16.51$$

Substituting for P in the supply equation the value of the equilibrium quantity is

$$Q^* = -5000 + 800(16.51) - 50(10) - 8(70) = 7148 \quad \text{units}$$

Note that the numerical values of the structural parameters of this model are

$$a = 2000, b = 400, c = 0.009, d = 50, e = 120, h = 5000, j = 800, k = 50, l = 8$$

and that of the reduced form parameters are

$$\alpha_1 = \frac{7000}{1200} \quad \alpha_2 = \frac{0.009}{1200} \quad \alpha_3 = \frac{50}{1200} \quad \alpha_4 = \frac{120}{1200} \quad \alpha_5 = \frac{50}{1200} \quad \alpha_6 = \frac{8}{1200}$$

4.2.1 Another Look at Single Commodity Linear Demand and Supply Model

$Q_{dt} = a - bP_t$ and $Q_{st} = -c + dP_t$ relate the quantity demanded and supplied of a commodity *contemporaneously* (during the same period of time) to its price. P_t, Q_{dt} and Q_{st} denote price, quantity demanded and quantity supplied at time period t, respectively. Both of these equations are "behavioral" equations. The demand equation indicates an inverse relation between quantity demanded and price, signifying the law of demand. Here, households, through a collective but uncoordinated action, reduce their purchase of the commodity when faced with a higher price for that commodity in the market. For supply, the relationship is direct, describing the collective but uncoordinated behavior of producers in the market as price fluctuates. Here we assume that we either arrive at the market supply and demand functions through an appropriate aggregation of firms' and households' supply and demand functions, or we present them simply as macroeconomic "analog" of their microeconomic counterparts. As we will see shortly, $-c$ in the supply function ensures that at or below a strictly positive price (here $\frac{c}{d}$ per unit), producers will not supply any units to the market.

To have a meaningful demand equation, we must have $b > 0$ and $a > 0$. If $P = 0$, then this good or service is "free". In this case 'a' signifies what is known as the *consumer's saturation point*. The saturation point is the maximum amount of a good that a consumer is willing to consume when it is offered for free. Since free goods are rare in any economy, this case seldom arises in the real world.

If $P_t = \dfrac{a}{b}$ then $Q_{dt} = 0$. At $P_t \geq \dfrac{a}{b}$ consumers are all *priced out of the market*, i.e. given their current levels of incomes they can't afford to buy any unit of this good or service. A general linear demand function should then be expressed as

$$Q_{dt} = \begin{cases} 0 & \text{for } P_t \geq a/b \\ a - bP_t & \text{for } 0 < P_t < a/b \\ a & \text{for } P_t = 0 \end{cases} \tag{4.16}$$

Similarly for producers, the price must be greater than c/d (where c and d are both positive numbers). If the market price falls to c/d or lower, the quantity supplied would be 0. At c/d or lower, producers simply shut down because the price cannot cover the unit variable cost of production.

In the short run, firms' maximum level of output is limited by their capacities and they cannot supply more than their collective capacity output Q_c no matter how high the price rises in the market. The short run supply function, therefore, should be presented as

$$Q_{st} = \begin{cases} 0 & \text{for } P_t \leq c/d \\ -c + dP_t & \text{for } c/d < P_t < e \\ Q_c & \text{for } P_t \geq e \end{cases} \tag{4.17}$$

It is clear from (4.16) and (4.17) that in order to a have a functioning and economically meaningful model, the price must be restricted in the range between c/d and a/b

$$\frac{c}{d} < P_t < \frac{a}{b} \quad \text{which implies that} \quad \frac{a}{b} \text{ must be greater than } \frac{c}{d} \tag{4.18}$$

In Fig. 4.3 demand and supply curve based on (4.16) and (4.17) are graphed.

A market reaches equilibrium at a price P^*, at which the quantity demanded by consumers would be equal to the quantity supplied by producers. At equilibrium $Q_{dt} = Q_{st}$ or $Q_{dt} - Q_{st} = 0$. This is the equilibrium condition of the model. The equilibrium condition implies that at the equilibrium excess demand $(Q_{dt} - Q_{st} > 0)$ or excess supply $(Q_{dt} - Q_{st} < 0)$ are eliminated. Utilizing the equilibrium condition for (4.16) and (4.17) we have,

$a - bP_t = -c + dPt$ or $(b + d)P_t = (a + c)$ and the equilibrium price is

$$P^* = \frac{a + c}{b + d} \tag{4.19}$$

By substituting for price in either the demand or supply equations, we get the equilibrium quantity

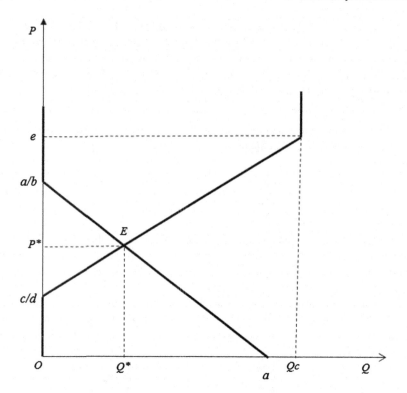

Fig. 4.3 Supply and demand curves

$$Q^* = \frac{ad - bc}{b + c}$$

Since Q^* cannot be negative, we must have another restriction on the parameters a, b, c, and d of the model, namely $ad - bc \geq 0$ or $ad \geq bc$ or $a/b \geq c/d$. Interestingly this is the same restriction that we introduced in (4.18). To see the consequence of violation of restriction (4.18), consider the following single commodity model

$$Q_{dt} = 50 - 10P_t$$
$$Q_{st} = -22 + 2P_t$$

Solving the model for P, we get

$$12P_t = 72 \quad \longrightarrow \quad P_t = 6$$

But at price $P_t = 6$, quantity demanded and supplied are -10 units, a totally absurd result!

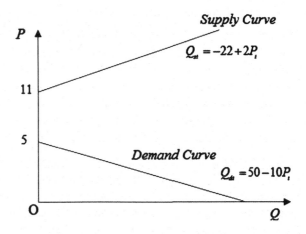

Fig. 4.4 Inconsistent market model

Here $a/b = 50/10 = 5 < c/d = 22/2 = 11$ and supply and demand curves intersect not in the first but in the second quadrants, leading to a negative quantity. This situation is depicted in Fig. 4.4.

If the two curves fail to intersect in the first quadrant, we say that the market is "inconsistent". This inconsistency generally arises from the breakdown in the process of "price discovery" due to lack of substantiative "information" about the market. As an example, this situation developed in the financial market after the crash of September 2008. The financial paper market (like the commercial paper market) froze because investors did not know how to price the papers. This was due to the tremendous uncertainty in the market and the possibility of bankruptcy and default by the issuing entities. The housing market experienced a similar situation, this time due to the massive default by homeowners and foreclosures.

Expressing quantity demanded and supplied as functions of price as in (4.16) and (4.17) is the conventional Walrasian formulation of market demand and supply equations. Here the argument is that households (firms) "adjust" their purchases (production) of this commodity in response to changes in the price. Changes in the price could be due to several exogenous factors, prominently changes in the prices of resources and technology—leading to changes in costs of production—and changes in the prices of other goods. Quantity adjustment is the mechanism for reaching equilibrium in the Walrasian system.

The Walrasian approach is also a *General Equilibrium* approach. An economy consists of a large number of interconnected markets (sectors or industries). The Walrasian model of an n-market economy consists of a system of $2n$ supply and demand equations with $2n$ endogenous variables (n prices and n quantities) and a number of exogenous variables. A market reaches equilibrium if all other markets reach equilibrium at the same time. Given a set of values for the exogenous variables and the parameters, the general equilibrium is achieved by solving the system of n equations for n unknown prices (a 2-market or 2-commodity model will be discussed in the next section). The quantity adjustment mechanism of the Walrasian model is readily adaptable to the long-run analysis, especially to the long run theory of production. In the long run firms are able to adjust their levels of output and production capacities. It also allows the economy more time to reach a general or a system-wide equilibrium.

The alternative Marshallian formulation expresses price as a function of quantity demanded and supplied; that is, price "adjusts" to changes in quantity. Changes in quantity could be due to changes in consumers' incomes and/or tastes. In the Marshallian formulation, price adjustment is the mechanism for reaching equilibrium. The Marshallian price adjustment mechanism is easily adaptable to *market period* and *short-run* analysis. Consumers cannot easily adjust their consumption patterns and firms cannot easily adjust their levels of outputs in the short-run, so a price adjustment is needed to eliminate excess demand or supply and clear the market. It is neither by accident nor by mistake that introductory economics books draw supply and demand curves with Q on the horizontal and P on the vertical axis. This is the graphical presentation of Marshallian supply and demand functions, even though, in many cases, they are mathematically expressed by their Walrasian version! The Marshallian approach is a *partial equilibrium* approach. Partial equilibrium examines the condition of equilibrium in an individual or isolated market.

A variant of our simple single market model expresses the supply function as a one period lagged equation

$$Q_{st} = -c + dP_{t-1}$$

This behavioral supply model implies that producers make their current supply decision based on the price they received last period. Utilizing the equilibrium condition

$$Q_{dt} = Q_{st} \quad \text{we have} \quad a - bP_t = -c + dP_{t-1} \quad \text{or}$$

$$P_t + \frac{d}{b}P_{t-1} = \frac{a+c}{b} \tag{4.20}$$

If this market reaches a stable equilibrium then the equilibrium price P^* would be the same for any time period, including periods t and $t-1$. In this case, we can write (4.20) as

$P^* + d/bP^* = (a+c)/b$ which leads to the same solution for P^* as in (4.19)
Equation (4.20) is an example of a *First Order Difference Equation with Constant Coefficients*. We will introduced and discuss Difference Equations and their applications in economics in Chap. 11.

The third variation of a single market price determination model is

$$P_{t+1} = P_t + \alpha(Q_{dt} - Q_{ds}) \tag{4.21}$$

where constant $\alpha > 0$ is called the *adjustment coefficient*. This model is sometimes called the *stock adjustment model*. Notice that:

♣ if $Q_{dt} - Q_{ds} > 0$ there is an excess demand or shortage in the market, leading to a rapid depletion of the stock of the good and the producers' inventories. In response to excess demand, price must adjust upward, that is $P_{t+1} > P_t$. As long as excess demand exists in the market, the process of price increases will continue. When excess demand is eliminated then in Eq. (4.21) $\alpha(Q_{dt} - Q_{ds}) = 0$ and price converges to the equilibrium price, i.e. $P_{t+1} = P_t = P^*$

♣ if $Q_{dt} - Q_{ds} < 0$, then there is an excess supply, or surplus, in the market generating additional unintended increase in the level of inventories. To eliminate excess supply, the price must adjust downward. Here $P_{t+1} < P_t$. The process of downward price adjustment will continue until excess supply is eliminated and the price converges to P^*.

♣ if $Q_{dt} - Q_{ds} = 0$, the market has already reached an equilibrium and no adjustment is required. In this case $P_{t+1} = P_t = P^*$

The following simple example demonstrates the dynamic nature of this model and the price/quantity adjustment process. Assume

$$Q_{dt} = 500 - 10P_t$$
$$Q_{st} = -112 + 8P_t$$

and adjustment coefficient $\alpha = 0.1$. With these assumptions, Eq. (4.21) can be written as

$$P_{t+1} = P_t + 0.1(Q_{dt} - Q_{ds})$$

Now assume that the market has not yet reached equilibrium and the price is $P_t = 35$. At this price

$$Q_{dt} - Q_{st} = 150 - 168 = -18$$

and the price adjusts to

$$P_{t+1} = 35 + 0.1 * (-18) = 33.2$$

But at this price $Q_{dt+1} - Q_{st+1} = 14.4$ and the price adjusts to

$$P_{t+2} = 33.2 + 0.1(14.4) = 34.64$$

In the next round of adjustment (iteration) excess demand changes to -11.52 and the price to $P_{t+3} = 33.4588$. After a series of adjustment, excess demand gradually reduces to zero and price converges to its equilibrium value of 34.

4.2.2 Two-Commodity Market Model

As it was discussed in the first section of this chapter, the demand for a product is not only a function of the price of the good itself, it is also a function of several other variables, including the prices of complementary P_c and/or substitute P_s goods. The market model for a single commodity analyses the market in isolation and treats P_c and P_s as exogenous variables whose values are somehow determined outside the model. A two-commodity market model is a first attempt to remedy this deficiency. This model connects markets for two related goods together and studies how the equilibrium quantities and prices are simultaneously determined. This is the first step in the direction of a more extensive general equilibrium n-commodity model.

The best way to understand a two-commodity model is by way of a numerical example. Consider two goods with prices P_1 and P_2. Assume demand and supply equations for these goods are given by

$$Q_{d1} = 15 - 4P_1 + 2P_2$$
$$Q_{s1} = -30 + 3P_1$$

and

$$Q_{d2} = 100 + 2P_1 - 2P_2$$
$$Q_{s2} = -20 + 3P_2$$

where Q_{d1}, Q_{d2}, Q_{s1}, and Q_{s2} denote quantity demanded and supplied of good 1 and 2. Note that the price of each good is entered with a positive coefficient in the demand equation for the other good, a clear indication that they are substitute goods. To determine the equilibrium prices and quantities, we utilize the standard market equilibrium condition that

$$Q_{d1} = Q_{s1} \quad \text{and} \quad Q_{d2} = Q_{s2}$$
$$15 - 4P_1 + 2P_2 = -30 + 3P_1 \quad \longrightarrow \quad -7P_1 + 2P_2 = -45$$
$$100 + 2P_1 - 2P_2 = -20 + 3P_2 \quad \longrightarrow \quad -2P_1 + 5P_2 = 120$$

We now have a system of two equations with two unknowns

$$\begin{cases} -7P_1 + 2P_2 = -45 \\ -2P_1 + 5P_2 = 120 \end{cases}$$

There are several methods for solving a system of equations. For a simple system of two equations with two unknowns, the most efficient method is often by elimination. By multiplying both sides of the first equation by -2 and both sides of the second equation by 7 and then adding them together, we can eliminate the first variable P_1 and easily find a solution for P_2.

$$-2(-7P_1 + 2P_2 = -45) \quad \longrightarrow \quad 14P_1 - 4P_2 = 90$$
$$7(-2P_1 + 5P_2 = 120) \quad \longrightarrow \quad -14P_1 + 35P_2 = 840$$

Adding the left hand side (LHS) and the right hand side (RHS) of the equations, we have

$$31P_2 = 930 \quad \longrightarrow \quad P_2^* = 930/31 = 30$$

Substituting for P_2 in the first equation $-7P_1 + 2P_2 = -45$, we get

$$-7P_1 + 2*30 = -45 \quad \longrightarrow \quad P_1^* = 15$$

By plugging in 15 and 30 for P_1 and P_2 in any of the supply or demand equations, we determine the equilibrium quantities in both markets as $Q_1^* = 15$ and $Q_2^* = 70$.

A two commodity market model can be easily extended to a 3, 4, ..., n market model. As it was mentioned earlier an n—market economy model consists of n supply and n demand equations with n prices and n quantities. The economy reaches equilibrium when all market simultaneously reach equilibrium. By utilizing the equilibrium condition and setting the quantity demanded equal to the quantity supplied in each market, we arrive at a system of n equations with n unknown prices. Solution to this system provides the set of equilibrium prices that could be in turn used for determining the equilibrium quantities.

4.3 US Economic Crisis and Return of Keynesian Economics

By December 2008 the United States financial crisis, which was rooted in two costly wars and emanated from the collapse of the housing market and freeze of the credit pipe lines (following the collapse of Lehman Brothers) had substantially deepened. What was once considered to be within the confines of the financial market had spread to the real economy. The impact on the real economy manifested itself in a sharp rise in the unemployment rate, a major drop in GDP, a dramatic increase in home foreclosure and business failure, a sharp decline in housing prices, and the near bankruptcy of industrial icons like General Motor Corporation. There is no shortage of 'adjectives' to characterize the severity of this crisis, but the most widely used phrases are 'the most severe slump since 1932 great depression' and the 'great recession'.

At the height of the depression the unemployment rate hit 25 % and real GDP drop by almost 33 %. With a 5 % decline in GDP and an unemployment rate of below 10 %, the 2008 crisis may not qualify as a depression, but the fear was that, if left unchecked, it may spin out of control and assume the designation.

To a great extent President Roosevelt's New Deal was an embrace of Keynesian economics. The British economist John Maynard Keynes was an advocate of increasing government spending to fight economic downturn and create jobs. In the following section we will introduce several versions of a simple Keynesian income determination model and discuss their solutions.

4.3.1 Macroeconomic Model and Equilibrium in National Income

A standard Keynesian macroeconomic model consists of the national income identity

$$Y \equiv C + I + G + X - M$$

and several equations explaining behavior of components of national income. Components of national income are consumption expenditure, C, investment expenditure, I, government expenditure, G, and export and import expenditures X and M. The simple version of the model assumes a closed economy. The model includes a consumption function, where consumption is expressed as a linear function of income

$$C = C_0 + bY \qquad C_0 > 0 \qquad 0 < b < 1 \tag{4.22}$$

C_0 is mostly interpreted as the *autonomous* consumption[9] and b is the *marginal propensity to consume*.[10] The simple version of the model assumes that investment and government expenditures are exogenously determined. The macroeconomic equilibrium condition is that the aggregate supply must equal aggregate demand, that is

$$Y = C + I_0 + G_0 \tag{4.23}$$

In the national income model consisting of (4.22) and (4.23) Y and C are the endogenous variables, I_0 and G_0 denote exogenously determined investment and government expenditures, C_0 is the autonomous or exogenous consumption, and b is a parameter. The consumption function (4.22) is a behavioral equation and (4.23) is an identity. Taken together (4.22) and (4.23) are called a *structural model*; they describe the structure of the economy. They are also called a *simultaneous equation model* since they simultaneously determine the values of the model's endogenous variables.

We arrive at what is known as the *reduced form* of the model by substituting for C in (4.23) from (4.22). This new equation yields the equilibrium national income.

$$Y = C_0 + bY + I_0 + G_0 \quad \longrightarrow \quad Y - bY = C_0 + I_0 + G_0$$
$$(1 - b)Y = C_0 + I_0 + G_0$$
$$Y^* = \frac{1}{1 - b}(C_0 + I_0 + G_0) \tag{4.24}$$

---

[9] As an analog of a household consumption function, C_0 is sometimes interpreted as the *subsistence* level of consumption.

[10] Some writers express (4.22) as $C = a + bY$ and treat the autonomous consumption as a parameter. Given the volatility of autonomous consumption, due to wealth and other non-income effects, it should not be considered a parameter. After all, based on the definition, a parameter changes very slowly and incrementally.

By putting Y^* from (4.24) into (4.22) we get the equilibrium level of consumption as

$$C^* = C_0 + b\left(\frac{C_0 + I_0 + G_0}{1 - b}\right) = \frac{C_0 + b(I_0 + G_0)}{1 - b}$$

Note that in (4.24) $\frac{1}{1-b} > 1$. This indicates that an increase in the exogenous variables increases national income by a 'multiple' larger than one. $\frac{1}{1-b}$ is, of course, the *multiplier* and plays a major role in *comparative-static analysis* of the economy. The comparative-static question in this model is the determination of new equilibrium values of the endogenous variables of the model, Y and C, when there is a change in the value of the exogenous variables, especially G_0.

If, for example, $C = 50 + 0.9Y$, $I_0 = 30$, and $G_0 = 20$ then the equilibrium values of output and consumption are

$$Y^* = \frac{1}{1-b}(C_0 + I_0 + G_0) = (\frac{1}{1-0.9})(50 + 30 + 20) = 10 * 100 = 1000$$
$$C^* = 50 + 0.9(1000) = 50 + 900 = 950$$

Here the value of the simple multiplier is $1/(1-0.9) = 1/0.1 = 10$. If government expenditure changes from 20 to 25, the economy moves to a new equilibrium where new Y^* and C^* are

$$Y^* = \frac{1}{1-b}(C_0 + I_0 + G_0) = (\frac{1}{1-0.9})(50 + 30 + 25) = 10 * 105 = 1050$$
$$C^* = 50 + 0.9(1050) = 50 + 945 = 995$$

Note that through the working of the multiplier, a 5-unit increase in government expenditure leads to a $10 * 5 = 50$ unit increase in income. A 5-unit increase in autonomous consumption C_0 from 50 to 55, or investment from 30 to 35, will have exactly the same impact on national income. If we denote the multiplier by m and the sum of the exogenous variables $C_0 + I_0 + G_0$ by E_0, we can write the change in the equilibrium national income resulting from a change in either, or all, of the exogenous variable(s) as $\Delta Y^* = m \, \Delta E_0$.

An expanded and more realistic version of the model includes a consumption function that is expressed as a function of *disposable* income, denoted by Y_D, and a tax equation:

$$Y = C + I_0 + G_0$$
$$C = C_0 + bY_D \qquad\qquad C_0 > 0 \quad 0 < b < 1$$
$$T = T_0 + tY \qquad\qquad\qquad 0 < t < 1$$

The above model has four exogenous variables (I_0, C_0, G_0, T_0), two parameters (b, t) and three endogenous variables (Y, C, T). In the tax equation T_0 is the autonomous tax and t is the *marginal tax rate*.

Given that $Y_D = Y - T$, by substituting for T from the tax function, we have

$$Y_D = Y - T = Y - T_0 - tY = (1 - t)Y - T_0$$

Substituting for Y_D into the consumption function, we have

$$C = C_0 + bY_D = C_0 + b[(1 - t)Y - T_0] = C_0 + b(1 - t)Y - bT_0$$

And finally, using the equilibrium condition for the national income, we have

$$Y = C + I_0 + G_0 = C_0 + b(1 - t)Y - bT_0 + I_0 + G_0$$
$$(1 - b + bt)Y = C_0 + I_0 + G_0 - bT_0 \quad \text{and}$$
$$Y* = \frac{1}{1 - b + bt} (C_0 + I_0 + G_0) - \frac{b}{1 - b + bt} T_0$$

By substituting for Y in the tax function, we determine T, and subsequently Y_D, which in turn is used for determining the equilibrium value of C from the consumption function. Note that for the model to have a meaningful solution bT_0 must be less than $C_0 + I_0 + G_0$.

In this model $m = \dfrac{1}{1 - b + bt} > 1$ is the autonomous consumption, investment and government expenditure multiplier and $m' = -\dfrac{b}{1 - b + bt} = -bm$ is the autonomous tax multiplier. Now we can write a change in the equilibrium national income resulting from a change in either, or all, of the exogenous variable(s) as

$$\Delta Y^* = m\, \Delta E_0 + m'\Delta T_0 = m\, \Delta E_0 - bm\Delta T_0$$

This expression clearly shows the impact of government fiscal policy on the equilibrium national income. An increase in government expenditure G_0 or a reduction in autonomous tax T_0, or a combination, will lead to higher income. Changes in the tax rate t, another fiscal policy instrument, has impact on the equilibrium income. If t is reduced, the denominator of both m and m' become smaller, leading to a larger m and m'. Since m is greater than m', a reduction in t will have a net positive impact on national income.

Examples

1. Consider the following national income model

$$Y = C + I_0 + G_0$$

$$C = 55 + 0.90Y_D$$
$$T = 0.15Y$$

If $I_0 = 60$ and $G_0 = 50$, what are the equilibrium values of national income, consumption, and tax? Is the budget of the government in this economy balanced?

For the first part of the problem

$$C = 55 + 0.90Y_D = 55 + 0.90(Y - T)$$
$$C = 55 + 0.90(Y - 0.15Y) = 55 + 0.765Y \quad \text{and}$$
$$Y = 55 + 0.765Y + 60 + 50 = 165 + 0.765Y$$
$$(1 - 0.765)Y = 165 \quad \longrightarrow \quad 0.235Y = 165$$
$$Y^* = \frac{165}{0.235} = 702.13$$
$$T^* = 0.15 * 702.13 = 105.32$$
$$C^* = 55 + 0.9(Y - T) = 55 + 0.9(702.13 - 105.32) = 592.13$$

Since $T = 105.32 > G_0 = 50$, the government in this economy has a 55.32 budget surplus.

What would be the impact of a 20 unit increase in government expenditure on output? By including government and taxation in the model, the multiplier changes from $\frac{1}{1 - b}$ to $\frac{1}{1 - b + bt}$, from 10 to 4.26. In this model an increase in autonomous investment or government expenditure by 10 adds only 42.6 to output.

Other versions of the macroeconomic model with different degree of complexity can be constructed. In the expanded open-economy version, for example, exports X and imports M are included. The following is a sample of a slightly more complicated version of national income determination model.

2. Consider an open economy with the following consumption, investment, and tax functions

$$C = 40 + 0.80Y_D$$
$$T = 10 + 0.10Y$$
$$I = 38 + 0.15Y$$

In this model investment is expressed as a linear function of income. The endogenous variables of the model are Y, C, I and T.

Assume $G_0 = 110$, $X_0 = 40$, and $M_0 = 90$. What are the equilibrium output, consumption, investment, and taxes in this economy?

$$C = 40 + 0.80Y_D = 40 + 0.80(Y - T) = 40 + 0.80[Y - (10 + 0.1Y)]$$
$$\quad = 40 + 0.80(0.9Y - 10) = 40 + 0.72Y - 8$$
$$C = 32 + 0.72Y$$
$$Y = C + I + G_0 + X_0 - M_0 = 32 + 0.72Y + 38 + 0.15Y + 110 + 40 - 90$$
$$Y = 130 + 0.87Y \quad \longrightarrow \quad Y - 0.87Y = 130 \quad \longrightarrow \quad 0.13Y = 130$$
$$Y^* = 130/0.13 = 1000$$
$$C^* = 32 + 0.72 * 1000 = 752$$
$$T^* = 10 + 0.1 * 1000 = 110$$
$$I^* = 38 + 0.15 * 1000 = 188$$

Readers should be able to verify that the exogenous expenditure multiplier, m, is 7.692 and the tax multiplier, m', is 6.154. Since $T^* = 110 = G_0$ the government budget in this economy is balanced.

Note that the economy's aggregate saving is

$$S = Y_D - C = (1000 - 110) - 752 = 138$$

while the aggregate investment is 188. The difference $188 - 138 = 50$ is balanced by the excess of imports over exports $90 - 40 = 50$.

3. As a practical real world example, consider the following model[11] base on aggregate annual data from 1959 to 1990

$$C = -228.78 + 0.832Y_D$$
$$I = -41.951 + 0.255\,Y_D - 11.511\,r$$
$$r = -0.178 + 0.010\,Y - 0.012\,M_0$$

where r is the corporate bond rate used as proxy for interest rate and M is the real money supply. By using $Y - T_0$ for Y_D and X_0 for net export, the implied reduce form of the model is obtained as

$$Y = \quad C + I + G_0 + X_0$$
$$Y = -228.78 + 0.832(Y - T_0) - 41.951 + 0.255(Y - T_0)$$
$$\quad -11.511(-0.178 + 0.010Y - 0.012M_0) + G_0 + X_0$$
$$\quad = -268.682 + 0.972Y + 0.138M_0 + G_0 + X_0 - 1.087T_0$$

and finally

$$Y = -9595.786 + 4.929M_0 + 35.714G_0 + 35.714X_0 - 38.821T_0$$

---

[11] From Ramu Ramanathan *Introductory Econometrics with Application* pp. 544–547.

Here Y is expressed in terms of exogenous variables M, G, X, and T. In 1
values of these variables were 3323.3, 820.8, -37.5, and 634.1, respectiv
models estimate of nominal GDP for 1990 is then 6211.74 or about \$6.2 trillion.

4. Consider the following model of national income determination

$$Y = C + I + G_0$$
$$C = C_0 + bY_D$$
$$T = T_0 + tY$$
$$I = I_0 + iY - k \, r_0$$

where r_0 is the exogenously determined interest rate expressed as percentage. To
determine the equilibrium income we write,

$$Y = C_0 + b(Y - T_0 - tY) + I_0 + iY - kr_0 + G_0$$
$$(1 - b + bt - i)Y = C_0 + I_0 + G_0 - bT_0 - kr_0$$
$$Y^* = \frac{1}{1 - b + bt - i}(C_0 + I_0 + G_0) - \frac{b}{1 - b + bt - i}T_0 - \frac{k}{1 - b + bt - i}r_0$$

where $\dfrac{1}{1 - b + bt - i}$, $-\dfrac{b}{1 - b + bt - i}$ and $-\dfrac{k}{1 - b + bt - i}$ are respectively
the expenditure, tax, and interest rate multiplier.
This formulation of national income determination model includes a monetary
policy instrument, namely the interest rate. The Federal Reserve Bank through the
Open Market Operations, change in the Discount rate, change in the Fed Fund rate,
or Quantitative Easing can influence interest rate in the economy and push the equilibrium national income toward a more desirable level.

4.4 Exercises

1. A market consists of 5000 identical households and 100 identical producers. The
   demand equation for a typical household over a week is given by

   $$Q_i^d = 30 - 2P + 0.001PCDI_i - 0.028P_c \qquad i = 1, 2, 3, \ldots, 5000$$

   And the supply equation for a typical firm over a week is given by

   $$Q_j^s = -50 + 10P - 0.5P_L - 0.1P_E \qquad j = 1, 2, 3, \ldots, 100$$

   (a) Write the market demand and supply equations.

(b) Assume a households' per capita disposable income PCDI is $8,000. Further assume that P_c, P_L, and P_E are $20, $100, and $80, respectively. Determine the market equilibrium price and quantity.

2. What would be the equilibrium price and quantity in problem (1) if the households per capita income increased to $8,500, ceteris paribus?

3. How would the equilibrium solution in problem (1) change if you doubled the number of households and producers?

4. In problem (1) assume that due to inflation, the cost of labor increases by 30% and price of energy by 40%. What is the new market equilibrium price and quantity? Measure the impact of the change in prices of labor and energy by comparing the new equilibrium values with values in problem (1) (You will be doing *comparative static analysis*)

5. In problem (1)

    (a) What is the weekly consumption of a household?

    (b) What is the weekly budget allocation of a household for this good?

6. In problem (1) assume that producers collectively employ 1000 units of labor and use 500 units of energy per week. Assuming no other production costs, what is the weekly profit of a producer?

7. Consider the following market demand and supply function for a commodity

$$Q_d = 20000 - 400P$$
$$Q_s = -4000 + 800P$$

Determine the equilibrium price and quantity.

8. Assume the demand and supply functions for a good are

$$Q_d = 200 - 5P + 0.002 \, INC_0$$
$$Q_s = -100 + 8P$$

where INC_0 is the exogenously determined average income.

    (a) Find the equilibrium price and quantity if the average income is 45,000 dollars.

    (b) Draw the demand and supply curves.

    (c) Assume average income rises to $50,000; write and graph the new demand function.

    (d) Find the new equilibrium solution. What is the impact of rise in income?

9. Find the equilibrium price and quantity for the following market model

    (a) $Q_d = 200 - 40P$ (b) $Q_d = 1400 - 60P$

          $Q_s = -40 + 80P$ $Q_s = -400 + 30P$

10. Graph the following demand and supply functions. Determine the equilibrium solution from the graph. Solve the equation to verify your answers.

    (a) $Q_d = 20 - 2P$ (b) $Q_d = 30 - 2P$

          $Q_s = -10 + 4P$ $Q_s = -50 + 3P$

    What is wrong with model (b)?

11. Find the equilibrium solution for the following models:

    (a) $Q_d = 30 - P^2$ (b) $Q_d = 95 - 3P^2$

          $Q_s = -2 + 4P$ $Q_s = -10 + 6P$

    {*Use the quadratic formula to solve the resultant quadratic equations.* This is an example of a *non-linear* market equilibrium models. We will discuss non-linear models in detail in Chap. 7.}

12. The demand and supply functions of two related goods are given by

$$Q_{d1} = 30 - 8P_1 + 4P_2$$
$$Q_{s1} = -60 + 6P_1$$
$$Q_{d2} = 200 + 4P_1 - 4P_2$$
$$Q_{s2} = -40 + 6P_2$$

    (a) What is the relationship between good one and good two?

    (b) Find the equilibrium prices and quantities.

13. The demand and supply function of two related goods are given by

$$Q_{d1} = 30 - 5P_1 + 2P_2$$
$$Q_{s1} = -20 + 4P_1$$
$$Q_{d2} = 70 + 4P_1 - 4P_2$$
$$Q_{s2} = -10 + 2P_2$$

    Find the equilibrium prices and quantities.

14. Assume that the national income model is specified as:

$$Y = C + I_0 + G_0$$
$$C = C_0 + bY_D$$
$$T = tY$$

(a) Identify the endogenous and exogenous variables and the parameters.

(b) Determine the equilibrium national income and the multiplier.

(c) What restrictions on the model's parameters are needed in order for the model to have a meaningful solution?

15. Assume that in problem (14) $C_0 = 30$, $b = 0.92$, $t = 0.12$, $I_0 = 70$, and $G_0 = 60$.

(a) Find the equilibrium values of the model's endogenous variables.

(b) What would be the change in national income if I_0 changes from 70 to 75?

(c) What would be the change in the equilibrium values of Y, C, and T if I_0 declines from 70 to 60 while G_0 increases from 60 to 70?

16. Consider an open economy with the following consumption and tax functions:

$$C = 5 + 0.75Y_D$$
$$T = 0.20Y$$

Assume $I_0 = 100$; $G_0 = 110$; $X_0 = 90$ and $M_0 = 105$

(a) Find the equilibrium output.

(b) Is the budget in this economy balanced?

(c) Is saving equal to investment?

(d) Determine the change in the equilibrium output if the autonomous imports decline by 20.

17. Consider the following model of income determination:

$$Y = C + I + G_0$$
$$C = C_0 + bY_D$$
$$I = I_0 + iY \qquad I_0 > 0 \qquad 0 < i < 1 \quad \text{and} \quad 0 < t < 1$$
$$T = tY$$

This model includes a new equation expressing investment as a function of income (*induced investment*) and an autonomous or exogenous component I_0.

(a) Solve the model for the equilibrium income. What is the model multiplier?

(b) What restrictions on the model's parameters are needed in order for the model to have a meaningful multiplier?

18. The following is a numerical version of the model in question (17)

$$C = 10 + 0.60Y_D$$
$$I = 20 + 0.20Y$$
$$T = 0.10Y$$

If $G_0 = 48$, solve the model for the equilibrium values of all the endogenous variables.

19. An expanded version of the national income determination model in problem (17) is formulated by adding an autonomous or exogenous tax component to the tax function:

$$Y = C + I + G_0$$
$$C = C_0 + bY_D$$
$$I = I_0 + iY \qquad I_0 > 0 \qquad 0 < i < 1 \quad \text{and} \quad 0 < t < 1$$
$$T = T_0 + tY \qquad T_0 > 0$$

Solve the model for the equilibrium income. Derive the model's multipliers.

20. The following is a numerical version of the model in question (19)

$$C = 10 + 0.85Y_D$$
$$I = 20 + 0.20Y$$
$$T = 30 + 0.15Y$$

(a) If $G_0 = 48$, solve the model for the equilibrium values of all the endogenous variables.
(b) If government gives a tax cut by reducing T_0 from 30 to 20, what would be the impact on the equilibrium income, consumption, and investment.
(c) What would be the impact on the equilibrium income, consumption, and investment if government reduces the average tax rate t from 15 to 12%?

21. Consider the following aggregate consumption and investment function for a closed economy

$$C = 15 + 0.70Y$$
$$I = 30 + 0.1Y - 7.5r$$

Here, investment is expressed as a function of income Y and interest rate r (expressed as percentage).

If $G_0 = 50$, write an equation relating Y and r and graph it. This graph is called the *IS curve*.

22. In problem (21) assume that the rate of interest r is 5.5% and that the government expenditure is 50. Determine the equilibrium income, consumption and investment.

23. In problem (21) assume that the Federal Reserve in order to stimulate the economy reduces the interest rate to 4.5%. Determine the new equilibrium income, consumption and investment, and compare them with the results in problem (22). Did the Fed achieve its objective? What is the rate of growth of the economy?

24. Below is a numerical version of the model presented in Example 4

$$C = 60 + 0.75Y_D$$
$$T = 20 + 0.20Y$$
$$I = 30 + 0.1Y - 7r$$

Assume that the rate of interest r is 5.0% and that the government expenditure is 80. Determine the equilibrium income, consumption and investment. What are the multipliers? What is the amount of budget deficit or surplus?

25. In problem (24) assume that the government expenditure is increased to 90 while the Federal Reserve reduces the interest rate to 3.5%. Determine the new equilibrium income, consumption and investment.

Chapter 5
Rates of Change and the Derivative

5.1 Linear Functions

A function of the form $y = a + mx$ is called a linear function because the graph of the function is a straight line. In $y = a + mx$ m is the slope of the line and a is its y-intercept ($y = a$ when $x = 0$). The main feature of a linear function is that the rate of change of the function is constant; it grows or declines at a steady rate. Figure 5.1 shows the graph of the function $y = f(x) = 3 + 0.5x$. Next to the graph is a table with some values of x and y.

Notice from the equation of the line and the accompanying table that as x increases by any value, y increases by 0.5 times that value. If we denote change in x by Δx and the corresponding change in the function by Δy, it is clear that $\Delta y = 0.5\Delta x$.

When $\Delta x = 1$, as x increases from 0 to 1 or from 1 to 2, the value of the function increases by $\Delta y = 0.5(1) = 0.5$ units, from 3 to 3.5 and then to 4. When $\Delta x = 0.5$, e.g. when x increases by 0.5 units from 2 to 2.5, the function changes by $\Delta y = 0.5(0.5) = 0.25$ units, from 4 to 4.25. When $\Delta x = 0.1$, e.g. when x changes from 2.5 to 2.6, y changes by $\Delta y = 0.5(0.1) = 0.05$ units, from 4.25 to 4.3. In general, $y = a + mx$ implies that

$$\Delta y = m\Delta x \quad \longrightarrow \quad m = \frac{\Delta y}{\Delta x}$$

This is the familiar expression for the slope of a line, which measures the rate of change of the function, or the change in the function per unit change in x. This rate in linear functions is constant and represented by m.

We can write the equation of a linear function, if we know

(a) the slope and intercept of the line,
(b) the coordinates of any two points on the line,
(c) the slope and coordinates of one point on the line.

S. Vali, *Principles of Mathematical Economics*,
Mathematics Textbooks for Science and Engineering 3,
DOI: 10.2991/978-94-6239-036-2_5, © Atlantis Press and the authors 2014

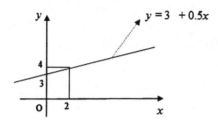

x	y	Δx	Δy	Δy/Δx
0	3			
1	3.5	1	0.5	0.5
2	4	1	0.5	0.5
2.5	4.25	0.5	0.25	0.5
2.6	4.3	0.1	0.05	0.5

Fig. 5.1 Graph of $y = 3 + 0.5x$

$y = a + mx$ is the slope-intercept form of the equation. To write a linear equation based on the coordinates of two points, lets assume that A and B are points on the line with the coordinates (x_1, y_1) and (x_2, y_2), as shown in Fig. 5.2.

It is clear from the graph that

$$\Delta x = x_2 - x_1 \qquad \text{and} \qquad \Delta y = y_2 - y_1$$

Therefore the slope of the line is

$$m = \frac{\Delta y}{\Delta x} = \left(\frac{y_2 - y_1}{x_2 - x_1} \right)$$

Since A is on the line then

$$y_1 = a + m\, x_1 = a + \left(\frac{y_2 - y_1}{x_2 - x_1} \right) x_1 \quad \longrightarrow \quad a = y_1 - \left(\frac{y_2 - y_1}{x_2 - x_1} \right) x_1$$

Substituting for the intercept and slope in the equation $y = a + mx$, we have

$$y = y_1 - \left(\frac{y_2 - y_1}{x_2 - x_1} \right) x_1 + \left(\frac{y_2 - y_1}{x_2 - x_1} \right) x$$

which can be simplified to

Fig. 5.2 Deriving the equation of a line from the coordinates of two points on the line

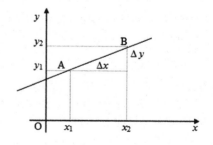

$$y - y_1 = \left(\frac{y_2 - y_1}{x_2 - x_1}\right)(x - x_1) \tag{5.1}$$

For case (c), we need to find the equation of a line with known slope m passing through a point with coordinates (x_1, y_1). This is an easy variation of the two-point case. If the slope is known then (5.1) can be written as

$$y - y_1 = m(x - x_1)$$

which is the point-slope form of the equation.

Let's reiterate the significant feature of linear functions: the slope of the line is the rate of change of the function which is constant over the entire domain of the function.

5.1.1 Rate of Change of Non-Linear Functions

Suppose $y = f(x)$ is a nonlinear function with its graph depicted in Fig. 5.3. Assume A and B are two points on this curve with coordinates $(x, f(x))$ and $(x + \Delta x, f(x + \Delta x))$. Here, the independent variable changes by a small increment Δx from x to $x + \Delta x$, and the corresponding change in the function is

$$\Delta y = f(x + \Delta x) - f(x)$$

From our discussion of linear functions, it is clear that

$$\frac{\Delta y}{\Delta x} = \frac{f(x + \Delta x) - f(x)}{x + \Delta x - x} = \frac{f(x + \Delta x) - f(x)}{\Delta x}$$

is the slope of the line intersecting the graph of the function $y = f(x)$ at the two points A and B. This slope measures the *average rate of change of the function* over the interval $[x, x + \Delta x]$.

Fig. 5.3 Rate of change of a non-linear function

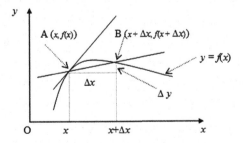

Now let's assume that B moves toward A along the curve. As B moves closer to A, Δx becomes smaller and smaller and $x + \Delta x$ approaches x. When B reaches A the line AB intersects the curve at only one point, point A. In this case the line becomes a *tangent* to the curve at point A. The slope of this tangent, m, is the limit of the ratio $\frac{\Delta y}{\Delta x}$ as Δx becomes smaller and smaller and approaches zero (which means it gets closer and closer to zero, but never becomes zero)

$$m = \lim_{\Delta x \to 0} \frac{\Delta y}{\Delta x} = \lim_{\Delta x \to 0} \frac{f(x + \Delta x) - f(x)}{\Delta x} \tag{5.2}$$

where $\lim_{\Delta x \to 0}$ is the symbolic expression for "the limit of ... as Δx approaches 0".[1] This slope measures the *instantaneous rate of change of the function at x*. In differential calculus, m is the derivative of the function $f(x)$ at x. There are a number of alternative notations for expressing the derivative of a function at x. The most commonly used notations are

$$\lim_{\Delta x \to 0} \frac{\Delta y}{\Delta x} = f'(x) = y' = \frac{dy}{dx} = \frac{d}{dx} f(x)$$

When this derivative (if it exists) is evaluated at $x = x_1$, the resulting value is the slope of the tangent to the curve at the point $(x_1, f(x_1))$. As we will see, the direct connection between the concept of the derivative and slope of a tangent plays a major role in determining the maximum or minimum of a function. Two commonly used notations representing the derivative a function evaluated at a specific value of x, like x_1, are $f'(x_1)$ and $\frac{dy}{dx}|_{x_1}$. In Fig. 5.3, the slope of the tangent to the curve at point A is $m = f'(x)$.

Examples

1. Find the slope of the tangent to function $f(x) = y = x^2 + 2$ at point P(1,3).
   We find the slope or the derivative of the function at $x = 1$ by using (5.2).

$$f(x + \Delta x) = (x + \Delta x)^2 + 2 = x^2 + (\Delta x)^2 + 2x\Delta x + 2$$

$$\frac{\Delta y}{\Delta x} = \frac{f(x + \Delta x) - f(x)}{\Delta x} = \frac{x^2 + (\Delta x)^2 + 2x\Delta x + 2 - x^2 - 2}{\Delta x}$$
$$= \Delta x + 2x$$

$$f'(x) = \frac{dy}{dx} = \lim_{\Delta x \to 0} \frac{\Delta y}{\Delta x} = \lim_{\Delta x \to 0} (\Delta x + 2x) = 2x \tag{5.3}$$

Then the slope of the tangent at $x = 1$ is

---

[1] More about limit in Sect. 5.2.1 below.

$$m = f'(1) = 2(1) = 2$$

The slope or the derivative of the function at $x = 1$ being equal to 2 indicates the instantaneous rate of change of the function in the close neighborhood of $x = 1$; that means if x changes from 1 by a small value, the function would change by approximately 2 times that value. The value of the function at $x = 1$ is $f(1) = (1)^2 + 2 = 3$. Assume x increases by a small amount from 1 to 1.001. The new value of y or $f(x)$ is

$$f(1.001) = (1.001)^2 + 2 = 3.002001$$

and the change in y is $3.002001 - 3 = 0.002001$ which is close to $2 * 0.001$.

2. What is the rate of change of the function at $x = 3$? From (5.3) we have

$$f'(x) = 2x \quad \text{Thus,} \quad f'(3) = 2 * 3 = 6$$

The rate of change of the function at $x = 3$ is 6. To show this, note that value of the function at $x = 3$ is

$$f(x) = x^2 + 2 \quad \longrightarrow \quad f(3) = (3)^2 + 2 = 11$$

If x decreases by a small increment of 0.001 from 3 to 2.999, the value of the function changes to

$$f(2.999) = (2.999)^2 + 2 = 10.994001$$

a decrease of -0.005999, which is close to $6 * (-0.001) = -0.006$.

It should be clear that the rate of change of the non linear function is different for different values of x. Only linear functions have a constant rate of change

3. Assume the market demand function for a good is $Q = f(p) = \dfrac{200}{p}$. What is the rate of change in quantity demanded if the price is \$10?

This rate of change is the slope of the tangent to the demand curve at point A with coordinates $p = 10$ and $Q = 200/10 = 20$, as shown in Fig. 5.4.

To determine the slope of the tangent we must evaluate the derivative of the function at $p = 10$. Given

$$f'(p) = \frac{dQ}{dp} = \lim_{\Delta p \to 0} \frac{\Delta Q}{\Delta p} = \lim_{\Delta p \to 0} \frac{f(p + \Delta p) - f(p)}{\Delta p}$$

$$f(p + \Delta p) = \frac{200}{p + \Delta p}$$

Fig. 5.4 Slope of the tangent as rate of change

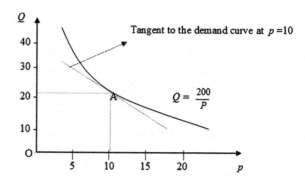

$$f'(p) = \frac{dQ}{dp} = \lim_{\Delta p \to 0} \frac{\dfrac{200}{p+\Delta p} - \dfrac{200}{p}}{\Delta p} = \lim_{\Delta p \to 0} \frac{\dfrac{200p - 200(p+\Delta p)}{p(p+\Delta p)}}{\Delta p}$$

$$= \lim_{\Delta p \to 0} \frac{\dfrac{200p - 200p - 200\Delta p}{p(p+\Delta p)}}{\Delta p} = \lim_{\Delta p \to 0} \frac{\dfrac{-200\Delta p}{p(p+\Delta p)}}{\Delta p}$$

$$= \lim_{\Delta p \to 0} \left(-\frac{200}{p^2 + p\Delta p} \right)$$

As Δp approaches zero, $p\Delta p$ goes to zero and

$$f'(p) = -\frac{200}{p^2}$$

Evaluating this derivative at p = 10, we have

$$f'(10) = -\frac{200}{10^2} = -2$$

At $p = 10$, the quantity demanded $Q = \dfrac{200}{10} = 20$. If the price increases by one cent to 10.01, the quantity demanded drops to $Q = (200/10.01) = 19.98$, a decrease of 0.02 units. This change in quantity demanded is -2 times the change in price $(-2 * 0.01 = -0.02)$.

The negative rate of change of the function signifies the inverse relationship that exists between price and quantity. At $p = 5$, the rate of change of the demand function is $f'(5) = -\dfrac{200}{5^2} = -8$. If price is reduced by one cent to \$4.99 the quantity demanded would be $Q = 40.08$, an increase of $(-8)(-0.01) = 0.08$ units.

5.2 Limits, Continuity, and Differentiability

5.2.1 Limits

Isaac Newton introduced the concept of a limit in the seventeenth century and labeled it as the "basic concept in calculus". We will try to provide a non-technical/non-threatening treatment of a limit.

Assume y is a function of x, $y = f(x)$. Assume x_1, an arbitrary value of x, is in the domain of the function. When we say 'the limit of $f(x)$, as x approaches x_1, is L' we mean that the values of $f(x)$ get closer and closer to L as x gets closer and closer to x_1, both *from below* and *from above* (or *from the left* and *from the right*.) Here, the objective is not to determine the value of the function when $x = x_1$ (where the function may be not even defined), but rather to study the behavior of the function in the vicinity of x_1. This limit is expressed symbolically as

$$\lim_{x \to x_1} f(x) = L$$

The following example should help in understanding this concept.

Let $y = f(x) = \dfrac{x - 5}{x^2 - 25}$ What is the limit of this function L as x approaches 5?

$$\lim_{x \to 5} \frac{x - 5}{x^2 - 25} = L$$

Notice that this function is not defined at $x = 5$. But, as stated above, we would like to see the pattern of values of the function when x is in the vicinity of 5. The following table gives the values of $f(x)$ for values of x approaching 5 from above and below.

It should be clear from Table 5.1 that the value of the function approaches 0.1 when x takes values closer and closer to 5; that is,

Table 5.1 Limit of $f(x) = \dfrac{x - 5}{x^2 - 25}$ as x approaches 5

$f(x)$ as x approaches 5 from below		$f(x)$ as x approaches 5 from above	
x	$f(x)$	x	$f(x)$
4.5	0.1052632	5.5	0.0952381
4.8	0.1020408	5.2	0.0980392
4.9	0.1010101	5.1	0.0990099
4.99	0.1001001	5.01	0.0999001
4.999	0.1000100	5.001	0.0999900
4.9999	0.1000010	5.0001	0.0999990
4.99999	0.1000001	5.00001	0.0999999

$$\lim_{x\to 5} \frac{x-5}{x^2-25} = 0.1$$

Approaching x_1 from the left or from below means that we start with values of x less than x_1 and incrementally approach x_1. In our example, we approached 5 from below or from the left starting at 4.5. This is the definition of the *left- hand limit*. The *right-hand limit* is similarly defined. For the right-hand limit, in our example, we started at 5.5 and gradually moved closer to 5.

$$\lim_{x\to x_1^-} f(x)$$

is the symbolic expression of the left-hand limit, and similarly

$$\lim_{x\to x_1^+} f(x)$$

is the symbolic expression of the right-hand limit of the function. The idea of the left-hand and right-hand side limits is depicted in Figs. 5.5 and 5.6.

In Fig. 5.5, as x approaches 5 from the left, i.e. x-values less than 5, $f(x)$ approaches 4. The left-hand limit of this function is 4. As x approaches 5 from the right, i.e. x-values greater than 5, $f(x)$ approaches 4 too. The right-hand limit of this function is also 4. Since the right-hand and left-hand limits of this function are equal, the limit of this function at 4 is 5.

Fig. 5.5 This function has a limit at 5

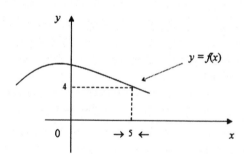

Fig. 5.6 This function limit's at 3 is not defined

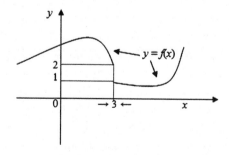

In Fig.5.6, as x approaches 3 from the left, i.e. x-values less than 3, $f(x)$ approaches 2. Thus, the left-hand limit of this function is 2. But as x approaches 3 from the right, $f(x)$ approaches 1. Thus, the right-hand limit of this function is 1. Since the right-hand and left-hand limits of this function are not equal, the limit of this function at 3 is not defined. By using the definition of one-sided limits, we can refine the definition of the limit by stating that a function has a limit at a certain point if it has the same left-hand and right-hand limit at that point.

$$\lim_{x \to x_1} f(x) = L \quad \text{if and only if} \quad \lim_{x \to x_1^-} f(x) = L \quad \text{and} \quad \lim_{x \to x_1^+} f(x) = L$$

Rules of Limits

If $f(x)$ is a constant function $f(x) = c$ where c is a constant, then it is intuitively obvious that

1. $\lim_{x \to x_1} f(x) = \lim_{x \to x_1} c = c$ the limit of a constant is the constant itself.

Assuming a constant c and two functions $f(x)$ and $g(x)$, then the following rules of limits should be also intuitively clear:

2. $\lim_{x \to x_1} [cf(x)] = c \lim_{x \to x_1} f(x)$

3. $\lim_{x \to x_1} [f(x) + g(x)] = \lim_{x \to x_1} f(x) + \lim_{x \to x_1} g(x)$

4. $\lim_{x \to x_1} [f(x) - g(x)] = \lim_{x \to x_1} f(x) - \lim_{x \to x_1} g(x)$

5. $\lim_{x \to x_1} [f(x).g(x)] = \lim_{x \to x_1} f(x). \lim_{x \to x_1} g(x)$

6. $\lim_{x \to x_1} \dfrac{f(x)}{g(x)} = \dfrac{\lim_{x \to x_1} f(x)}{\lim_{x \to x_1} g(x)}$ Assuming $\lim_{x \to x_1} g(x) \neq 0$

7. $\lim_{x \to x_1} [f(x)]^n = [\lim_{x \to x_1} f(x)]^n$

5.2.2 Continuity

If the limit of a function $f(x)$ as x approaches x_1 is $f(x_1)$, we say the function is *continuous* at x_1. This is formally expressed as

$$\text{If} \quad \lim_{x \to x_1} f(x) = f(x_1) \quad \text{then} \quad f(x) \text{ is continuous at } x_1$$

It is, of course, implicitly understood that x_1 is in the domain of the function $f(x)$ and $\lim f(x)$ as $x \to x_1$ exists.

A function is continuous over an interval if it is continuous at every value in that interval. This means that the graph of the function over the interval has no gap or break, and we can draw the curve without removing our pen from the paper. The graph of $f(x)$ in Fig.5.5 is smooth and without gaps and breaks, so the function is continuous over the interval shown. On the contrary, as the break in the graph of the function in Fig. 5.6 indicates, the function is not continuous (or it is *discontinuous*) at $x = 3$.

The function $f(x) = y = x^2 + 2$ is continuous at $x = 2$. Using rules 3, 7, and 1 for limits in sequence, we have

$$\lim_{x \to 2} (x^2 + 2) = \lim_{x \to 2} x^2 + \lim_{x \to 2} 2 = [\lim_{x \to 2} x]^2 + 2 = 2^2 + 2 = 6$$

As a matter of fact this quadratic function is continuous over the entire set of real numbers R. In general, all polynomials are continuous over their domains.

Function $Q = f(p) = \dfrac{200}{p}$ is not continuous (or it is discontinuous) at $p = 0$. Applying rule 6, we have

$$\lim_{p \to 0} \frac{200}{p} = \frac{\lim_{p \to 0} 200}{\lim_{p \to 0} p} = \frac{200}{0}$$

So, the function is undefined and is discontinuous when the price is equal to zero. If p is zero, then this good is a free good and must be treated outside of the market supply and demand analysis.

5.2.3 Differentiability

In mathematics, *differentiation* is the act or process of finding the derivative of a function. A function $f(x)$ is *differentiable* at x_1 if the derivative of the function at x_1, $f'(x_1)$, exists. As we saw earlier, there is a direct connection between the derivative of a function and the tangent to a curve. Geometrically, the derivative of $f(x)$ at $x_1, f'(x_1)$, is the slope of the tangent line to the graph of the function at the point $(x_1, f(x_1))$.

For a function to be differentiable at a point, it must be continuous at that point. This is a *necessary* condition. But continuity of a function at a point is not *sufficient* for differentiability.

The fact that continuity of the function at a point does not guarantee differentiability is exhibited by Fig. 5.7. In Fig. 5.7a, the function is discontinuous at x_1 and one can virtually draw an infinite number of tangents to the curve at the two ends of the gap in the curve (2 are drawn as a sample). The graph of the function in Fig. 5.7b has a sharp point or corner at x_1, and one can draw an infinite number of tangent lines to the curve at this point too, including a vertical line with an indeterminate slope (the slope of a vertical line is ∞). Note that the function in Fig. 5.7b is continuous, just not differentiable. In Fig. 5.7c, the graph of the function has a vertical bend, so the

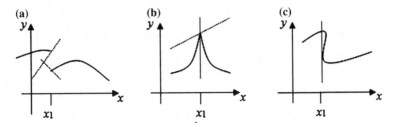

Fig. 5.7 Non differentiable functions

tangent to the graph at this point is again a vertical line, and the function is similarly not differentiable at x_1.[2]

5.3 Rules of Differentiation

In this section we discuss techniques of finding derivatives of various algebraic univariate functions (functions with only one independent variable). The rules of differentiating a bi- or multi- variate function are discussed later in the chapter. It is possible to calculate derivatives directly from the definition, but this approach is tedious and time consuming. The following rules of differentiation greatly simplify the process of finding derivatives.

1. Derivative of a constant function:

The derivative of a constant function $y = f(x) = c$ is always 0. The graph of a constant function is a horizontal line parallel to the x-axis. Since the slope of a horizontal line is 0, then $f'(x) = 0$. We can also prove this result by using the definition:

$$f'(x) = \frac{dy}{dx} = \lim_{\Delta x \to 0} \frac{\Delta y}{\Delta x} = \lim_{\Delta x \to 0} \frac{f(x + \Delta x) - f(x)}{\Delta x} = \lim_{\Delta x \to 0} \frac{c - c}{\Delta x} = 0$$

2. Derivative of a power function $y = f(x) = x^n$:

We first find the derivative of simpler power functions $f(x) = x$, $f(x) = x^2$, and $f(x) = x^3$ with the hope of recognizing a pattern that could be generalized to $f(x) = x^n$.

Applying the definition to $f(x) = x$, we have

$$f'(x) = \frac{dy}{dx} = \lim_{\Delta x \to 0} \frac{f(x + \Delta x) - f(x)}{\Delta x} = \lim_{\Delta x \to 0} \frac{x + \Delta x - x}{\Delta x} = \lim_{\Delta x \to 0} \frac{\Delta x}{\Delta x} = 1$$

---

[2] The curve shown in Fig. 5.7c is technically not a function since there are multiple y-values for a given x-value.

Therefore, $\dfrac{d}{dx}x = 1$. Note that the graph of function $y = x$ is a ray through the origin with a slope of 1.

What is the derivative of $f(x) = x^2$? Applying the definition, we have

$$f'(x) = \lim_{\Delta x \to 0} \frac{f(x + \Delta x) - f(x)}{\Delta x} = \lim_{\Delta x \to 0} \frac{(x + \Delta x)^2 - x^2}{\Delta x}$$

$$= \lim_{\Delta x \to 0} \frac{x^2 + (\Delta x)^2 + 2x\Delta x - x^2}{\Delta x} = \lim_{\Delta x \to 0} \frac{(\Delta x)^2 + 2x\Delta x}{\Delta x}$$

$$= \lim_{\Delta x \to 0} \frac{\Delta x(\Delta x + 2x)}{\Delta x} = \lim_{\Delta x \to 0} (\Delta x + 2x) = 2x$$

Therefore, $\dfrac{d}{dx}x^2 = 2x$

Determining the derivative of $y = f(x) = x^3$ is a little more involved but similarly straightforward

$$f'(x) = \lim_{\Delta x \to 0} \frac{f(x + \Delta x) - f(x)}{\Delta x} = \lim_{\Delta x \to 0} \frac{(x + \Delta x)^3 - x^3}{\Delta x} \qquad (5.4)$$

After expanding $(x + \Delta x)^3$, we can write

$$\frac{(x + \Delta x)^3 - x^3}{\Delta x} = \frac{x^3 + 3x^2\Delta x + 3x(\Delta x)^2 + (\Delta x)^3 - x^3}{\Delta x}$$

$$= \frac{\Delta x[3x^2 + 3x\Delta x + (\Delta x)^2]}{\Delta x} = 3x^2 + 3x\Delta x + (\Delta x)^2$$

And going back to (5.4), we have

$$\lim_{\Delta x \to 0} \frac{(x + \Delta x)^3 - x^3}{\Delta x} = \lim_{\Delta x \to 0} [3x^2 + 3x\Delta x + (\Delta x)^2]$$

As Δx approaches 0, $3x\Delta x$ and $(\Delta x)^2$ drop out, therefore

$$f'(x) = \frac{d}{dx}x^3 = 3x^2$$

Summarizing the results of our exercise so far

$$
\begin{array}{llll}
f(x) = x^1 & f'(x) = 1 & \text{or} & f'(x) = 1x^{1-1} = 1x^0 = 1 \\
f(x) = x^2 & f'(x) = 2x & \text{or} & f'(x) = 2x^{2-1} = 2x^1 = 2x \\
f(x) = x^3 & f'(x) = 3x^2 & \text{or} & f'(x) = 3x^{3-1} = 3x^2
\end{array}
$$

The pattern emerging leads to the derivative of a general power function

$$y = f(x) = x^n \qquad\qquad y' = f'(x) = nx^{n-1}$$

3. Derivative of $cf(x)$, where c is constant:

Let $\varphi(x) = cf(x)$. Then by using the definition of derivatives

$$\varphi'(x) = \lim_{\Delta x \to 0} \frac{\varphi(x + \Delta x) - \varphi(x)}{\Delta x} = \lim_{\Delta x \to 0} \frac{cf(x + \Delta x) - cf(x)}{\Delta x}$$

$$= \lim_{\Delta x \to 0} c \left[\frac{f(x + \Delta x) - f(x)}{\Delta x} \right]$$

And by rule 2 of limits

$$\lim_{\Delta x \to 0} c \left[\frac{f(x + \Delta x) - f(x)}{\Delta x} \right] = c \lim_{\Delta x \to 0} \frac{f(x + \Delta x) - f(x)}{\Delta x} = cf'(x)$$

4. Derivative of $f(x) \pm g(x)$, the sum or difference rule:

We find the derivative of the sum of two functions and then extend the result to their difference. If $f(x)$ and $g(x)$ are two differentiable functions of x, then by definition

$$\frac{d}{dx}[f(x) + g(x)] = \lim_{\Delta x \to 0} \frac{[f(x + \Delta x) + g(x + \Delta x)] - [f(x) + g(x)]}{\Delta x}$$

$$= \lim_{\Delta x \to 0} \frac{[f(x + \Delta x) - f(x)] + [g(x + \Delta x) - g(x)]}{\Delta x}$$

$$= \lim_{\Delta x \to 0} \left[\frac{f(x + \Delta x) - f(x)}{\Delta x} + \frac{g(x + \Delta x) - g(x)}{\Delta x} \right]$$

and by using rule 3 of limits

$$\frac{d}{dx}[f(x) + g(x)] = \lim_{\Delta x \to 0} \frac{f(x + \Delta x) - f(x)}{\Delta x} + \lim_{\Delta x \to 0} \frac{g(x + \Delta x) - g(x)}{\Delta x}$$

$$= f'(x) + g'(x)$$

We can easily extend this result to the sum of more than two functions.
If $y_i = f_i(x) \qquad i = 1, 2, 3, ..., n$ are n different functions of x, then

$$\frac{d}{dx} \sum_{i=1}^{n} f_i(x) = \sum_{i=1}^{n} f_i'(x)$$

It is also easy to see that the derivative of the difference of two functions $f(x)$ and $g(x)$ is the difference of their derivatives, that is

$$\frac{d}{dx}[f(x) - g(x)] = f'(x) - g'(x)$$

5. Derivative of $y = f(x)g(x)$, the product rule:

Assume $f(x)$ and $g(x)$ are two differentiable functions of x, then by definition

$$\frac{dy}{dx} = \frac{d}{dx}[f(x)\, g(x)] = \lim_{\Delta x \to 0} \frac{f(x + \Delta x)\, g(x + \Delta x) - f(x)\, g(x)}{\Delta x}$$

By adding and subtracting $f(x + \Delta x)\, g(x)$ to the numerator, we have

$$\frac{d}{dx}[f(x)\, g(x)] = \lim_{\Delta x \to 0} \frac{f(x + \Delta x)\, g(x + \Delta x) - f(x)\, g(x) + f(x + \Delta x)\, g(x) - f(x + \Delta x)\, g(x)}{\Delta x}$$

$$= \lim_{\Delta x \to 0} \frac{f(x + \Delta x)\, [g(x + \Delta x) - g(x)] + g(x)[f(x + \Delta x) - f(x)]}{\Delta x}$$

$$= \lim_{\Delta x \to 0} \left[f(x + \Delta x)\, \frac{g(x + \Delta x) - g(x)}{\Delta x} + g(x)\, \frac{f(x + \Delta x) - f(x)}{\Delta x} \right]$$

Applying rules 3 and 5 of limits

$$\frac{d}{dx}[f(x)\, g(x)] = \lim_{\Delta x \to 0} f(x + \Delta x)\, \lim_{\Delta x \to 0} \frac{g(x + \Delta x) - g(x)}{\Delta x}$$

$$+ \lim_{\Delta x \to 0} g(x)\, \lim_{\Delta x \to 0} \frac{f(x + \Delta x) - f(x)}{\Delta x}$$

In the above expression, as Δx approaches 0, $f(x + \Delta x) = f(x)$. Note also that $\lim_{\Delta x \to 0} g(x)$ is simply $g(x)$ because there is no change in the function. Therefore,

$$\frac{d}{dx}\, [f(x)\, g(x)] = f(x)\, g'(x) + g(x)\, f'(x)$$

The five rules we have learned so far can be summarized:

1. $\dfrac{d}{dx} c = 0$

2. $\dfrac{d}{dx} x^n = n x^{n-1}$

3. $\dfrac{d}{dx} cf(x) = cf'(x)$

4. $\dfrac{d}{dx}[f(x) \pm g(x)] = f'(x) \pm g'(x)$

5. $\dfrac{d}{dx}[f(x)\, g(x)] = f(x)\, g'(x) + f'(x)\, g(x)$

Using these 5 rules we can generate a series of new rules, such as

6. Derivative of the reciprocal of a power function, $f(x) = \dfrac{1}{x^n}$:

We write $f(x) = \dfrac{1}{x^n} = x^{-n}$ and then apply rule 2

$$\frac{d}{dx}x^{-n} = -nx^{-n-1} = -nx^{-(n+1)} = -\frac{n}{x^{n+1}}$$

7. Derivative of $y = [f(x)]^n$:

We can express $y = [f(x)]^2 = f(x)f(x)$ and apply rule 5, the product rule

$$\frac{d}{dx}[f(x)]^2 = \frac{d}{dx}[f(x)\,f(x)] = f(x)f'(x) + f'(x)f(x) = 2f(x)f'(x)$$

We can also express $y = [f(x)]^3 = [f(x)]^2 f(x)$ and apply rule 5

$$\frac{d}{dx}[f(x)]^3 = \frac{d}{dx}[f(x)]^2 f(x) = [f(x)]^2 f'(x) + 2f(x)f'(x)f(x)$$

$$= 3[f(x)]^2 f'(x)$$

Recognizing the pattern, we can safely infer that

$$\frac{d}{dx}[f(x)]^n = nf'(x)\,[f(x)]^{n-1}$$

8. Derivative of the mth root of a function $y = \sqrt[m]{f(x)}$:

Let's first find the derivative of $y = \sqrt{x}$. We can write $y = \sqrt{x} = x^{\frac{1}{2}}$ and then apply rule 2

$$\frac{d}{dx}\sqrt{x} = \frac{d}{dx}x^{\frac{1}{2}} = \frac{1}{2}x^{\frac{1}{2}-1} = \frac{1}{2}x^{-\frac{1}{2}} = \frac{1}{2}\frac{1}{x^{\frac{1}{2}}} = \frac{1}{2\sqrt{x}}$$

We determine the derivative of a general root function $y = \sqrt[m]{f(x)}$ by writing $y = f(x)^{\frac{1}{m}}$ and then applying rule 7

$$\frac{d}{dx}\sqrt[m]{f(x)} = \frac{d}{dx}f(x)^{\frac{1}{m}} = \frac{1}{m}f'(x)\left[f(x)^{\frac{1}{m}-1}\right] = \frac{1}{m}f'(x)\left[f(x)^{-\frac{m-1}{m}}\right]$$

$$\frac{d}{dx}\sqrt[m]{f(x)} = \frac{f'(x)}{m\sqrt[m]{f(x)^{m-1}}}$$

9. Derivative of $y = \dfrac{f(x)}{g(x)}$ the quotient rule:

We write $y = \dfrac{f(x)}{g(x)} = f(x)[g(x)]^{-1}$ and then apply rules 5 and 7. By rule 5

$$\frac{d}{dx} f(x)[g(x)]^{-1} = f(x)\frac{d}{dx}[g(x)]^{-1} + f'(x)[g(x)]^{-1} \qquad \text{and by rule 7}$$

$$\frac{d}{dx}[g(x)]^{-1} = -g'(x)[g(x)]^{-2} = -\frac{g'(x)}{[g(x)]^2} \qquad \text{So we have}$$

$$\frac{d}{dx}\frac{f(x)}{g(x)} = -f(x)\frac{g'(x)}{[g(x)]^2} + \frac{f'(x)}{g(x)} = \frac{f'(x)g(x) - f(x)g'(x)}{[g(x)]^2}$$

10. Function of a function or chain rule:

Assume z is a function of y, $z = f(y)$, and y is a function of x, $y = g(x)$, then the instantaneous rate of change of z with respect to x, $\dfrac{dz}{dx}$, is determined as

$$\frac{dz}{dx} = \frac{dz}{dy}\frac{dy}{dx}$$

By using the ten rules developed above we can differentiate polynomial-based functions with any degree of complexity. Other rules are needed for trigonometric, exponential, and logarithmic functions. We will cover the derivatives of exponential and logarithmic functions in Sect. 9.4.

───────────

Examples

Find the derivative of the following functions

1. $y = 5x^3 + 10x^2 - 25x + 5$

   We first apply rule 4 and then a combination of rules 2, 3, and 1

$$\frac{d}{dx}(5x^3 + 10x^2 - 25x + 5) = \frac{d}{dx}5x^3 + \frac{d}{dx}10x^2 - \frac{d}{dx}25x + \frac{d}{dx}5$$

$$y' = \frac{dy}{dx} = 15x^2 + 20x - 25$$

2. $y = (3x^2 + 5)(2x - 1)$

   Let $f(x) = (3x^2 + 5)$ and $g(x) = (2x - 1)$, then

$$y = (3x^2 + 5)(2x - 1) = f(x)\,g(x)$$

Applying rule 5, the product rule, we have

$$\frac{dy}{dx} = f(x)g'(x) + f'(x)g(x) = (3x^2 + 5)(2) + (6x)(2x - 1)$$
$$= 6x^2 + 10 + 12x^2 - 6x$$
$$= 18x^2 - 6x + 10$$

3. $f(u) = \sqrt{u}(u - 1)$

$$\frac{d}{du}f(u) = \frac{u - 1}{2\sqrt{u}} + \sqrt{u}$$

4. $\varphi(t) = \dfrac{t^2 - 2}{t + 5}$

$$\frac{d}{dt}\varphi(t) = \frac{2t(t + 5) - (t^2 - 2)}{(t + 5)^2} = \frac{t^2 + 10t + 2}{(t + 5)^2}$$

5. Find the slope of the tangent to the graph of $y = 3x^2 - 10x + 25$ at $x = 4$.
   The slope of this tangent is the value of the derivative of the function at $x = 4$.

$$f'(x) = 6x - 10 \quad \longrightarrow \quad f'(4) = 6(4) - 10 = 14$$

6. What is the rate of change of $f(w) = \dfrac{5w - 2}{w^2 + 3}$ at the point where w is equal 3.

$$f'(w) = \frac{5(w^2 + 3) - 2w(5w - 2)}{(w^2 + 3)^2} = \frac{-5w^2 + 4w + 15}{(w^2 + 3)^2}$$

$$f(3) = \frac{-5(3)^2 + 4(3) + 15}{(3^2 + 3)^2} = \frac{-18}{144} = -0.125$$

7. An object dropped from the top of a tall building falls a distance which is a function of time the object is in free fall. The Galileo law, in honor of Galileo Galilei who discovered it, is written as

$$D(t) = 4.9t^2$$

where D is distance traveled in meters and t is time in seconds since the object was released. For example, an object travels about $4.9(5)^2 = 122.5$ meters in its first 5 seconds of free fall. After 10 seconds the distance traveled is 490 meters, indicating that the object covers 3 times more distance in its "second" 5 seconds of free fall compared to the "first" 5 seconds. This is due to the earth's gravitational force causing free falling objects to pick up speed as they fall. Determining the speed of a falling object at various points or over intervals of time frequently arises in physics. The average speed (*velocity*) of an object is generally computed as the ratio of distance traveled to the length of the time interval, that is

$$\text{Average Speed} = \frac{\text{Distant Traveled}}{\text{Elapsed Time}}$$

For example, if a car travels 50 miles from 9:30 to 10:10, the average speed of this car is

$$\text{Average Speed} = \frac{50}{40} = 1.25 \quad \text{miles/min or 75 miles/hours}$$

What is the average speed of a falling object between 5 and 6 seconds after its release?

The distance traveled by the object in the interval of 5 to 6 seconds is

$$D(6) - D(5) = 4.9(6)^2 - 4.9(5)^2 = 53.9 \quad \text{meters}$$

$$\text{Average Speed} = \frac{53.9}{6 - 5} = 53.9 \quad \text{meters/second}$$

What is speed of an object 5 seconds into free fall?

Speed or velocity is the instantaneous rate of change of an object's distance traveled with respect to time. Therefore S, speed, is

$$S = \frac{d}{dt}D(t) = \frac{d}{dt}4.9t^2 = 9.8t$$

and after 5 seconds an object moves at $S = 9.8(5) = 49$ meters/second.

8. A firm has the following total cost (TC) function

$$TC(Q) = Q^3 - 24Q^2 + 200Q + 500$$

where Q is the level of output in 100 units and TC is in \$100. What is the rate of increase in production cost if this firm is currently producing 1000 units?

The rate of change in production cost is the *marginal cost (MC)*, that is

$$MC = \lim_{\Delta Q \to 0} \frac{\Delta TC}{\Delta Q} = \frac{d}{dQ}TC(Q) = TC'(Q)$$

In our example, MC is

$$MC = TC'(Q) = 3Q^2 - 48Q + 200, \text{ therefore}$$

$$\text{At } Q = 10, \quad MC = TC'(10) = 3(10)^2 - 48(10) + 200 = 20$$

$MC = 20$ indicates that at the current level of output the rate of change in total cost is 20; that is, a small increase in production will cost additional 20 times that increase in cost. Note that the total cost of producing 1000 units is

$$TC(10) = (10)^3 - 24(10)^2 + 200(10) + 500 = 1100 \quad \text{or } \$110,000$$

If this firm decides to produce one extra unit, 1001st unit, the total cost would be

$$TC(10.01) = (10.01)^3 - 24(10.01)^2 + 200(10.01) + 500$$
$$= 1100.20 \quad (\text{or } \$110,020)$$

indicating the additional $20 cost for the 1001st unit.

9. A firm has the following total cost function

$$TC = Q^3 - 30Q^2 + 400Q + 500$$

At what level of output is the firm's marginal cost equal to $100?
The marginal cost function is the derivative of the total cost function with respect to Q, therefore,

$$MC = \frac{d}{dQ}TC(Q) = 3Q^2 - 60Q + 400$$

To determine the level of output where MC is 100

$$3Q^2 - 60Q + 400 = 100 \quad \longrightarrow \quad 3Q^2 - 60Q + 300 = 0 \quad \text{or} \quad Q^2 - 20Q + 100 = 0$$

Using the quadratic formula, solution to the above quadratic equation is

$$Q = \frac{-(-20) \pm \sqrt{(-20)^2 - 4*100}}{2} = \frac{20}{2} = 10$$

Then at $Q = 10$ the rate of change in the cost of producing one additional unit of output is $100.

10. If demand for the product of a monopolistically competitive firm is given by

$$Q = 215 - 5P$$

what are the firm's total and marginal revenue functions?
Total revenue (TR) is $P * Q$, where Q is the number of units demanded by the household and sold by the firm. To express total revenue as a function of Q we must first solve the demand equation in terms of price, that is

$$5P = 215 - Q \quad \text{or} \quad P = 43 - 0.2Q \quad \text{then we have,}$$
$$TR = TR(Q) = P * Q = (43 - 0.2Q) * Q = 43Q - 0.2Q^2$$

Since the marginal revenue (*MR*) is the rate of change of the total revenue, then in this example

$$MR = \frac{d}{dQ}TR(Q) = TR'(Q) = 43 - 0.4Q$$

In general, if a demand equation is linear $P = a - bQ$ then total revenue function is

$$TR = PQ = (a - bQ)Q = aQ - bQ^2$$

Differentiating *TR* with respect to *Q* we get the marginal revenue function

$$MR = \frac{dTR}{dQ} = a - 2bQ$$

Compare *MR* with the demand function. They both have the same *y*-intercept *a*, but slope of the *MR* function is twice that of the demand function. The *x*-intercept of the demand curve is $\frac{a}{b}$. The *x*-intercept of *MR* is $\frac{a}{2b}$, that is the *MR* curve cuts the horizontal axis at a point half the distance between the origin and the point where the demand curve cuts the axis. This is very helpful in graphing *MR* curves. Figure 5.8 shows the graph of a linear demand function and the associated marginal revenue function.

11. A monopolist faces the following demand function for its product in the market

$$Q = \frac{9335}{P^{1.25}} \quad \text{or} \quad Q = 9335P^{-1.25}$$

Write the monopolist's marginal revenue function and graph it. What is the monopolist's additional revenue from selling the 100th unit of output?
The firm's total revenue is $TR = P * Q$. We first solve the demand function for *P* in terms of *Q*.

$$Q = \frac{9335}{P^{1.25}} \quad \longrightarrow \quad P^{1.25} = \frac{9335}{Q}$$

Fig. 5.8 Linear demand and marginal revenue curves

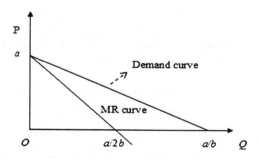

and after raising both sides of the equation to a power 0.8, the reciprocal of 1.25, we have

$$(P^{1.25})^{0.8} = \left(\frac{9335}{Q}\right)^{0.8} \quad \longrightarrow \quad P = \frac{1500}{Q^{0.8}}$$

then the total revenue function is

$$TR = \frac{1500}{Q^{0.8}}Q = 1500\,Q^{0.2}$$

The marginal revenue function is

$$MR = \frac{dTR}{dQ} = 0.2 * 1500Q^{0.2-1} = 300Q^{-0.8} \quad \text{or} \quad MR = \frac{300}{Q^{0.8}}$$

Using the marginal revenue function we can approximately determine the additional revenue from selling 100th unit of output as

$$MR = \frac{300}{Q^{0.8}}|_{Q=100} = \frac{300}{(100)^{0.8}} = \frac{300}{39.81} = 7.54$$

Figure 5.9 depicts the non-linear demand and marginal revenue curves.

12. A monopolistically competitive firm faces the following demand function for its product

$$P = \sqrt{10 - Q}$$

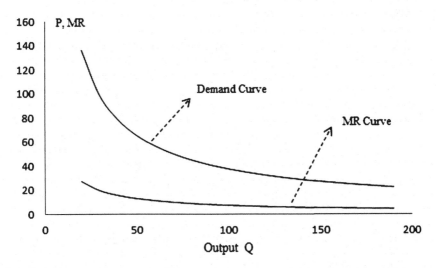

Fig. 5.9 Non-linear demand and marginal revenue curves

What is the firm's marginal revenue function?

$$TR = P * Q = Q\sqrt{10 - Q}$$

$$MR = \frac{dTR}{dQ} = \sqrt{10 - Q} - \frac{Q}{2\sqrt{10 - Q}}$$

$$MR = \frac{20 - 3Q}{2\sqrt{10 - Q}}$$

13. A firm has the following quadratic cost function

$$TC = 4Q^2 - 30Q + 400$$

Determine the point where the marginal cost curve intersects the average total cost curve.

$$MC = 8Q - 30$$

$$ATC = \frac{TC}{Q} = \frac{4Q^2 - 30Q + 400}{Q} = 4Q - 30 + \frac{400}{Q}$$

At the point of intersection, $MC = ATC$:

$8Q - 30 = 4Q - 30 + \frac{400}{Q}$ $\longrightarrow$ $4Q = \frac{400}{Q}$ leading to $Q^2 = 100$ or $Q = 10$
At $Q = 10$, $MC = ATC = 50$

5.4 Exercises

1. What are the y- intercept(s) and x-intercept(s) of graphs of the following functions?

   (a) $y = 3x - 10$ (b) $y = -5x + 10$

   (c) $f(x) = 12 + \frac{1}{3}x$ (d) $g(x) = 12 - \frac{1}{3}x$

   (e) $f(x) = x^2 + 5x + 6$ (f) $y = x^2 + x - 6$

   (g) $y = x^2 - 8x + 16$ (h) $y - 3x + 8 = 0$

   (i) $y = -0.005x + 0.01$ (j) $y = 10$

   (k) $x = -4$ (l) $y = x$

   (m) $Q = 20 - 0.5P$ (n) $Q = -10 + 2P$

2. Write the equation of the line joining points $P(5, 10)$ and $Q(20, 30)$.

3. A line joins point $A(2, 12)$ to point $B(4, 22)$ and point B to point $C(8, 38)$. Is this line a straight line?

4. Find the point of intersection of the lines with equations

$$y = 3x + 4$$
$$y = 5x - 10$$

5. Write the equation of the line that passes through the point $A(4, 6)$ and is parallel to a line with equation

$$y = 2x - 5$$

6. Use the rules of differentiation to find the derivative of the following functions

(a) $y = 3x - 10$ (b) $y = -5x + 10$

(c) $f(x) = 12x^2 + 1/3x$ (d) $g(x) = 12 - 1/3x^2$

(e) $f(x) = 1/3x^3 + 5x^2 - 6x$ (f) $y = x^2 + x - 6$

(g) $y = (x^2 - 8x + 16)^2$ (h) $y = \frac{3x-5}{x^2+10}$

(i) $y = 10$ (j) $\varphi(w) = \frac{3w^2-2}{w-1}$

7. Find the rate of change of the following functions at the given value

(a) $y = 3x - 10$ at $x = 5$

(b) $y = -5x + 10$ at $x = -10$

(c) $f(x) = 12x^2 + 1/3x$ at $x = 3$

(d) $f(x) = 1/3x^3 + 5x^2 - 6x$ at $x = -2$

(g) $y = (w^2 - 8w + 16)^2$ at $w = 4$

(h) $f(t) = \frac{3t-5}{t^2+10}$ at $t = 2$

(d) $g(x) = 12\sqrt{x} - 1/3x^2$ at $x = 9$

8. Differentiate

(a) $f(t) = \frac{3t}{t^2-4}$ (b) $y = ax^2 + bx + c$

(c) $f(x) = ax^3 + bx^2 + cx + d$ (d) $g(z) = \frac{z^2}{3+\sqrt{z}}$

(e) $y = (3x^2 + 2)^4$ (f) $y = 2x\sqrt[4]{x^2 - 3}$

(g) $z = \frac{ax-\beta x^2}{x-1}$ (h) $w = 5u^{\frac{1}{3}} + 2u$

9. The total cost function of a steel company is

$$TC(Q) = 0.0003Q^3 - 0.02Q^2 + 0.5Q + 300$$

What is the firm's marginal cost at the level of output 50? At the level of output 100?

10. A firm has the following total cost function

$$TC(Q) = Q^3 - 24Q^2 + 200Q + 500$$

   (a) What is this firm's fixed cost?

   (b) At what level(s) of output does the firm's marginal cost equal $50?

11. A monopolist faces the following demand function in the market

$$P = 15500 - 100\sqrt{Q}$$

   find its marginal revenue if it produces 10000 units.

12. A firm's profit, π, is the difference between its total revenue TR and its total costs TC. If $TR = f(Q)$ and $TC = g(Q)$ write the profit function and find the firm's marginal profit. What is the economic interpretation of marginal profit?

13. Market demand for a good produced by a monopolist is

$$P = \frac{1500}{Q^{0.8}}$$

   What is the monopolist's marginal revenue from selling the 100th unit?

14. A firm has the following total cost TC function

$$TC = Q^3 - 30Q^2 + 400Q + 500$$

   At what level of output the firm's marginal cost MC is $100?

15. A monopolistically competitive firm is facing the following demand for its product

$$Q = \frac{1000}{P^2}$$

   What is the firm's marginal revenue from selling the 100th unit?

16. A firm has the following quadratic cost function

$$TC = 0.025Q^2 + 12Q + 500$$

   Find the coordinates of the point where the marginal cost curve intersects the average total cost curve.

5.5 Optimization: Determining the Maximum and/or Minimum of a Function

Consider the set of two graphs in Fig. 5.10. The top one is the graph of the cubic function $y = (1/3)x^3 - 9x$ over the domain $D = \{x \in R \mid -5 \le x \le 5\}$. The bottom one is the graph of the first derivative of the function $y' = x^2 - 9$. As is generally the case with cubic functions, the graph of $y = (1/3)x^3 - 9x$ has a hill and a valley. As x increases from -5 the function increases, reaching its highest value of $y = 18$ at $x = -3$. The value of the function decreases as x moves toward zero and into positive values. At $x = 3$ the function reaches it's lowest value of $y = -18$.

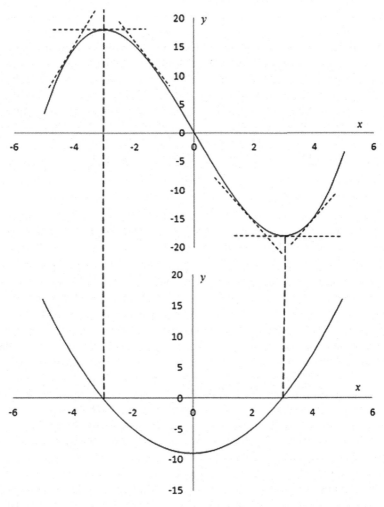

Fig. 5.10 Graphs of $y = (1/3)x^3 - 9x$ (*top*) and $y' = x^2 - 9$ (*bottom*)

Fig. 5.11 Local and global
extreme points

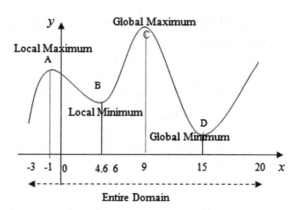

Expressed differently, this function reaches it's *peak* or it's *maximum* when $x = -3$
and it's *trough* or it's *minimum* at $x = 3$, over the specified domain $-5 \leq x \leq 5$. If
we extend the domain to, for example, $-7 \leq x \leq 7$ then the lowest and the highest
values of the function would change to $y = -51.333$ and $y = 51.333$. The maximum
and minimum values of a function are called its *extreme values* or its *extrema*.

We must distinguish between a *local* or *relative extremum* (local minimum or
maximum) and a *global* or *absolute extremum* (global minimum or maximum). The
function depicted in Fig. 5.11 has several extreme points. Assume that

$$D = \{x \in R \mid -3 \leq x \leq 20\}$$

is the entire domain of the function, then points C and D are the maximum and
minimum of the function over the entire domain. They are called global extreme
points.

If a function has global maximum at point x^*, then $f(x^*) \geq f(x)$ for all x in the
domain. Similarly, if a function has a global minimum at x^*, then $f(x^*) \leq f(x)$ for
all x in the domain. In Fig. 5.11 the function has a global maximum at $x^* = 9$ and a
global minimum at $x^* = 15$. On the other hand consider a subset of the entire domain
where x takes values between -3 and 6. Over this limited interval the function has
a local or relative maximum point A at $x^* = -1$ and a local or relative minimum
point B at $x^* = 4.6$.

Table 5.2 gives values of the cubic function $y = \frac{1}{3}x^3 - 9x$, its first derivative
$y' = x^2 - 9$, and its second derivative $y'' = 2x$ for selected values of x.

It is clear from Fig. 5.10 and Table 5.2 that as x increases from -5 to -3 the
value of the function increases from 3.33 to a maximum of 18.00, while values of
the derivative of the function decreases from 16 to 0.

Recall that the derivate of the function evaluated at any specific point is the
slope of the tangent to the function or the rate of change of the function at that
point. As we move uphill from the altitude of 3.33 toward the summit (which has

Table 5.2 $f(x), f'(x)$ and $f''(x)$ for selected values of x

x	$y = 1/3x^3 - 9x$	$y' = x^2 - 9$	$y'' = 2x$
-5	3.33	16	-10
-4.5	10.13	11.25	-9
-4	14.67	7	-8
-3.5	17.21	3.25	-7
-3	18.00	0	-6
-2	15.33	-5	-4
-1	8.67	-8	-2
0	0.00	-9	0
1	-8.67	-8	2
2	-15.33	-5	4
3	-18.00	0	6
3.5	-17.21	3.25	7
4	-14.67	7	8
4.5	-10.13	11.25	9
5	-3.33	16	10

a height of 18.00) the rate of change of the function decreases. This is measured by the first derivative or slope of the tangent to the graph of the function. Note that $f'(-5) = 16$, $f'(-4.5) = 11.25$, $f'(-4) = 7$, and $f'(-3.5) = 3.25$. At the summit or maximum the tangent to the graph is parallel to the x-axis with slope equal to zero, $f'(-3) = 0$. Moving to the right of the maximum, the slope of the tangent changes direction and becomes negative, e.g. $f'(-2) = -5$. Note that at the maximum point $f''(-3) = -6 < 0$.

We observe a similar pattern as we move to the right of $x = -3$. This time the value of the function decreases and reaches a minimum of -18 at $x = 3$. Over the interval $-3 < x < 3$ the slopes of the tangents to the function are negative. At $x = 3$ the first derivative of the function is zero. That is, the tangent to the graph of the function at the minimum point is again parallel to the x-axis with slope equal to zero. To the right of the minimum the slope of the tangent changes direction and becomes positive. Also note that in the case of a minimum $f''(3) = 6 > 0$.

The above description leads to two important conclusions about local extreme points:

(1) At the function's extrema (maximum or minimum points) its derivative *vanishes*, i.e. the rate of change of the function becomes zero. For this reason the extrema of a function are called the *stationary points* of the function. $f'(x) = 0$ is the *necessary* or the *first-order condition* (*FOC*) for a <u>local</u> extremum.[3]
(2) If $f'(x)$ goes from positive to zero and then negative, the extremum is a maximum. Otherwise, if it changes from negative to zero and then positive, the extremum is a minimum. This is the *first derivative test* for identifying whether an extremum

---

[3] Global minimum or maximum can occur at endpoints of a domain, where $f'(x)$ does not necessarily equal zero.

is a maximum or a minimum. Alternatively, if at x, $f'(x) = 0$ and $f''(x) < 0$ then the extremum is a maximum. But if at x, $f'(x) = 0$ and $f''(x) > 0$, then the extremum is a minimum. Checking the *sign* of the second derivative of the function at its extreme point(s) is the *second derivative test*. The first or second derivative tests are the *sufficient* or the *second-order condition (SOC)* for identifying an extremum of a function as a maximum or minimum.

The above conclusions provide a blueprint for determining a function's maximum or minimum values. Assume the function $y = f(x)$ is continuous and twice differentiable, meaning $f(x)$, $f'(x)$, and $f''(x)$ exist for all x in the domain of the function. To determine the stationary point(s) of the function we take the first derivative of the function, set it equal to zero and solve for x. Let's denote the solution by x_0. At x_0, where $f'(x_0) = 0$, the necessary or first-order condition for an extremum is satisfied and $[x_0, f(x_0)]$ are the coordinates of the extreme point.

Next we check the second-order condition *(SOC)* by examining $f''(x_0)$. If $f''(x_0) < 0$ then the stationary point at x_0 is a maximum, and if $f''(x_0) > 0$ then the stationary point is a minimum. The second derivative test may not be conclusive if $f''(x_0) = 0$. In this case, at x_0, the function might have a maximum, minimum, or an *inflection point*. In such cases the first derivative test must be used. By using the first derivative test we can avoid the complication of using the n-th derivative test discussed in some mathematical economics texts. Next we solve several numerical examples to show how stationary points of a function are identified.

Examples

1. Find the relative maximum or minimum of the function

$$y = f(x) = 5x^2 - 40x + 10$$

The first order or necessary condition for the function's stationary point(s) is that $\dfrac{dy}{dx} = f'(x) = 0$.

$$y' = f'(x) = 10x - 40 = 0 \longrightarrow x = 4$$

At $x = 4$ value of the function is

$$y = f(4) = 5(4^2) - 40 * 4 + 10 = 80 - 160 + 10 = -70$$

Therefore, $(4, -70)$ is the function's stationary point. To determine whether the function has a maximum or a minimum at this point, we apply the second order or sufficient condition.

$$\frac{d^2y}{dx^2} = f''(x) = 10 > 0 \quad \text{indicating the function has a minimum at } x = 4$$

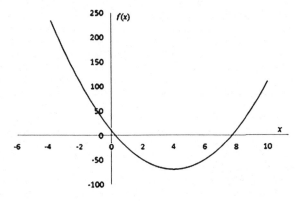

Fig. 5.12 Graph of $y = f(x) = 5x^2 - 40x + 10$

Table 5.3 $f(x)$, $f'(x)$ and $f''(x)$ for selected values of x

x	$f(x)$	$f'(x)$	$f''(x)$
−3	175	−70	10
−2	110	−60	10
−1	55	−50	10
0	10	−40	10
1	−25	−30	10
2	−50	−20	10
3	−65	−10	10
4	−70	0	10
5	−65	10	10
6	−50	20	10
7	−25	30	10

This result is confirmed by the graph of the function $f(x) = 5x^2 - 40x + 10$ in Fig. 5.12 and by the figures in Table 5.3, which are the values of $f(x)$, $f'(x)$, and $f''(x)$ for selected values of x.

Table 5.3 shows that as x increases from −3 to 0 then to 4 and beyond, values of $f'(x)$ change from −70 to −40 to 0 and then to positive numbers. Given that $f'(x) = 0$ at $x = 4$ the function has an extremum at this point. Since $f'(x)$ changes sign from *negative* to 0 and then *positive* in the vicinity of $x = 4$, the function has a minimum at this point. The conclusion of the first derivative test is verified by the second order condition *SOC*; the second derivative of the function is positive 10 at every value of x, including $x = 4$

2. Find the relative maximum or minimum of the function

$$y = f(x) = -5x^2 + 40x + 10$$

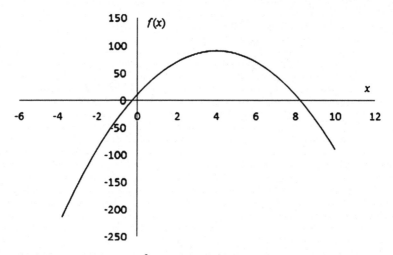

Fig. 5.13 Graph of $y = f(x) = -5x^2 + 40x + 10$

$$\frac{dy}{dx} = f'(x) = -10x + 40 = 0 \quad \longrightarrow \quad 10x = 40 \quad \text{and} \quad x = 4$$

At $x = 4$, the value of the function is

$$y = f(4) = -5(4^2) + 40 * 4 + 10 = -80 + 160 + 10 = 90$$

Therefore $(4, 90)$ is the function's stationary point. To determine whether the function has a maximum or a minimum at this point, we apply the *SOC* (Fig. 5.13)

$$f''(x) = -10 < 0 \quad \text{indicating the function has a maximum at} \quad x = 4$$

3. Find the relative maximum or minimum of the function

$$y = f(x) = x^3 - 6x^2 + 9x$$

The *FOC* is

$$f'(x) = 3x^2 - 12x + 9 = 0 \quad \longrightarrow \quad x^2 - 4x + 3 = 0$$

and solution to this quadratic equation

$$x = \frac{-(-4) \pm \sqrt{(-4)^2 - 4 * 3}}{2} = \frac{4 \pm \sqrt{16 - 12}}{2} = \frac{4 \pm 2}{2}$$

leading to $x_1 = 3$ and $x_2 = 1$. To determine which stationary point is a local maximum and which is a local minimum, we must check the *SOC*,

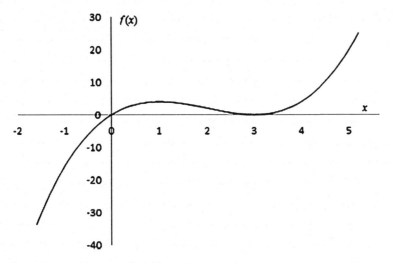

Fig. 5.14 Graph of $y = f(x) = x^3 - 6x^2 + 9x$

$$f''(x) = 6x - 12 \quad \longrightarrow \quad f''(1) = 6(1) - 12 = -6 < 0$$

so at $x = 1$ this function has a local maximum, and

$$f''(x) = 6x - 12 \quad \longrightarrow \quad f''(3) = 6(3) - 12 = 6 > 0$$

indicating that the function has a local minimum at $x = 3$. Given that $f(1) = 1^3 - 6(1)^2 + 9*1 = 4$ and $f(3) = 3^3 - 6(3)^2 + 9*3 = 0$ then the function's local maximum and minimum are 4 and 0, respectively.

This results are confirmed by the graph of the function $f(x) = x^3 - 6x^2 + 9x$, shown in Fig. 5.14, and by Table 5.4, which shows values of $f(x)$, $f'(x)$, and $f''(x)$ for selected values of x.

4. Find the stationary values of the function

$$y = f(x) = 3x^4 - 10x^3 + 6x^2 + 1$$

The function's relative extrema occur where

$$f'(x) = 12x^3 - 30x^2 + 12x = 0 \quad \text{which leads to three possible solutions}$$
$$12x^3 - 30x^2 + 12x = 0 \quad \longrightarrow \quad 12x(x^2 - 2.5x + 1) = 0$$

leading to $x = 0$ or $x^2 - 2.5x + 1 = 0$. This quadratic equation has two solutions $x = 0.5$ and $x = 2$. To check the SOC, we must evaluate the second derivate of the function at stationary values

Table 5.4 $f(x)$, $f'(x)$ and $f''(x)$ for selected values of x

x	$f(x)$	$f'(x)$	$f''(x)$
−0.9	−13.689	22.23	−17.4
−0.3	−3.267	12.78	−13.8
0	0	9	−12
0.3	2.187	5.67	−10.2
0.6	3.456	2.88	−8.4
1.0	4	0	−6
2.1	1.701	−2.25	0.6
2.7	0.243	−1.53	4.2
3.0	0	0	6
3.3	0.297	2.07	7.8
3.6	1.296	4.68	9.6

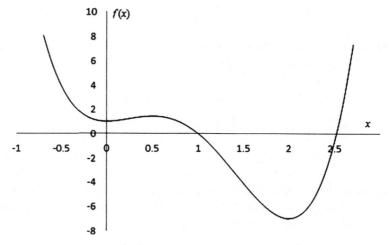

Fig. 5.15 Graph of $y = 3x^4 - 10x^3 + 6x^2 + 1$

$$f''(x) = 36x^2 - 60x + 12$$
$$f''(0) = 12 > 0 \quad \text{the function has a local minimum at} \quad x = 0$$

$f''(0.5) = 36(0.25) - 60(0.5) + 12 = 9 - 30 + 12 = -9 < 0$ the function has a local maximum at $x = 0.5$

$f''(2) = 36(4) - 60(2) + 12 = 144 - 120 + 12 = 36 > 0$ the function has a local

minimum at $x = 2$. Readers are urged to carefully check these results using the graph of the function in Fig. 5.15 and Table 5.5 which provides values of $f(x)$, $f'(x)$, and $f''(x)$ for selected values of x.

Table 5.5 $f(x)$, $f'(x)$ and $f''(x)$ for selected values of x

x	$f(x)$	$f'(x)$	$f''(x)$
−0.4	2.68	−10.37	41.76
−0.2	1.32	−3.70	25.44
0.0	1.00	0.00	12.00
0.2	1.16	1.30	1.44
0.5	1.44	0.00	−9.00
0.7	1.23	−2.18	−12.36
1.5	−4.06	−9.00	3.00
2.0	−7.00	0.00	36.00
2.5	−0.56	30.00	87.00

Fig. 5.16 Graph of fuel consumption and speed

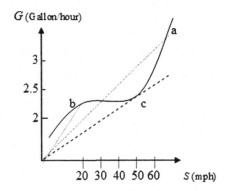

5. Assume that fuel consumption of a car G (gallons per hour) is a function of the speed of the car S (miles per hour–mph), as depicted in Fig. 5.16.[4]

What speed maximizes fuel efficiency? Fuel efficiency is achieved when gas consumption per mile (g) is minimized. Fuel consumption per mile is the ratio of G to S, that is

$$g = \frac{G}{S} = \frac{\text{gallons per hour}}{\text{miles per hour}} = \text{gallons per mile}$$

The reciprocal of G is, of course, *miles per gallon*.

Note that gas consumption per mile at any speed is the slope of the line drawn from the origin to the point on the graph corresponding to that speed. For example, point 'a' on the graph corresponds to 70 miles/hour speed and 3.5 gallons of fuel consumed. Therefore, g is $\frac{3.5}{70}$ = 0.05 gallons per mile (or saying it differently 70/3.5 = 20 miles/gallon). Similarly point 'b' corresponds to 20 miles per hour speed and 2.3 gallons of fuel, with 2.3/20 = 0.115 gallon/mile. To minimize g we must determine the point on the graph such that the slope of the ray from the origin to that point is

---

[4] Adopted from P.D.Taylor, *Calculus: The Analysis of Functions*, Wall & Emerson, Toronto, 1992.

Fig. 5.17 Graphs of convex
and concave functions

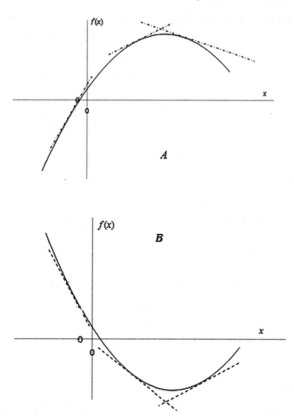

the smallest. That point is 'c'. At point c, a speed of 55 miles/hour corresponds to
2.4 gallons of fuel so gallons per mile is 2.4/55 = 0.044, the lowest possible value.

Inflection Point

Before defining the inflection point of a function we must first define a *Concave*
or *Convex* function. If the graph of a function over an interval lies below all of
its tangents, as in Fig. 5.17a, it is called *Concave* or *Concave downward* over that
interval. If the graph lies above all of its tangents, as in Fig. 5.17b, it is called *Convex*
or *Concave upward*.

Inflection point: a point on the graph of a function $y = f(x)$ is an inflection point
(or point of inflection) if at that point the curve changes from concave to convex or
from convex to concave. At the point of inflection the *curvature* of the graph of the
function changes from concave (positive curvature) to convex (negative curvature)
or vice versa.

Consider the function $y = \frac{1}{3}x^3 - 9x$ discussed at the beginning of this section
and graphed in the top part of Fig. 5.10. As x increases, the curvature of the curve
changes from concave to convex. This function has an inflection point at the origin.

Fig. 5.18 Graph of $TC(Q) = Q^3 - 24Q^2 + 200Q + 500$

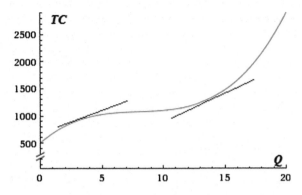

Table 5.6 Change of sign of second derivative

x	$TC(Q)$	$TC'(Q)$	$TC''(Q)$
4	980	56	−24
5	1025	35	−18
6	1052	20	−12
7	1067	11	−6
8	1076	8	0
9	1085	11	6
10	1100	20	12
11	1127	35	18
12	1172	56	24

At the point of inflection the second derivative of the function changes sign. Change in the sign of the second derivative of a function is a reliable test for identifying an inflection point. It is possible for an inflection point to be a stationary point.

As an example, consider the total cost function

$$TC(Q) = Q^3 - 24Q^2 + 200Q + 500$$

with the graph depicted in Fig. 5.18.

This function's curvature changes from concave to convex and, therefore, must have an inflection point. The function's inflection point occurs at $Q = 8$, where the sign of the second derivative of the function $TC''(Q) = 6Q - 48$ changes from negative to positive (see Table 5.6).

5.5.1 Constraint Optimization

In *constraint optimization* we must maximize or minimize a function subject to certain condition(s). Examples of constraint optimizations in economics are minimizing the cost of producing a specific level of output or maximizing the level of output for a specified cost. To get a flavor of constraint optimization, consider the following problem. What is the maximum area of a rectangle if its perimeter is 20 inches?

Let x and y denote the sides of the rectangle. Denote the rectangle's area by A. Then we have

$$\text{Maximize} \qquad A = x.y$$
$$\text{Subject to} \qquad 2x + 2y = 20$$

From the *constraint* $2x + 2y = 20$ we solve for y in terms of x.

$$2x + 2y = 20 \quad \longrightarrow \quad 2y = 20 - 2x \text{ and } y = 10 - x$$

Substituting for y in the area function, we have

$$A = x(10 - x) = 10x - x^2$$

The *FOC* for maximum is

$$\frac{dA}{dx} = 10 - 2x = 0 \quad \longrightarrow \quad x = 5$$

The second-order condition for a maximum is satisfied since $\dfrac{d^2A}{dx^2} = -2 < 0$. The other side of the rectangle is also 5 inches, making it a square, and the maximum area is 25 square inches.

A wide range of economics application of optimizations mentioned in this chapter will appear in the next several chapters.

5.6 Exercises

1. Find the relative maximum and/or minimum values of the following functions
   (a) $y = 2x^2 - 4x + 10$
   (b) $y = -2x^2 - 4x + 10$
   (c) $y = x^3 - 3x + 10$
   (d) $y = x^3 - 9x^2 - 48x + 50$

2. A baseball player hits a home run. The height, in feet, of the ball t second after it is hit is given by
   $$h = -15t^2 + 96t + 4$$

   (a) How high does the ball goes before returning to the ground?
   (b) How many seconds is the ball in the air?

3. A travel agency offers a one week Caribbean tour for $1,000 per person. If more than 50 passengers join the tour, the price is reduced by $5 per additional passenger. The tour can accommodate 150 passengers. What number of passengers maximizes the travel agency's total revenue? What price does each passenger pays?

4. A cruise line offers a one week tour of Alaska for $2,400 per person. If more than 200 passengers join the tour, the price is reduced by $3 per additional passenger. The tour can accommodate 1000 passengers. What number of passengers maximizes the cruise line's total revenue? What price does each passenger pays?

5. Find the relative maximum or minimum of the function

$$y = x(1 - x)^4$$

5.7 Partial Derivatives and Extreme Values of Bivariate Functions

In the previous sections we discussed the necessary and sufficient conditions for the local maximum or minimum of a univariate function, the function of a single independent variable, $y = f(x)$. In this section we want to derive similar conditions for a function of two variables $y = f(x_1, x_2)$, a bivariate function, where the variables x_1 and x_2 are independent of one another.

If in the multivariate function $y = f(x_1, x_2, \ldots, x_n)$, variable x_1 changes by Δx_1 while all the other variables $x_2, \ldots, x_n$ remain unchanged, then the change in y is

$$\Delta y = f(x_1 + \Delta x_1, x_2, \ldots, x_n) - f(x_1, x_2, \cdots, x_n)$$

and the rate of change of the function can be expressed as

$$\frac{\Delta y}{\Delta x} = \frac{f(x_1 + \Delta x_1, x_2, \ldots, x_n) - f(x_1, x_2, \ldots, x_n)}{\Delta x_1}$$

The limit of $\frac{\Delta y}{\Delta x_1}$ as $\Delta x_1 \to 0$ constitutes a derivative. This derivative is called the *partial derivative* of y with respect to x_1. We can similarly define the partial derivative of the function with respect to the other variables $x_2, \ldots, x_n$. The process of taking a partial derivative is called *partial differentiation*.

The partial derivative of $y = f(x_1, x_2, \ldots, x_n)$ with respect to the ith variable x_i $i = 1, 2 \ldots, n$ can be denoted by $\frac{\partial y}{\partial x_i}$, $\frac{\partial f}{\partial x_i}$, f_{x_i} or f_i. The following examples should help readers to learn the techniques of partial differentiation.

Examples

1. Assume z is a function of x and y such that $z = f(x, y) = 3x^2 + 4y^3 - xy$. Find the partial derivatives. First we partially differentiate the function with respect to x, while we treat y as a constant

$$\frac{\partial z}{\partial x} = f_x = 6x - y$$

Next we partially differentiate the function with respect to y, while treating x as a constant

$$\frac{\partial z}{\partial y} = f_y = 12y^2 - x$$

Find the partial derivatives of the following functions:

2. $f(x, y) = 3x - 2y^4$

$$\frac{\partial f}{\partial x} = 3 \qquad\qquad \frac{\partial f}{\partial y} = -8y^3$$

3. $f(x, y) = x^4 - 3x^3y^2 + 3xy^3$.

$$\frac{\partial f}{\partial x} = 4x^3 - 9x^2y^2 + 3y^3 \qquad\qquad \frac{\partial f}{\partial y} = -6x^3y + 9xy^2$$

4. $f(x, y) = \dfrac{x - y}{x + y}$.

$$f_x = \frac{(x+y) - (x-y)}{(x+y)^2} = \frac{2y}{(x+y)^2}$$
$$f_y = \frac{-(x+y) - (x-y)}{(x+y)^2} = -\frac{2x}{(x+y)^2}$$

5. $f(x, y, z) = x^2y^3z - 3yz + 3xz^2$

$$\frac{\partial f}{\partial x} = 2xy^3z + 3z^2 \qquad \frac{\partial f}{\partial y} = 3x^2y^2z - 3z \qquad \frac{\partial f}{\partial z} = x^2y^3 - 3y + 6xz$$

6. $w = \sqrt{r^2 + 3rs^3}$

$$\frac{\partial w}{\partial r} = \frac{2r + 3s^3}{2\sqrt{r^2 + 3rs^3}} \qquad\qquad \frac{\partial w}{\partial s} = \frac{9rs^2}{2\sqrt{r^2 + 3rs^3}}$$

7. $y = \dfrac{3x_1x_2 + 5x_1^2 - 5x_2^2}{3x_1x_2 + 5}$

$$\frac{\partial y}{\partial x_1} = \frac{(3x_2 + 10x_1)(3x_1x_2 + 5) - 3x_2(3x_1x_2 + 5x_1^2 - 5x_2^2)}{(3x_1x_2 + 5)^2}$$

$$= \frac{15x_2(x_1^2 + x_2^2 + 1) + 50x_1}{(3x_1x_2 + 5)^2}$$

$$\frac{\partial y}{\partial x_2} = \frac{(3x_1 - 10x_2)(3x_1x_2 + 5) - 3x_1(3x_1x_2 + 5x_1^2 - 5x_2^2)}{(3x_1x_2 + 5)^2}$$

$$= -\frac{15x_1(x_1^2 + x_2^2 - 1) + 50x_2}{(3x_1x_2 + 5)^2}$$

The graph of a bivariate function $z = f(x, y)$ is a surface. If we cut this surface at a point $z = f(a, b)$ with a plane parallel to the zx coordinate plane, the intersection is a curve. The partial derivative of $z = f(x, y)$ with respect to x, $f_x(x, y)$, evaluated at the point (a, b), $f_x(a, b)$, is the slope of the tangent to the curve at the point $(a, b, f(a, b))$ in the x direction. Similarly, If we cut this surface at a point $z = f(a, b)$ by a plane parallel to the zy coordinate plane, the intersection is also a curve. The partial derivative of $z = f(x, y)$ with respect to y, $f_y(x, y)$, evaluated at point (a, b), $f_y(a, b)$, is the slope of the tangent to the curve at the point $(a, b, f(a, b))$ in the y direction.

Figure 5.19 (taken from Google's "images for graphs of 3d functions" and modified) should help with the geometric visualization of the above statements. As the figure shows, the partial derivatives of f evaluated at $(3, 2)$ are the slopes of the tangents to the surface of f at point $(3, 2, f(3, 2) = 4)$, the red dot. A positive $f_x(a, b)$ indicates that the surface "rises" in the direction of the x-axis; a negative $f_x(a, b)$ indicates that the surface "falls" in the direction of the x-axis.

5.7.1 Higher Order Partial Derivatives

If a function $f(x, y)$ has partial derivatives f_x and f_y at each point (x, y) in a region, then these partial derivatives are themselves functions of x and y, which may also have partial derivatives. The partial derivatives of f_x and f_y with respect to x and y are called the *second order partial derivatives* and are denoted by

$$\frac{\partial}{\partial x}\left(\frac{\partial f}{\partial x}\right) = \frac{\partial^2 f}{\partial x^2} = f_{xx} \qquad\qquad \frac{\partial}{\partial y}\left(\frac{\partial f}{\partial y}\right) = \frac{\partial^2 f}{\partial y^2} = f_{yy}$$

$$\frac{\partial}{\partial y}\left(\frac{\partial f}{\partial x}\right) = \frac{\partial^2 f}{\partial y\partial x} = f_{xy} \qquad\qquad \frac{\partial}{\partial x}\left(\frac{\partial f}{\partial y}\right) = \frac{\partial^2 f}{\partial x\partial y} = f_{yx}$$

Fig. 5.19 The partial derivatives of f evaluated at $(3, 2)$
are the slopes of the tangents
to the surface of f at point
$(3, 2, f(3, 2) = 4)$

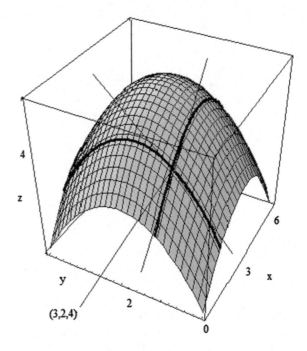

If the *cross partial derivatives* f_{xy} and f_{yx} are continuous, then they are equal, which
means that the order of cross partial differentiation is immaterial.

Examples

Find the second order partial derivatives of the function for examples 1 through 4 in
the previous section

1. We found the first order partial derivatives of the function

$$z = f(x, y) = 3x^2 + 4y^3 - xy \quad \text{as}$$

$$\frac{\partial z}{\partial x} = f_x = 6x - y \qquad\qquad \frac{\partial z}{\partial y} = f_y = 12y^2 - x$$

Thus the second order partial derivatives are:

$$\frac{\partial^2 z}{\partial x^2} = f_{xx} = 6 \qquad\qquad \frac{\partial^2 z}{\partial y^2} = f_{yy} = 24y$$

$$\frac{\partial^2 z}{\partial y \partial x} = f_{yx} = -1 \qquad\qquad \frac{\partial^2 z}{\partial x \partial y} = f_{xy} = -1$$

2. $f(x, y) = 3x - 2y^4$

$$\frac{\partial f}{\partial x} = 3 \qquad\qquad \frac{\partial f}{\partial y} = -8y^3$$

$$\frac{\partial^2 f}{\partial x^2} = f_{xx} = 0 \qquad\qquad \frac{\partial^2 f}{\partial y^2} = f_{yy} = -24y^2$$

$$\frac{\partial^2 f}{\partial y \partial x} = f_{xy} = 0 \qquad\qquad \frac{\partial^2 f}{\partial x \partial y} = f_{yx} = 0$$

3. $f(x, y) = x^4 - 3x^3 y^2 + 3xy^3$.

$$\frac{\partial f}{\partial x} = 4x^3 - 9x^2 y^2 + 3y^3 \qquad\qquad \frac{\partial f}{\partial y} = -6x^3 y + 9xy^2$$

$$\frac{\partial^2 f}{\partial x^2} = f_{xx} = 12x^2 - 18xy^2 \qquad\qquad \frac{\partial^2 f}{\partial y^2} = f_{yy} = -6x^3 + 18xy$$

$$\frac{\partial^2 f}{\partial y \partial x} = f_{yx} = -18x^2 y + 9y^2 \qquad\qquad \frac{\partial^2 f}{\partial x \partial y} = f_{xy} = -18x^2 y + 9y^2$$

4. $f(x, y) = \dfrac{x - y}{x + y}$.

$$f_x = \frac{2y}{(x+y)^2} \qquad\qquad f_y = -\frac{2x}{(x+y)^2}$$

$$f_{xx} = \frac{-4y(x+y)}{(x+y)^4} = -\frac{4y}{(x+y)^3} \qquad\qquad f_{yy} = \frac{4x(x+y)}{(x+y)^4} = \frac{4x}{(x+y)^3}$$

$$f_{yx} = \frac{2(x+y)^2 - 4y(x+y)}{(x+y)^4} = \frac{2x - 2y}{(x+y)^3}$$

$$f_{xy} = -\frac{2(x+y)^2 - 4x(x+y)}{(x+y)^4} = -\frac{2y - 2x}{(x+y)^3} = \frac{2x - 2y}{(x+y)^3}$$

Note that in all of the 4 examples we have $f_{xy} = f_{yx}$.

5.7.2 Maximum and Minimum of a Bivariate Function

As we saw earlier, an important application of the derivative is finding maximum or minimum values of a univariate function $y = f(x)$. We can apply the same idea and use partial derivatives to identify the maximum or minimum of bivariate functions. A necessary condition for $f(x, y)$ to have a maximum or minimum at point $(a, b, f(a, b))$ is that $f_x(a, b) = 0$ and $f_y(a, b) = 0$. If the first partial derivatives both equal zero at $(a, b, f(a, b))$ then the function has a critical, stationary, or extreme

Fig. 5.20 A saddle point

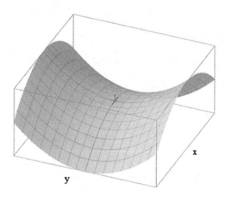

point at $(a, b, f(a, b))$. We now need the second order conditions. If this function is continuous over the region containing (a, b) such that the second order partial derivatives exist, we form

$$\Delta = f_{xx}(a, b) f_{yy}(a, b) - \left[f_{xy}(a, b) \right]^2$$

and

> If $\Delta > 0$ and $f_{xx}(a, b) > 0$ then $f(a, b)$ is a local minimum.
>
> If $\Delta > 0$ and $f_{xx}(a, b) < 0$ then $f(a, b)$ is a local maximum.

If $\Delta < 0$ then $(a, b, f(a, b))$ is neither a local minimim or a local maximum. In the case of $\Delta < 0$, $(a, b, f(a, b))$ is a *saddle point*. Figure 5.20 (taken from Wikipedia and modified) is an illustration of a saddle point. At the point identified by a red dot, the function has a local maximum in the direction of the x axis and at the same time a local minimum in the direction of the y axis. If $\Delta = 0$, more information is needed to determine the function's behavior at the point.

Examples

1. Find the relative maximum and/or minimum of the function

$$f(x, y) = 2x^3 + 3y^2 - 6x - 6y + 24$$

$$\frac{\partial f}{\partial x} = 6x^2 - 6 = 0 \quad \longrightarrow \quad x = \pm 1$$

$$\frac{\partial f}{\partial y} = 6y - 6 = 0 \quad \longrightarrow \quad y = 1$$

$$f(1, 1) = 2 + 3 - 6 - 6 + 24 = 17 \quad f(-1, 1) = -2 + 3 + 6 - 6 + 24 = 25$$

Then the function has extrema at the points $(1, 1, 17)$ and $(-1, 1, 25)$. The second order partial derivatives are

Fig. 5.21 Graph of
$f(x, y) = 2x^3 + 3y^2 - 6x - 6y + 24$

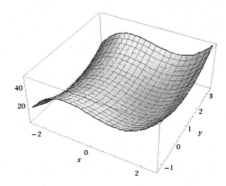

$$f_{xx} = 12x \quad f_{xx}(1, 1) = 12 \quad f_{xx}(-1, 1) = -12 \quad f_{yy} = 6 \quad f_{yx} = f_{xy} = 0$$

The values of Δ at $(1, 1, 17)$ and $(-1, 1, 25)$ are

$$\Delta = 12 * 6 - 0 = 72 > 0 \quad \text{and} \quad \Delta = -12 * 6 - 0 = -72 < 0$$

Thus this function has a local minimum at point $(1, 1, 17)$ and a saddle point at point $(-1, 1, 25)$. Figure 5.21 is the graph of the function.

2. Find the local maximum, minimum, or saddle point of the function

$$f(x, y) = 2x^2y + 2y^2 - 2x^2 - 8y$$
$$f_x = 4xy - 4x = 0 \quad \longrightarrow \quad x(y - 1) = 0 \tag{5.5}$$

$$f_y = 2x^2 + 4y - 8 = 0 \tag{5.6}$$

To find the extreme points we must solve Eqs. (5.5) and (5.6) simultaneously. In Eq. 5.5 we have either $x = 0$ or $y - 1 = 0$, leading to $y = 1$. In the case of $x = 0$, Eq. (5.6) becomes $4y - 8 = 0 \longrightarrow y = 2$ and we have a extreme point at $(0, 2, -8)$. In the case of $y = 1$ Eq. (5.6) becomes $2x^2 - 4 = 0$, which yields the pair of solutions $x = \pm\sqrt{2}$. Therefore, we have two more critical points at $(\sqrt{2}, 1, -6)$ and $(-\sqrt{2}, 1, -6)$. To check the second order condition, we have

$$f_{xx} = 4y - 4|_{y=2} = 4 \quad \text{and} \quad f_{xx} = 4y - 4|_{y=1} = 0$$
$$f_{yy} = 4 \quad \text{and} \quad f_{xy} = f_{yx} = 4x|_{x=0} = 0$$

Subsequently $\Delta = 4 * 4 - 0 = 16$. When $x = \sqrt{2}$,

$$f_{xy} = f_{yx} = 4x = 4\sqrt{2}, \quad \text{and} \quad \Delta = 0 * 4 - [4\sqrt{2}]^2 = -32$$

Fig. 5.22 Graph of
$f(x, y) = 2x^2y + 2y^2 - 2x^2 - 8y$

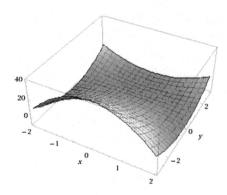

and Δ is the same when $x = -\sqrt{(2)}$. These results are summarized in the following table

Extreme point	f_{xx}	f_{yy}	f_{xy}	Δ	Conclusion
$(0, 2, -8)$	4	4	0	16	Local minimum
$(\sqrt{2}, 1, -6)$	0	4	$4\sqrt{2}$	-32	Saddle point
$(-\sqrt{2}, 1, -6)$	0	4	$-4\sqrt{2}$	-32	Saddle point

This function has two saddle points. Figure 5.22 is the graph of the function.

3. A manufacturer produces two models of pumps; standard and deluxe. It costs $40 to manufacture standard pump and $60 to make its deluxe version. The firm's marketing department estimates that if standard pump is priced at P_1 dollars and the deluxe at P_2 dollars, then the manufacturer will sell $500(P_2 - P_1)$ units of the standard and $45,000 + 500(P_1 - 2P_2)$ units of the deluxe each year. How should the items be priced to maximize profit?

$$TR = P_1Q_1 + P_2Q_2 = P_1[500(P_2 - P_1)] + P_2[45,000 + 500(P_1 - 2P_2)]$$

$$TR = 500P_1P_2 - 500P_1^2 + 45000P_2 + 500P_1P_2 - 1000P_2^2$$

$$TR = 1000P_1P_2 - 500P_1^2 - 1000P_2^2 + 45000P_2$$

$$TC = 40 * [500(P_2 - P_1)] + 60 * [45000 + 500(P_1 - 2P_2)]$$

$$TC = 2700000 + 10000P_1 - 40000P_2$$

$$\pi = TR - TC$$

$$\pi = 1000P_1P_2 - 500P_1^2 - 1000P_2^2 + 45000P_2 - 2700000 - 10000P_1$$
$$+ 40000P_2$$

$$\pi = -500P_1^2 - 1000P_2^2 + 1000P_1P_2 - 10000P_1 + 85000P_2 - 2700000$$

$$\frac{\partial\pi}{\partial P_1} = -1000P_1 + 1000P_2 - 10000 = 0$$

$$\frac{\partial\pi}{\partial P_2} = -2000P_2 + 1000P_1 + 85000 = 0$$

The solution to this system of two equations with two unknowns is $P_1 = 65$ and $P_2 = 75$. To check the second order conditions, we have

$$\frac{\partial^2\pi}{\partial P_1^2} = -1000 \qquad \frac{\partial^2\pi}{\partial P_2^2} = -2000 \quad \text{and} \quad \frac{\partial^2\pi}{\partial P_1 \partial P_2} = 1000$$

Therefore

$$\Delta = f_{xx}(a, b)f_{yy}(a, b) - \left[f_{xy}(a, b)\right]^2 = (-1000)(-2000) - (1000)^2$$
$$= 1000000 > 0$$

With $\Delta > 0$ and $f_{xx} = -1000 < 0$, the second order condition for a maximum is satisfied. The manufacturer's profit maximizing combination of the two products is $(P_1, Q_1) = (65, 5000)$ and $(P_2, Q_2) = (75, 2500)$, with a maximum profit of \$162,500.

5.8 Exercises

1. If $f(x, y) = 3x^2 - 2xy^3 + 10$
   find *a)* $f(1, 2)$ *b)* $f(-1, 3)$ *c)* $f(x + 1, 2y)$

2. If $g(w, z) = \dfrac{2w + z}{5 - wz}$ find $g(2, -2)$

3. Find the partial derivatives of the following functions
   (a) $f(x, y) = 3x^3 - 2x^2y^3 + y^4 - 6xy^2 + 10$
   (b) $g(w, z) = -2w^3 + 5z^2 - 5wz + 10$
   (c) $z = \dfrac{x^2 - 2y}{3x + y^2} + 3x^2 - 2y$
   (d) $f(x, y) = 2\sqrt{x^2 - 5y} + 5\sqrt{x + 2y^2}$
   (e) $z = (w^2 + x^2)^3$

4. The graph of a bivariate function $f(x, y)$ is a surface in three-dimensional space. The *level curves* or the *contour plot* for $f(x, y)$ are curves in the xy plane defined by $f(x, y) = c$, where c is a constant. Graph the level curves for $f(x, y) = x^2 + y^2 - 2y$ for $c = 2$ and $c = 10$.

5. Find the extreme values of the following functions

   (a) $f(x, y) = x^2 + y^2 - 2y$

   (b) $z = 3x^3 + 3x^2y - y^2 + 5$

   (c) $z = \sqrt{x^2 + y^2}$

Chapter 6
Optimal Level of Output and Long Run Price

6.1 Golden Rule of Profit Maximization

The cost structure of a firm is reflected in its costs functions: total cost TC; average cost AC; and marginal cost MC. Total cost, the sum of total variable and total fixed costs, is generally expressed as a function of the level of output Q. While a firm's production function is often a bivariate or multivariate function relating its output to various inputs, the cost functions are usually univariate functions relating different costs to only one variable, output. It is therefore much easier to specify and estimate various cost functions for a firm than specify and estimate its production function.[1]
A typical firm may faces three types of *optimization* (maximization or minimization) problems. A firm may decide its level of output Q_0 and focus on minimizing the costs of producing it, i.e. minimizing its operating cost. This is an example of a *constraint optimization*. Here the constraint is the exogenously determined level of output. In the second class of optimization problems, a firm may decide its total expenditure on various inputs—or its operating cost—and try to determine its profit maximizing level of output. This is another constraint optimization problem. Here the constraint is the exogenously determined total cost TC_0. As we will see, if the total cost function of a firm is known then these two types of constraint optimizations are rather trivial problems.

The third type of optimization problem arises when a firm attempts to maximize its profit (or minimize its loss) without any constraint on output or cost. This is an example of *unconstrained optimization*. In this case the main objective of a firm with cost function $TC(Q)$ is to find the level of output Q^* that maximizes its profit π. Q^* is the firm's *optimal* or equilibrium level of output.

We began by exploring the unconstrained optimization. An unconstrained profit-maximizing firm expands its output as long as its marginal cost MC is less than its marginal revenue MR. The optimal level of output is achieved when firm's marginal cost becomes equal to its marginal revenue, that is $MC = MR$. Denoting the firm's

---

[1] A detailed discussion of production functions and related topics is presented in Chap. 10.

S. Vali, *Principles of Mathematical Economics*,
Mathematics Textbooks for Science and Engineering 3,
DOI: 10.2991/978-94-6239-036-2_6, © Atlantis Press and the authors 2014

total revenue function as $TR(Q)$, the profit function of the firm could be written as

$$\pi(Q) = TR(Q) - TC(Q) \tag{6.1}$$

The firm reaches its profit maximizing level of output when the first order condition (*FOC*) for a maximum, $\dfrac{d\pi}{dQ} = 0$, is satisfied. By setting the derivative of (6.1) with respect to Q equal to zero we have

$$\pi'(Q) = \frac{d\pi}{dQ} = \frac{dTR}{dQ} - \frac{dTC}{dQ} = TR'(Q) - TC'(Q) = 0$$

$$TR'(Q) - TC'(Q) = 0 \quad \longrightarrow \quad TR'(Q) = TC'(Q) \tag{6.2}$$

Given that $TR'(Q)$ and $TC'(Q)$ are marginal revenue and marginal cost, the *FOC* in (6.2) is a mathematical reiteration of the 'golden rule' of profit maximization, the condition $MC = MR$.

The second order condition (*SOC*) for maximization requires that the second derivative of the profit function evaluated at the level of output Q^*, a solution to (6.2), be negative, that is

$$\pi''(Q^*) = \frac{d^2\pi}{dQ^2}\bigg|_{Q^*} = TR''(Q^*) - TC''(Q^*) < 0 \quad \text{or}$$

$$TR''(Q^*) < TC''(Q^*) \tag{6.3}$$

The Economic implication of (6.3) is that the firm maximizes its profit at the level of output where the rate of change of marginal cost is greater than the rate of change of marginal revenue. To better understand this condition let us consider two cases.

If the firm operates in a competitive market then its marginal revenue equals the price, P, determined in the market, and the *FOC* can be expressed as

$$\frac{d\pi}{dQ} = P - TC'(Q) = 0 \quad \text{or} \quad P = MC$$

The *SOC* requirement for maximum profit is

$$\frac{d^2\pi}{dQ^2} = \frac{d}{dQ}[P - TC'(Q)] = 0 - TC''(Q^*) < 0 \quad \text{or} \quad TC''(Q^*) > 0$$

This means that the maximum occurs at the intersection of the price line with the upward sloping segment of the marginal cost curve. This situation is depicted in Fig. 6.1.

In Fig. 6.1 the price line intersects the MC curve at points A and B, corresponding to levels of output Q_1 and Q_2, at which price is equal to MC. Point A is on the

Fig. 6.1 A Competitive firm's marginal revenue and marginal cost curves

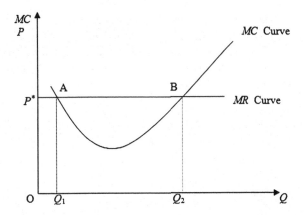

downward sloping portion of *MC* curve where $\dfrac{dMC}{dQ} = TC''(Q^*) < 0$. This is the *SOC* for a minimum (at Q_1 profit is actually minimized). Point B is on the upward sloping segment of *MC*, where

$$\frac{dMC}{dQ} = TC''(Q^*) > 0$$

At Q_2 the *SOC* for a maximum is satisfied and profit is maximized.

If the market in which this firm operates is not competitive then the graph of the firm's marginal revenue function should be a downward sloping curve.[2] This scenario is represented by Fig. 6.2. In Fig. 6.2 point A corresponds to the level of output leading to a minimum profit. The firm's profit is maximized at the output level corresponding to point B, lying on the upward swing of *MC* curve.

Fig. 6.2 A noncompetitive firm's marginal revenue and marginal cost curves

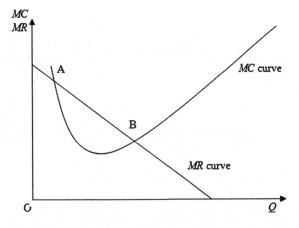

---

[2] I am ignoring the possibility of an upward sloping marginal revenue curve here.

The total cost function TC plays a major role in study of firms' behavior in the market. Often quadratic, and most often cubic, cost functions are used in mathematical modeling of firms' cost structure and study of their operations in both competitive and non-competitive markets. We first discuss the application of quadratic cost functions and then proceed to a more detailed presentation of the cubic cost functions.

6.1.1 Firms with Quadratic Cost Functions in a Competitive Market

Examples

Assume the variable cost function of a perfectly competitive firm is given by

$$VC = VC(Q) = 0.5Q^2 - 25Q$$

Assume the firm's fixed cost FC is 100. If the price of a unit of output is $20 in the market, what is the firm's profit maximizing level of output? What is the maximum profit?

The firm's total cost, total revenue, and profit functions are:

$$TC = TC(Q) = VC(Q) + FC = 0.5Q^2 - 25Q + 100$$
$$TR = TR(Q) = P * Q = 20Q$$
$$\pi(Q) = TR(Q) - TC(Q) = 20Q - 0.5Q^2 + 25Q - 100$$
$$= -0.5Q^2 + 45Q - 100$$

The stationary value of the profit function is

$$\frac{d\pi}{dQ} = \pi'(Q) = -Q + 45 = 0 \quad \longrightarrow \quad Q^* = 45$$

Since $\pi''(Q^*) = -1 < 0$ the *SOC* for a maximum is satisfied. At 45 units of output the firm's profit is

$$\pi(45) = -0.5(45)^2 + 45(45) - 100 = 912.5$$

Are these figures "economically" legitimate? Let explore.

In the above example we used the specified cost functions and market price and applied the first and second order conditions for profit maximization and determined that our firm makes the maximum profit of $912.5 at the level of 45 units of output.

From a 'mathematical' perspective, the specified functions and the results look fine. But from an 'economic' perspective they have major flaws. These functions are in violation of a number of economic 'ordinances'. The most critical violation is that at the "optimal" level of output the variable cost is actually negative,

$$VC(45) = (0.5Q^2 - 25Q)\Big|_{Q=45} = 0.5(45)^2 - 25(45) = -112.5$$

This firm produces 45 units of output with variable costs of -112.5!!

This numerical example was introduced to highlight possible pitfalls of mechanical applications of mathematics in economics, particularly in optimization problems. With an eye on requirements which will be explored in the next section, we can offer a different formulation of the problem with the total cost function expressed as

$$TC(Q) = 5Q^2 - 30Q + 400$$

and the market price set at $P = \$100$ per units. With these revisions, the firm's profit function is

$$\pi(Q) = TR - TC = 100Q - 5Q^2 + 30Q - 400 = 130Q - 5Q^2 - 400$$

FOC for a maximum is

$$\frac{d\pi}{dQ} = \pi'(Q) = 130 - 10Q = 0 \quad \longrightarrow \quad Q^* = 13$$

And the *SOC* for a maximum $\pi''(Q^*) = -10 < 0$ is satisfied. With the optimal output at 13 units, the variable and total costs are \$455 and \$855, respectively, and the firm's maximum profit is \$445.

Figure 6.3 is graphic representation of profit maximizing output and maximum profit. As the top graph shows, the profit maximizing level of output is reached when $MR = P = MC$ at $Q = 13$ units. At this level of output the average total cost or unit cost is \$65.77 and profit per unit is $100 - 65.77 = 34.23$, leading to a total profit of $13 * 34.77 = \$445$. The bottom graph confirms that profit reaches it maximum value at $Q = 13$ units.

6.1.2 Economically Legitimate Quadratic Cost Functions

The graph of a quadratic function of the form $y = ax^2 + bx + c$ is a *parabola*. If a is positive the *vertex* of the parabola is its lowest point, or the function's minimum (Fig. 6.4). If a is negative the vertex of the parabola is its highest point or the function's maximum (Fig. 6.5).

The *FOC* for an extreme point of a quadratic function is

$$\frac{dy}{dx} = 2ax + b = 0$$

Fig. 6.3 Profit maximizing
output and maximum profit

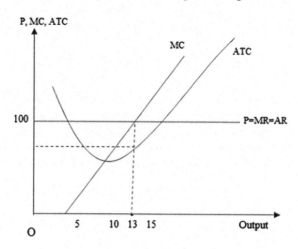

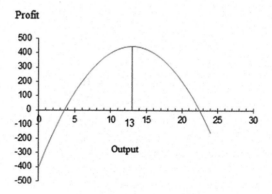

leading to $x = -\dfrac{b}{2a}$. The *SOC* for a minimum or maximum is

$$\frac{d^2y}{dx^2} = 2a > 0 \ \text{ if } \ a > 0, \quad \text{otherwise} \quad \frac{d^2y}{dx^2} = 2a < 0 \ \text{ if } \ a < 0$$

It is clear from the *SOC* that the parabola has a minimum if a is positive and a maximum if a is negative.

It is possible to represent the cost structure of a firm by quadratic variable cost and total cost function $VC(Q) = aQ^2 + bQ$ and $TC(Q) = aQ^2 + bQ + c$. But in order to have economically legitimate cost functions, we must impose certain restrictions on the variable Q and the values of the parameters a, b, and c. The obvious restriction on Q is that it cannot be a negative number. And obviously the variable and total costs cannot be negative either. This requires that we work with only the segment or branch of the parabola that is located in the first quadrant of the coordinate system. Next, we examine the necessary restrictions on the parameters.

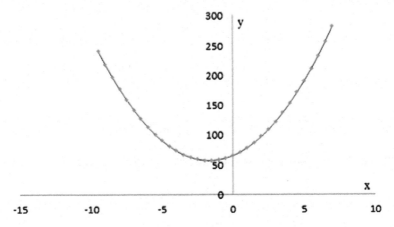

Fig. 6.4   Graph of $y = 3x^2 + 10x + 65$

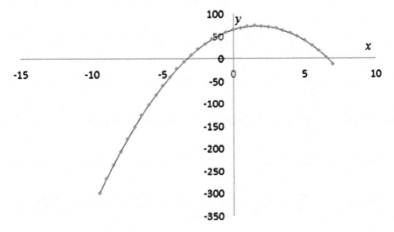

Fig. 6.5   Graph of $y = -3x^2 + 10x + 65$

Writing a competitive firm's profit function as

$$\pi(Q) = TR(Q) - TC(Q) = PQ - aQ^2 - bQ - c,$$

the *FOC* and *SOC* for maximizing profit become

$$\frac{d\pi}{dQ} = \pi'(Q) = TR'(Q) - TC'(Q) = MR(Q) - MC(Q) = 0$$

$$P - 2aQ - b = 0 \quad \longrightarrow \quad Q^* = \frac{P - b}{2a} \tag{6.4}$$

$$\frac{d^2\pi}{dQ^2} = -2a < 0 \tag{6.5}$$

It is clear that in order to satisfy the SOC for a maximum a must be positive. Let's designate this restriction as R1

$$a > 0 \tag{R1}$$

R1 immediately eliminates a family of quadratic functions with negative a, the coefficient of the quadratic term Q^2, from consideration.

At $Q = 0$ $TC = c$. Then c must be the fixed cost and non-negative.

$$c \geq 0 \tag{R2}$$

For an operating firm, VC must be positive. This must be true at the profit maximizing level of output Q^*. By substituting from Eq. (6.4) into the VC function $aQ^2 + bQ$, we have

$$a\left[\frac{(P-b)}{2a}\right]^2 + b\left[\frac{(P-b)}{2a}\right] > 0$$

$$\frac{P^2 + b^2 - 2bP}{4a} + \frac{bP - b^2}{2a} > 0 \quad \longrightarrow \quad \frac{P^2 - b^2}{4a} > 0 \tag{6.6}$$

We know from (R1) that a must be positive, so for (6.6) to be positive the numerator must be positive, that is

$$P^2 - b^2 > 0 \quad \longrightarrow \quad (P+b)(P-b) > 0 \tag{6.7}$$

If b is positive then $P + b$ is positive. Thus for (6.7) to hold $P - b$ must also be greater than zero: that is, $P - b > 0$, leading to $P > b$. If, on the other hand, b is negative then $P - b$ in (6.7) is positive and for (6.7) to hold we must have $P + b > 0$, leading to $P > -b$. Combination of $P > b$ or $P > -b$ can be expressed as

$$P > |b| \tag{R3}$$

It should be clear from (6.4) that if in the firm's quadratic cost function b is positive, then the market price $P = b$ will be the firm's shutdown price. At any price greater than b, the marginal cost $2aQ + b$ will be greater than the average variable cost $aQ + b$ and the firm can recover part or all of its fixed cost.

At any output level $Q > 0$, total variable costs TVC must be positive

$$TVC > 0 \quad \longrightarrow \quad aQ^2 + bQ > 0 \quad \longrightarrow \quad aQ^2 > -bQ \quad \longrightarrow \quad Q > -(b/a)$$

This condition, of course, is satisfied if (R3) holds true.

6.1.3 Quadratic Cost Function and Firm's Break-Even Point

A firm's *break-even point* is the level of output at which the firm's total cost is equal to its total sales revenue and consequently its profit is zero. If Q is a break-even level of output, then

$$\pi(Q) = TR(Q) - TC(Q) = 0 \rightarrow PQ - (TVC + TFC) = 0$$
$$\frac{PQ}{Q} = \frac{TVC + TFC}{Q} = \frac{TVC}{Q} + \frac{TFC}{Q}$$
$$P = AVC + AFC$$

It is clear that at a break-even point the market price must cover the sum of average variable and average fixed costs, otherwise the firm incurs losses. For a firm to reach profitability, the market price must be greater than $AVC + AFC$. A firm continues operation as long as the sales revenue exceeds the total variable cost. But if the business environment deteriorates and the market price falls below the minimum average variable cost then the firm loss is more than its fixed cost and there is no reason for continued operation.

Graphically the break-even point(s) occur at the intersection(s) of the total cost curve and the total revenue curve, which is a straight line through the origin called a ray.

Break-even analysis of a firm with quadratic cost function leads to additional restriction on price and coefficients of the cost function. If firm's total cost is $aQ^2 + bQ + c$ and market price is P, then at break-even

$$\pi(Q) = 0 \quad \longrightarrow \quad PQ - aQ^2 - bQ - c = 0 \quad \longrightarrow \quad aQ^2 - (P - b)Q + c = 0$$

The above quadratic equation has real solution(s) if its discriminant is non-negative, that is

$$(P - b)^2 \geq 4ac \quad \text{or} \quad P - b \geq \sqrt{4ac}$$

leading to

$$P \geq b + 2\sqrt{ac} \qquad (R4)$$

If in (R4) the strict inequality $P > b + 2\sqrt{ac}$ holds then the firm will have two break-even points

$$Q_1 = \frac{(P - b) - \sqrt{(P - b)^2 - 4ac}}{2a} \quad \text{and} \quad Q_2 = \frac{(P - b) + \sqrt{(P - b)^2 - 4ac}}{2a}$$

In this case the firm can operate in the range $Q_1 < Q < Q_2$ and make a profit. Any level of output below Q_1 or above Q_2 results in losses. Interestingly the profit maximizing level of output is the average of these two break-even points (in problem number 10 at the end of this section, you are asked to show this).

If in (R4) $P = b + 2\sqrt{ac}$, the firm will have one break-even point

$$Q = \frac{P - b}{2a}$$

At the level of output below or above the break-even point the firm incur losses. Note that $Q = \dfrac{P - b}{2a}$ is exactly the same as (5), which provides the profit maximizing/loss minimizing level of output. When $P = b + 2\sqrt{ac}$, therefore, the break-even point is the optimal level of output which is

$$Q = \frac{P - b}{2a} = \frac{b + 2\sqrt{ac} - b}{2a} = \sqrt{\frac{c}{a}}$$

Graphically this is the case where the total revenue line is tangent to the total cost curve. At the point of tangency not only is the total cost equal to total revenue but the slope of the total cost curve, which is the marginal cost, is equal to the slope of the total revenue line, which is the price. This means that the following two conditions $TC = TR$ and $MC = P$ are simultaneously satisfied. For a firm with quadratic cost function these conditions are:

$$\begin{cases} aQ^2 + bQ + c = PQ \\ 2aQ + b = P \end{cases}$$

Students are encouraged to verify that solution to the above system of two equations with two unknowns consists of $P = b + 2\sqrt{ac}$ and $Q = \sqrt{\dfrac{c}{a}}$. As we will see shortly, this single break-even point is the firm's long run normal profit price and level of output.

If the first three requirements (R1), (R2), and (R3) are satisfied, we have an economically meaningful quadratic total cost function and a legitimate optimization model. But for a firm to operate *profitably* we need the strict inequality in (R4). Next we examine these restrictions by way of several numerical examples.

Examples

1. $TC = 3Q^2 + 10Q + 60.75$ for $Q \geq 0$ is a legitimate total cost function. A firm with this total cost function can maximize profit (minimize loss) as long as (R3) is satisfied; the prevailing price in the market must be more than $10 per unit, $P > 10$. Let see why we need this restriction. Assume the market price is $8 per unit. The profit is maximized when MC is equal P. Here the marginal cost function is $MC = 6Q + 10$, therefore $MC = P$ leads to $6Q + 10 = 8$ which in turn leads to $6Q = -2$ and $Q = -1/3$, violating the basic requirement $Q > 0$ [we can get the same result by using (6.4), $Q^* = (P - b)/2a = (8 - 10)/2 * 3 = -1/3$]. So

when $P \leq 10$ the optimization model is inconsistent: the graph of the marginal cost function, which is a straight line, does not intersect the price line in the first quadrant.

For price $P = b + 2\sqrt{ac} = 10 + 2\sqrt{3 * 60.75} = 37$ the firm has a single break-even point of $Q = 4.5$ units:

$$3Q^2 + 10Q + 60.75 = 37Q \rightarrow -3Q^2 + 27Q - 60.75 = 0 \rightarrow Q = 4.5$$

At this price total revenue line is tangent to the total cost curve at $Q = 4.5$ and $TR = TC = 166.5$ (point C on the graph in Fig. 6.6). At $P = 37$ any other output level leads to a loss.

For prices greater than 37, the firm has two break-even points. If, for example, the market price is \$52 then

$$3Q^2 + 10Q + 60.75 = 52Q \quad \longrightarrow \quad -3Q^2 + 42Q - 60.75 = 0 \quad \text{leading to}$$

$Q_1 = 1.64$; and $TR = TC = 85.28$ (point A on the graph in Fig. 6.6) and $Q_2 = 12.36$; and $TR = TC = 642.2$ (point B on the graph in Fig. 6.6). If the firm chooses any output level in the range of 1.64 to 12.36 units, it makes a profit. Any output level outside this range leads to losses. The profit maximizing level of output is $6Q + 10 = 52 \rightarrow Q = 7$ units, which is the average of the two break-even points.

2. We would always have a consistent optimization model if the quadratic cost function did not have a linear term, i.e. when b is zero and $TC = aQ^2 + c$. In this case, from (6.4) we have $Q^* = \dfrac{P}{2a}$, which provides a nontrivial solution as long as $P > 0$.

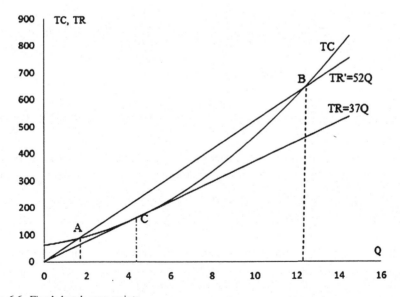

Fig. 6.6 Firm's break-even points

Fig. 6.7 Graph of linear marginal cost $MC = 6Q$ and $P = 30$

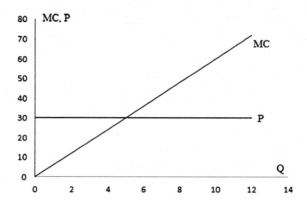

Since $P = MC = 2aQ$ is always greater than $AVC = aQ$ for any $Q > 0$, then the firm recovers its variable cost at any non-zero price. From (R4), $P = 2\sqrt{ac}$ is the break-even price and $Q = \sqrt{\dfrac{c}{a}}$ is the break-even level of output. As a numerical example consider a firm with the total cost function $TC = 3Q^2 + 60.75$. As shown in Fig. 6.7, the graph of the marginal cost function $MC = 6Q$ is a line through the origin (a ray), guaranteeing an intersection with the price line. If, for instance, the market price is \$30 per unit, the optimal level of output is $Q^* = (30/6) = 5$ units. Note that $P = 30$ is greater than $2\sqrt{ac} = 2\sqrt{3 * 60.75} = 27$, generating a profit for the firm.

3. A quadratic total cost function may have a negative linear term, i.e. $b < 0$. In this case (R3) plays a crucial role. Let's consider a firm with the following total cost function

$$TC = 2.5Q^2 - 30Q + 80$$

The associated variable and marginal cost functions are

$$VC = 2.5Q^2 - 30Q$$
$$MC = 5Q - 30$$

VC and MC for a firm that operates must be positive. $MC > 0$ leads to the condition that Q must be greater than 6. But a more binding constraint originates form $VC > 0$.

$$VC > 0 \quad \longrightarrow \quad 2.5Q^2 - 30Q > 0 \quad \longrightarrow \quad 2.5Q^2 > 30Q \quad \longrightarrow \quad Q > 12$$

This means that the appropriate specification of the total cost function must be

$$TC = 2.5Q^2 - 30Q + 80 \qquad Q > 12$$

In order to have a consistent optimization model, (R3) must be satisfied; that is, $P > |-30|$. If the market price is more than \$30 per unit, then we have a working model. Otherwise the model will generate nonsensical results. If for example $P = 20$, by following the profit maximizing procedure, we have

$$\pi = TR - TC = 20Q - 2.5Q^2 + 30Q - 80 = -2.5Q^2 + 50Q - 80$$

$$\frac{d\pi}{dQ} = -5Q + 50 = 0 \quad \longrightarrow \quad Q = 10$$

generating a profit of $-2.5(10)^2 + 50 * 10 - 80 = 170$. But this is a nonsensical result, because at $Q = 10$ the variable cost is -50. If price is \$40

$$\pi = TR - TC = 40Q - 2.5Q^2 + 30Q - 80 = -2.5Q^2 + 70Q - 80$$

$$\frac{d\pi}{dQ} = -5Q + 70 = 0 \quad \longrightarrow \quad Q = 14$$

generating a profit of $-2.5(14)^2 + 50 * 14 - 80 = 130$. At $Q = 14$ the variable cost is $2.5(14)^2 - 30 * 14 = 70$.

Note that the two constraints $P > 30$ and $Q > 12$ are binding and must be satisfied under any condition, including break-even analysis. If, for example, price is \$35 then the break-even output is determined by solving

$$35Q - 2.5Q^2 + 30Q - 80 = 0 \quad \longrightarrow \quad -2.5Q^2 + 65Q - 80 = 0$$

This equation has two solutions: $Q = 1.3$ and $Q = 24.7$. But the first solution must be discarded since it violates the constraint $Q > 12$. Therefore the range of output that generates positive profit is $12 < Q < 24.7$, with profit maximizing level equal to 13 units.

6.1.4 Quadratic Cost Function and Long-Run Price and Output

Consider a firm with $TC(Q) = 5Q^2 - 30Q + 400$ in a competitive market. If the market price is \$100, it is easy to verify that the firm's profit maximizing level of out is 13 units and the firm's profit is \$445. This is a positive or above normal profit, a short run situation. The equilibrium of this competitive industry in the long run would be different from the short run equilibrium. Firms in this industry are making above normal profit and this is a powerful incentive for existing firms to expand their output and/or for new firms to enter this market, thus increasing the total quantity of the good supplied and deriving the market price down. This process continues as long as above normal profit is not eliminated.

In the long run a firm may adapt a new and more efficient technology and/or change its scale of operation, leading to a different cost structure captured by a different total cost and average cost curve. The *Long-Run Average Cost (LRAC)* curve is the *envelope* of the firm's short run average cost curves, and the *Long-Run Total Cost (LRTC)* curve is the envelope of the short-run curves. Points on the *LRAC* curve are the minimum costs of producing different levels of output associated with different techniques or scales of production, and in the long run the equilibrium price in the market will be established at a point on the long-run average cost *LRAC* curve.

We seldom know the shape or equation of a firm's *LRAC* curve, which not only depends on adaptation of new technologies but also the type of return to scale—constant, increasing, or decreasing—the firm operates under. Therefore to provide a general answer to questions about the long-run equilibrium requires that we make certain simplifying assumptions. To start with, let us assume that all firms operating in a competitive industry have access to the same technology and subsequently have the same cost structure,[3] captured at specific point in time by the following quadratic total cost function

$$TC(Q) = aQ^2 + bQ + c$$

The second assumption is that a firm arrives at its current total cost function by adapting the most efficient technology in the short run and the most efficient scale of operation or the *optimal scale of operation* in the long-run. With these assumptions in the background, let P denote the long run price established in this market. Thus the profit function of a typical firm is

$$\pi(Q) = TR(Q) - TC(Q) = PQ - aQ^2 - bQ - c$$

The *FOC* for maximizing profit is

$$\frac{d\pi}{dQ} = \pi'(Q) = P - 2aQ - b = 0 \quad \longrightarrow \quad Q^* = \frac{P-b}{2a} \qquad (6.8)$$

But if P is the market long run price, then the profit of the firm at Q^* must be zero, that is

$$\pi(Q^*) = PQ^* - a(Q^*)^2 - bQ^* - c = 0$$

Substituting for Q^* from (6.8) in the above equation, we have

$$P\left(\frac{P-b}{2a}\right) - a\left(\frac{P-b}{2a}\right)^2 - b\left(\frac{P-b}{2a}\right) - c = 0$$

After multiplying both sides of the equation by $4a^2$ and some algebraic manipulation, we have

$$P^2 - 2bP + b^2 - 4ac = 0$$

---

[3] This is not actually an assumption but rather the implication of the assumptions needed for the existence of perfect competition.

This quadratic equation has two solutions

$$P = \frac{2b \pm \sqrt{4b^2 - 4b^2 + 16ac}}{2} = b \pm 2\sqrt{ac}$$

Since $2\sqrt{ac}$ is positive—based on (R1) and (R2)—then the only acceptable solution satisfying (R3) that P is greater than $|b|$ is

$$P = b + 2\sqrt{ac}$$

And after substituting for P in (8) we have

$$Q^* = \frac{[b + 2\sqrt{ac} - b]}{2a} = \sqrt{\frac{c}{a}}$$

which is exactly the solution for a single break-even point discussed earlier.

We can obtain the same results by determining the minimum average total cost function $ATC(Q)$. First we provide a general proof of this statement.

Let $TVC = f(Q)$ denote the total variable cost function. With the market price P, the total revenue function is $TR = P * Q$. If at the point $[Q, f(Q) + FC]$ the total revenue line TR is tangent to TC curve, then the following two conditions must hold

$$\begin{cases} f(Q) + FC = PQ \\ f'(Q) = P \end{cases}$$

By multiplying both sides of the second equation by $-Q$ and adding it to the first equation we get

$$f(Q) + FC - Qf'(Q) = 0$$

Now, the average total cost function is $ATC(Q) = \dfrac{f(Q) + FC}{Q}$. The FOC for a minimum is

$$\frac{dATC}{dQ} = \frac{f'(Q)Q - [f(Q) + FC]}{Q^2} = 0 \quad \longrightarrow \quad f'(Q)Q - [f(Q) + FC] = 0$$

leading to exactly the same equation as $f(Q) + FC - Qf'(Q) = 0$, which we just derived above.

The $ATC(Q)$ for our general quadratic cost function is

$$ATC(Q) = \frac{TC(Q)}{Q} = \frac{aQ^2 + bQ + c}{Q}$$

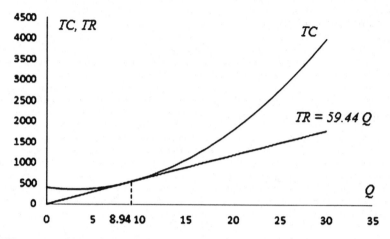

Fig. 6.8 Log-run Situation; *TR* tangent to *TC*

The first order condition for its minimum is

$$\frac{dATC}{dQ} = \frac{(2aQ + b)Q - aQ^2 - bQ - c}{Q^2} = 0$$

$$= \frac{2aQ^2 + bQ - aQ - bQ - c}{Q^2} = 0 \quad \longrightarrow \quad aQ^2 - c = 0,$$

leading to the same answer $Q^* = \sqrt{\dfrac{c}{a}}$ and $P = b + 2\sqrt{ac}$

Using these results for our numerical example given by the total cost function

$$TC(Q) = 5Q^2 - 30Q + 400$$

we have the market long run price and a firm's long run output (or equivalently, the firm's long run break-even price and output) as

$$P^* = -30 + 2(5 * 400)^{1/2} = 59.44 \quad \text{and} \quad Q^* = (400/5)^{1/2} = 8.94 \ \text{units}$$

Figures 6.8 and 6.9 are the graphic presentations of the two methods.

Exercises

1. Find the stationary points of the following quadratic functions.

    (a) $y = 3x^2 + 65$
    (b) $y = 3x^2$

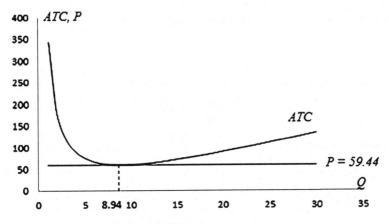

Fig. 6.9 Log-run situation; minimum of ATC curve

(c) $y = 3x^2 - 10x + 65$

(d) $y = -2.5x^2 - 30x + 80$

2. Determine whether the following quadratic functions could be employed as legitimate total cost functions. Specify any restrictions needed for these functions for legitimate profit optimization.

(a) $TC = 2Q^2 + 15Q + 50$

(b) $TC = -2Q^2 + 15Q + 250$

(c) $TC = -Q^2 - 15Q + 50$

(d) $TC = 5Q^2 + 15Q$

(e) $TC = 10Q^2$

(f) $TC = 10Q^2 - 40Q + 150$

(g) $TC = 8Q^2 + 25Q + 60$

3. A competitive firm has the following quadratic total cost function

$$TC = aQ^2 + 10Q + 50$$

determine a if the market price is \$50 per unit and the firm's profit maximizing level of output is 10 units.

4. A competitive firm has the following quadratic total cost function:

$$TC = 3Q^2 + bQ + 50$$

Determine b if the market price is \$70 per unit and the firms profit maximizing level of output is 10 units.

5. A competitive firm has the following quadratic total cost function:

$$TC = 5Q^2 - 25Q + 300$$

(a) Determine the firm's break-even points if the market price is $P = \$80$.

(b) Show that the profit maximizing level of output is the average of the two break-even points.

6. To generalize from the last problem, show that for a firm with a quadratic cost function the optimal level of output is the average of its break-even points, as long as the market price is greater than b.

7. Assume a firm has the following total cost function

$$TC = 10Q^2 - 40Q + 150$$

(a) Specify the bounds of output and price for the function.

(b) Find the break-even points of output if the market price is $55.

(c) Find the profit maximizing level of output and show that it is the average of the two break-even points in part (b).

(d) Find the price and output when the total revenue line is tangent to the TC curve.

(e) Show that the market equilibrium price and the firm's output in the long run are the same as in part (d).

8. Assume all firms in an industry have the following total cost function

$$TC = aQ^2 + 25Q + 60$$

Determine a if the market equilibrium price and a single firm's output in the long run are 49 and 5, respectively.

6.2 Firms with Cubic Cost Functions in Competitive Industry

In microeconomics a cubic function more appropriately represents a total cost function than a quadratic function. In general a cubic function, if properly specified, has all the nice properties expected from a well-behaved total cost function. Most importantly, it generates a marginal cost curve that is, to the great liking of economists, U-shaped. A U-shaped MC curve exhibits the *law of increasing marginal cost*, which is the main factor behind the *law of increasing cost* and the *law of diminishing return*. For this reason cubic cost functions are preferred to quadratic functions and are more widely used by economists for analyzing firms profit maximizing (loss minimizing) behavior.

Example

Assume all firms in a competitive industry have the following cubic total cost function $TC = TC(Q) = \frac{1}{12}Q^3 - 2.5Q^2 + 30Q + 100$. If the market price is $P = 21$, we want to find the profit maximizing output of a typical firm and its total profit.

A firm's marginal cost function is

$$MC = \frac{dTC}{dQ} = \frac{1}{4}Q^2 - 5Q + 30$$

We know that firm's profit is maximized at $MC = P$, therefore

$$\frac{1}{4}Q^2 - 5Q + 30 = 21 \quad \longrightarrow \quad \frac{1}{4}Q^2 - 5Q + 9 = 0 \quad \text{or} \quad Q^2 - 20Q + 36 = 0$$

This quadratic equation has two solutions

$$(Q - 18)(Q - 2) = 0 \quad \longrightarrow \quad Q = 18 \quad \text{and} \quad Q = 2$$

We must now check the *SOC* for a maximum

$$\frac{dMC}{dQ} = \frac{1}{2}Q - 5 \Big|_{Q=2} = -4 < 0 \qquad \text{Minimum profit}$$

$$\frac{dMC}{dQ} = \frac{1}{2}Q - 5 \Big|_{Q=18} = 4 > 0 \qquad \text{Maximum profit}$$

Thus, the profit maximizing output is $Q^* = 18$ units and maximum profit is $\pi = 62$. Recall from Sect. 6.1 that the *profit maximizing level of output* occurs at the intersection of price line with the *upward sloping* segment of the marginal cost curve. The intersection of price with the *downward sloping* part of the marginal cost curve determines the *profit minimizing level of output* (the *SOC* reiterates this fact).

Figure 6.10a represents the *MC* curves and price $P = 21$ of the above example. Figure 6.10b represents the profit function of the same example. The dash lines joining the two Figures highlights the points of intersection between *MC* and the price line where profit is maximized ($Q^* = 18$) and minimized ($Q = 2$).

6.2.1 Restrictions on Coefficients of a Cubic Cost Function

Similar to a quadratic cost function, parameters of an economically legitimate cubic cost function must satisfy certain conditions. In this section we examine the types of restrictions on parameters of a cubic function that are needed in order for it to be a viable total cost function.

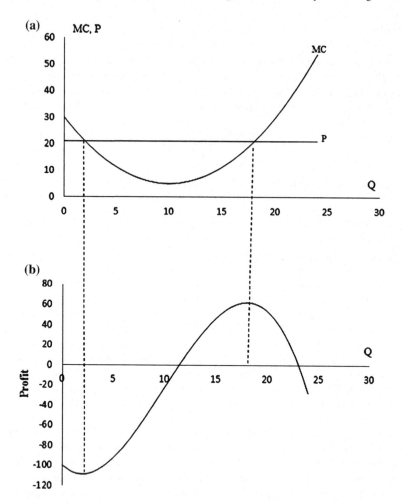

Fig. 6.10 Optimal level of output and profit for a firm with cubic cost function

Assume the following cubic total cost function

$$TC = aQ^3 + bQ^2 + cQ + d \qquad (6.9)$$

At $Q = 0$ $TC = d$. Therefore, 'd' must be the fixed cost. In the short run $d > 0$. In the long run $d = 0$.

Whether the short run or the long run, variable costs are always non-negative. A firm must incur additional cost for producing additional units of output. This means that the entire variable cost curve must be above the Q axis unless $Q = 0$. The MC function, derived from the variable cost function, must be also positive at all levels of output, including at its minimum point. Writing MC as

$$MC = \frac{dTC}{dQ} = 3aQ^2 + 2bQ + c$$

the *FOC* for minimum is

$$\frac{dMC}{dQ} = 6aQ + 2b = 0 \quad \longrightarrow \quad Q = -\frac{b}{3a} \tag{6.10}$$

Since Q must be positive (unless the firm ceases operation, which in that case $Q = 0$) then in (6.10) $\frac{-b}{3a} > 0$.

The *SOC* for a minimum requires that

$$\frac{d^2 MC}{dQ^2} = 6a > 0 \quad \text{leading to condition} \quad a > 0 \tag{R1}$$

With $a > 0$ for $\frac{-b}{3a}$ in (6.10) to be positive b must be negative

$$b < 0 \tag{R2}$$

Even at its minimum *MC* must be positive, therefore

$$MC(-b/3a) = 3a(-b/3a)^2 + 2b(-b/3a) + c > 0$$

which after simplification yields

$$b^2 < 3ac \tag{R3}$$

Since b^2 and a are both positive, for $R3$ to hold c must be also positive

$$c > 0 \tag{R4}$$

With variable cost and fixed cost positive at all levels of output, total cost must be positive at all range of outputs, including $Q = 1$, that is

$$TC(1) = a + b + c + d > 0 \tag{R5}$$

If parameters of a cubic function satisfy all four restrictions,[4] the function can be appropriately used to represent a total cost function.

Unfortunately some authors of Mathematical Economics textbooks carelessly specify costs functions that 'mathematically' exhibit nice properties in optimization problems, while violating meaningful 'economic' properties. As an example the following cubic total cost function appears in a widely used textbook

---

[4] Can you show that in light of other restrictions ($R5$) might be redundant?

$$TC(Q) = 3Q^3 - 7.5Q^2 + 2Q + 5$$

This function satisfies R1, R2, and R4, but fails to satisfy R3. The result is that this firm can produce one unit of output at variable cost of -2.5 and two units at variable cost of negative 2!

6.2.2 Profit Maximization with Cubic Cost Functions: Additional Restrictions

Use of cubic cost function for a profit-maximizing firm necessitates additional restrictions. These restrictions arise from the profit maximization process. Let us first consider the case of a competitive firm with cubic total cost function $TC = aQ^3 + bQ^2 + cQ + d$. Let P denote the price established in the market. The profit function for this firm is

$$\pi = TR - TC = PQ - aQ^3 - bQ^2 - cQ - d$$

The FOC for the profit maximizing level of output is

$$d\pi/dQ = P - 3aQ^2 - 2bQ - c = 0$$

or alternatively
$$3aQ^2 + 2bQ + (c - P) = 0$$

For the above quadratic equation to have a real solution the *discriminant* of the quadratic formula must be non-negative, that is

$$4b^2 - 12a(c - P) \geq 0 \quad \text{or} \quad b^2 - 3ac + 3aP \geq 0 \qquad (6.11)$$

After simplifying (6.11) we arrive at a new restriction relating price and the coefficients of the cubic cost function, namely

$$P \geq c - \frac{b^2}{3a}$$

When the above inequality holds, we have a mathematically consistent model; that is our optimization routine leads to real solution(s). But as it was mentioned before, mathematical niceties are not always sufficient for economically meaningful results. To demonstrate, consider a firm with the following cubic total cost function

$$TC = 0.2Q^3 - 6Q^2 + 80Q + 100$$

Assume the price established in the market is $25 per unit. Since this price satisfies the above inequality

$$25 > 80 - \frac{(-6)^2}{3 * 0.2} \quad \longrightarrow \quad 25 > 20$$

we are assured that the profit maximization routine generates real solution(s), as shown below

$$\pi = TR - TC = 25Q - 0.2Q^3 + 6Q^2 - 80Q - 100 = -0.2Q^3 + 6Q^2 - 55Q - 100$$

$$\frac{d\pi}{dQ} = -0.6Q^2 + 12Q - 55 = 0$$

leading to two solutions 7.1 and 12.9, of which the second one $Q = 12.9$ satisfies the second order condition for a maximum. With this price/quantity combination the total revenue is 322.5 and the total cost is 562.88, a loss of 240.38. Here the price specified for the model does not cover even the average variable cost, let alone the average fixed cost. It is easy to verify that at $Q = 12.9$ the AVC is 35.88 and the AFC is 7.75. For the firm to operate, the market price must at least covers the average variable cost.

6.2.3 Cubic Cost Function and Break-Even Points

Recall that the level of output where the total cost is equal to total revenue is the break-even point for a firm. Break-even analysis for a firm with a cubic cost function is similar to the quadratic case, with the minor exception of being computationally more involved.

Consider a firm with the cost function $TC = 0.2Q^3 - 6Q^2 + 80Q + 300$. If market price is $70, the firm's break-even level of output is found by setting

$$TC = TR \quad \longrightarrow \quad 0.2Q^3 - 6Q^2 + 80Q + 300 = 70Q$$

and solving the resulting cubic equation

$$0.2Q^3 - 6Q^2 + 10Q + 300 = 0$$

This equation has three real roots, one negative and two positive.[5] Discarding the negative root, the remaining two roots yield the break-even points as $Q_1 = 10.0$ and $Q_2 = 25.81$. The profit maximizing level of output is $Q = 19.13$, which is between the two break-even points.

---

[5] See the chapter's Appendix II for application of a simple numerical method using Microsoft Excel and use of other tools in solving cubic and other nonlinear functions.

When the market price is such that the revenue line is tangent to the total cost curve then there is only one break-even point. This break-even point and market price are the normal profit optimal combination. Similar to the case of a firm with a quadratic cost function, to determine this price we must solve the following system of two equations ($TR = TC$ and $MC = P$) with two unknowns,

$$\begin{cases} 0.2Q^3 - 6Q^2 + 80Q + 300 = PQ \\ 0.6Q^2 - 12Q + 80 = P \end{cases}$$

By multiplying both sides of the second equation by Q and subtracting it from the first equation we get

$$-0.4Q^3 + 6Q^2 + 300 = 0$$

The real root of this equation is $Q = 17.46$. By plugging this value into the second equation, we find the market price $P = \$53.39$. For market prices greater than $\$53.39$, this firm makes a profit. For prices below $\$53.39$, it loses money but stays in business as long as revenue exceeds the total variable cost and the firm is able to recover part of its fixed cost. But if the downward slide of price continues and crosses the shutdown threshold, the firm must discontinue production. The shutdown price threshold is the price that makes revenue line tangent to the total variable cost curve, where $TR = TVC$ and $MC = P$. In this example this point is determined by the following system

$$\begin{cases} 0.2Q^3 - 6Q^2 + 80Q = PQ \\ 0.6Q^2 - 12Q + 80 = P \end{cases}$$

By eliminating P, we have $-0.4Q^3 + 6Q^2 = 0$ leading to $Q = 15$. We determine P from the second equation, $P = 35$.

The graphs in Fig. 6.11 depict both the long-run break-even and shutdown situations. In the graph, $TR = 53.39Q$ is tangent to the total cost curve at $Q = 17.46$ units and $TR' = 35Q$ is tangent to the total variable cost at $Q = 15$.

6.2.4 Cubic Cost Functions and Market Equilibrium Price and Firm's Output in the Long Run

When the market price equal to $\$70$ the profit maximizing level of output for the firm with the total cost function $TC = 0.2Q^3 - 6Q^2 + 80Q + 300$ is $Q = 19.13$ units and the firm generates a profit of $\$304.29$, an above normal profit. As it was noted before, an above normal profit in a competitive market is a short run situation. In the long-run, equilibrium price in the market will be determined such that firms' make only normal profit. To determine the long-run equilibrium price and quantity, we use the same assumptions we made for the case of firms with quadratic cost

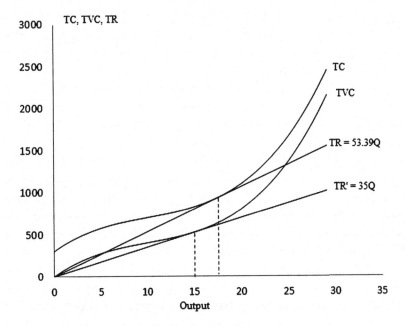

Fig. 6.11 Long-run normal profit and shutdown situation

function, namely that the long-run market price will be established at the lowest possible average total cost *ATC*, which is a point on the firm's *LRAC* curve. At this price all firms earn only normal profit.

In the long-run equilibrium price P^* is the minimum of *ATC* function. There average total cost function is:

$$ATC(Q) = \frac{TC(Q)}{Q} = \frac{0.2Q^3 - 6Q^2 + 80Q + 300}{Q} \tag{6.12}$$

The FOC for a minimum is:

$$\frac{dATC}{dQ} = \frac{(0.6Q^2 - 12Q + 80)Q - (0.2Q^3 - 6Q^2 + 80Q + 300)}{Q^2} = 0$$

$$\frac{dATC}{dQ} = \frac{0.6Q^3 - 12Q^2 + 80Q - 0.2Q^3 + 6Q^2 - 80Q - 300}{Q^2} = 0$$

$$= \frac{0.4Q^3 - 6Q^2 - 300}{Q^2} = 0$$

$$0.4Q^3 - 6Q^2 - 300 = 0$$

The positive rational solution to this cubic equation is $Q = 17.46$. To make sure that the function reaches a minimum at $Q = 17.46$ we must check the second order condition:

$$\frac{d^2 ATC}{dQ^2} = \frac{(1.2Q^2 - 12Q)Q^2 - 2Q(0.4Q^3 - 6Q^2 - 300)}{Q^4}$$

$$\frac{d^2 ATC}{dQ^2} = \frac{0.4Q^4 + 600Q}{Q^4} > 0 \quad \text{for all values} \quad Q > 0$$

Thus the SOC for a minimum is satisfied.

Using Eq. (6.12) above, the value of the ATC function at $Q = 17.46$ is 53.39. Therefore, the long-run market equilibrium price is $P^* = 53.39$. At this price this firm produces $Q^* = 17.46$ units and its economic profit is zero. This is the same result we derived in Sect. 6.3, when we demonstrated that for a single break-even point the revenue line is tangent to the total cost curve.

HW 7 6.3 Profit Maximization in a Non-Competitive Market

In a profit maximization model of a firm operating in a perfectly competitive market price is an exogenous variable. Since a firm has no market power to influence the outcome of interaction between forces of supply and demand, it takes the price as it establishes in the market. For firms operating in a non-competitive environment—a monopoly, or monopolistic competition—price is not an exogenous variable. Firms in noncompetitive markets must search for a combination of price and quantity that maximizes their profit.

Regardless of the market structure, profit is simply total revenue minus total cost. The difference between a competitive and a noncompetitive firm is the way total revenue is expressed. Under perfect competition price is the same as marginal revenue MR, the contribution to revenue by an additional unit sold, and firms sell as many units as they can produce at the same price, so total revenue TR is simply price times quantity $TR = P * Q$. However, if there is imperfection in the market marginal revenue would deviate from price. A monopolist or monopolistically competitive firm faces a downward sloping demand curve for its product. This has an obvious implication: for a firm to sell more it must lower its price. Then marginal revenue will differ from price. The demand curve faced by a noncompetitive firm represents all possible combinations of price-quantity available to the firm. The firm's challenge is to find the combination that maximizes its profit.

Single-period[6] profit maximizing noncompetitive firms use the same search mechanism for finding their profit maximizing price-quantity combination as firms

---

[6] We should carefully distinguish between a *single-period* and *multi-period* profit maximization strategy. A noncompetitive firm, especially a monopoly with substantial market power, may follow a strategy of maximizing profit over several production cycles rather than over a single cycle. This

in a competitive market. That is, as long as the firm's marginal revenue MR is more than its marginal cost MC, it expands its output. The firm reaches its equilibrium at a combination of output-price where $MR = MC$.

The objective function for noncompetitive firm is to maximize

$$\pi(Q) = TR(Q) - TC(Q)$$

By expressing the demand function faced by the noncompetitive firm as $P = f(Q)$ the total revenue function would be

$$TR(Q) = PQ = f(Q) * Q$$

We can rewrite the profit function as

$$\pi(Q) = f(Q) * Q - TC(Q)$$

The first and second order conditions for the maximization of π at Q^* is the same as for the competitive case, namely

$$\pi'(Q^*) = 0 \quad \text{and} \quad \pi''(Q^*) < 0$$

which are equivalent to $MR = MC$ and $TR''(Q^*) < TC''(Q^*)$.

Figure 6.12 depicts the process for a firm with linear demand function.

Fig. 6.12 A noncompetitive firm maximizes profit at the level of output where $MC = MR$

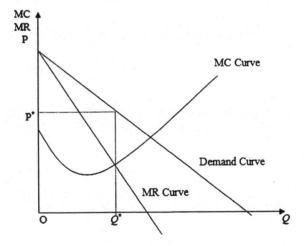

is dynamic optimization strategy and is beyond the scope of this book. Here, and in other parts of the book, by "profit maximization" we simply mean single period maximization.

Example

A firm operates in a *monopolistically competitive* market. This noncompetitive firm
has the following cubic total cost function

$$TC(Q) = \frac{1}{3}Q^3 - 7Q^2 + 111Q + 50 \tag{6.13}$$

The equation of demand function faced by the firm in the market is

$$P = f(Q) = 100 - Q$$

We wish to determine firm's profit maximizing combination of output and price.
This firm's total revenue function is

$$TR(Q) = PQ = f(Q) * Q = (100 - Q)Q = 100Q - Q^2 \tag{6.14}$$

Writing the profit function as

$$\pi(Q) = TR(Q) - TC(Q) = f(Q) * Q - TC(Q)$$

we have,

$$\pi(Q) = 100Q - Q^2 - (\frac{1}{3}Q^3 - 7Q^2 + 111Q + 50)$$

$$\pi(Q) = -\frac{1}{3}Q^3 + 6Q^2 - 11Q - 50$$

The *FOC* for profit maximization requires that

$$\frac{d\pi}{dQ} = \pi'(Q) = -Q^2 + 12Q - 11 = 0$$

This quadratic equation has two solutions, $Q = 1$ and $Q = 11$. We must next check
the second order condition. The second derivative of the profit function is

$$\frac{d^2\pi}{dQ^2} = -2Q + 12$$

Evaluating the second derivative at $Q = 1$ and $Q = 11$, we have

$$\frac{d^2\pi}{dQ^2}\bigg|_{Q=1} = -2Q + 12\bigg|_{Q=1} = 10 > 0$$

$$\frac{d^2\pi}{dQ^2}\bigg|_{Q=11} = -2Q + 12\bigg|_{Q=11} = -10 < 0$$

Therefore, profit is maximized at $Q^* = 11$. By substituting 11 for Q in the demand function, the firm's profit maximizing price is determined as

$$P^* = 100 - Q^* = 100 - 11 = 89$$

The optimal combination $(Q^*, P^*) = (11, \$89)$ generate the maximum profit of $111.4.

$$\pi(11) = -1/3(11)^3 + 6(11)^2 - 11(11) - 50 = 111.4$$

Notice that we get exactly the same results by using the firm's equilibrium condition, $MR = MC$, directly. From total cost function (6.13) and total revenue function (6.14), we derive the marginal cost and marginal revenue functions as

$$MC = Q^2 - 14Q + 111 \quad \text{and} \quad MR = 100 - 2Q$$

Setting marginal revenue equal to marginal cost yields the following:

$$100 - 2Q = Q^2 - 14Q + 111 \quad \longrightarrow \quad Q^2 - 12Q + 11 = 0$$

This expression leads to the same profit maximizing level of output. Figure 6.13 is a graphic illustration of the model.

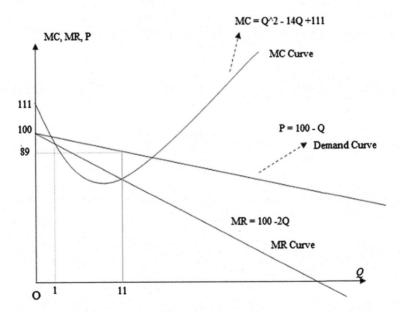

Fig. 6.13 Optimal combination $(Q^*, P^*) = (11, \$89)$ maximizes profit for the monopolistically competitive firm

6.4 Long Run Price-Output Combination for a Monopolistically Competitive Firm

The firm in the above numerical example operating in monopolistically competitive market produces 11 units of output priced at $89 a unit and makes $111.3 profit. This is a short-run situation, very similar to the short-run state of a competitive firm discussed in previous sections. Lured by above normal profit, new firms enter this market. As the new firms, which produce and supply a close substitute, enter the market demand for the good or service produced by this firm declines. This decline manifests itself in a downward slide in the demand curve and the associated marginal revenue curve for this firm. Each downward shift reduces the firm's profit. This process continues until opportunity for profit is eliminated. This occurs when price becomes equal to average total cost ATC. Graphically, this occurs at the point of tangency between the *dislocated* demand curve and the ATC curve. Figure 6.14 is a graphic depiction of the long run equilibrium of the firm. But what is the firm's long run optimal output-price combination?

The original demand equation facing the firm is $P = 100 - Q$, with the associated marginal revenue equation $MR = 100 - 2Q$. In the process of market adjustment the demand curve shifts to the left. Since these shifts are parallel, the demand equation at the end of the long-run equilibrium process will have the same slope but a different intercept. Let's denote this demand equation by $P = a - Q$. At equilibrium the demand curve is tangent to the ATC curve. At the point of tangency the slopes of the ATC curve and demand curve are the same. The long-run equilibrium price and quantity occurs at this point. We use this fact and write

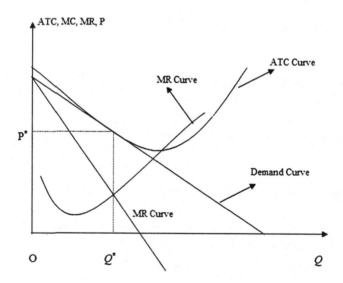

Fig. 6.14 Long run equilibrium of a monopolistically competitive firm

$$ATC(Q) = \frac{TC(Q)}{Q} = \frac{\left(\frac{1}{3}Q^3 - 7Q^2 + 111Q + 50\right)}{Q}$$

The slope of *ATC* at any point is

$$\frac{dATC}{dQ} = \frac{(Q^2 - 14Q + 111)Q - \left(\frac{1}{3}Q^3 - 7Q^2 + 111Q + 50\right)}{Q^2}$$

$$\frac{dATC}{dQ} = \frac{\frac{2}{3}Q^3 - 7Q^2 - 50}{Q^2}$$

At the point of tangency between the *ATC* and demand curves, the slope of *ATC* curve is equal to the slope of the demand curve, which in our example is -1. That is

$$\frac{d\overset{\bullet}{ATC}}{dQ} = \frac{dP}{dQ} = -1 \quad \text{therefore}$$

$$\frac{\frac{2}{3}Q^3 - 7Q^2 - 50}{Q^2} = -1 \quad \text{or} \quad \frac{2}{3}Q^3 - 7Q^2 - 50 = -Q^2$$

After simplifying the expression, we have

$$Q^3 - 9Q^2 - 75 = 0$$

An acceptable solution to this cubic equation is $Q^* = 9.7836 \approx 9.784$.

Since at equilibrium price is equal to *ATC*, we can determine the long run price by calculating the value of *ATC* at $Q = 9.784$

$$ATC(9.784) = \frac{1}{3}(9.784)^2 - 7(9.784) + 111 + \frac{50}{9.784} = 79.53$$

Therefore, the long run equilibrium price is $P = \$79.53$.

To find the long-run dislocated demand equation we must find its new intercept. Since we already know the long-run price and quantity, then

$$P = a - Q \quad \longrightarrow \quad 79.53 = a - 9.784 \quad \longrightarrow \quad a = 89.314$$

and the dislocated demand equation is

$$P = 89.314 - Q$$

Note that at the new price-quantity combination (79.53, 9.784) total revenue and total cost are both 778.06 and profit is zero.

If the demand function in the model is nonlinear, the above procedure in rare occasion may fail. In the next example we offer an alternative method for determining the long run price-output combination of a firm operating in a non-competitive industry that is certain to work.

Example

Assume a monopolistically competitive firm has the following quadratic cost function

$$TC = 5Q^2 + 30Q + 400$$

Also assume that this firm faces the following demand function for its product

$$P = 500 - 11Q$$

This firm achieves its profit maximizing goal at a level of output where marginal revenue is equal to marginal cost, i.e. when

$$\frac{dTR}{dQ} = \frac{dTC}{dQ}$$
$$MC = 10Q + 30$$
$$TR = P * Q = (500 - 11Q)Q = 500Q - 11Q^2 \text{ and } MR = 500 - 22Q$$

By equating MR and MC, we have

$$500 - 22Q = 10Q + 30$$

leading to $32Q = 470$ and $Q^* = 14.688$. By plugging 14.688 for Q in the demand function, we have the profit maximizing price

$$P^* = 500 - 11 * 14.688 = 338.44$$

This quantity-price combination maximizes the firm's profit at $\pi = 3051.56$ Lured by potential of making above normal profit, new firms enter this market. As new competitors enter the market, demand for goods or services produced by this firm declines, causing a downward slide in the demand curve and its associated marginal revenue curve. The long-run equilibrium is reached when following two conditions are simultaneously satisfied

$$\begin{cases} TC = TR \\ MC = MR \end{cases} \tag{6.15}$$

Graphically, this happens when the point of intersection of MR and MC curves vertically aligns with the point of tangency between the dislocated demand curve and the ATC curve (see Fig. 6.14). As it was noted before, the dislocated demand function has the same slope as the original demand curve but different intercept. Expressing this function as $P = a - 11Q$, we have

$$TR = PQ = (a - 11Q)Q = aQ - 11Q^2 \quad \rightarrow \quad MR = \frac{dTR}{dQ} = a - 22Q$$

By substituting for $TC, TR, MC,$ and MR in (6.15), we have

$$\begin{cases} 5Q^2 + 30Q + 400 = aQ - 11Q^2 \\ 10Q + 30 = a - 22Q \end{cases}$$

We can solve this system of two equations with two unknowns by multiplying both sides of the second equation by $-Q$ and adding it to the first equation:

$$-5Q^2 + 400 = 11Q^2 \quad \rightarrow \quad 16Q^2 = 400 \quad \rightarrow \quad Q^2 = 25 \text{ and } Q = 5$$

Substituting 5 for Q in the second equation we get $a = 190$. The equation of dislocated demand curve is then

$$P = 190 - 11Q$$

At the long-run level of out $Q_{LR} = 5$, the long-run price is $P_{LR} = 135$.

6.5 Short-Run Production Function and Profit Maximization for a Noncompetitive Firm

Consider a noncompetitive firm with the short-run production function[7] $Q = L^{\frac{1}{2}}$, where L is the labor input. Assume the wage rate w is \$5 per unit of labor and firm fixed cost is \$50. What is the firm profit maximizing level of output if demand for its product is given by $P = 144 - 7Q$?

Since the firm's only variable input is labor, its total cost function TC is

$$TC = VC + FC = wL + FC = 5L + 50$$

Substituting for $L = Q^2$ from the production function into the total cost function, we have

$$TC = 5Q^2 + 50$$

The firm's profit function is

$$\pi(Q) = TR - TC = PQ - TC = (144 - 7Q)Q - 5Q^2 - 50$$
$$\pi(Q) = -12Q^2 + 144Q - 50$$

---

[7] Again, see Chap. 10 for a comprehensive discussion of the production function.

The FOC for a local maximum is

$$d\pi/dQ = -24Q + 144 = \quad \longrightarrow \quad Q^* = 6 \text{ units}$$

Substituting for Q in the production function and demand equation we find the equilibrium price and employment for the firm

$$L^* = (Q^*)^2 = 36 \text{ units} \quad P^* = 144 - 7(6) = \$102$$

The firm total profit is $\pi^* = \pi(6) = -432 + 864 - 50 = \$382$.

Now assume that due to exogenous factors the market demand for the noncompetitive firm's product declines to the point that its above normal profit is eliminated. What would be the firm's new Q, L, P, and π?

This situation is similar to Fig. 6.14. The noncompetitive firm's ATC function is,

$$ATC = TC/Q = (5Q^2 + 50)/Q$$

At the point of tangency between the ATC and the dislocated demand curve,

$$\frac{dATC}{dQ} = \frac{dP}{dQ} \quad \longrightarrow \quad \frac{10Q^2 - (5Q^2 + 50)}{Q^2} = -7$$

$$5Q^2 - 50 = -7Q^2 \quad \longrightarrow \quad 12Q^2 = 50 \quad \longrightarrow \quad Q = 2.04 \text{ units}$$

leading to the price and labor employment of

$$P = ATC(2.04) = 34.71 \text{ an } \quad L = Q^2 = 2.04^2 = 4.16 \text{ units}$$

To find the dislocated demand equation we must find its new intercept. Writing the new demand equation as

$$P = a - 7Q$$

we have

$$34.71 = a - 7(2.04) \quad \longrightarrow \quad a = 49 \quad \text{so} \quad P = 49 - 7Q$$

is the dislocated demand function. With this demand equation, the new profit function is

$$\pi(Q) = TR - TC = PQ - TC = (49 - 7Q)Q - 5Q^2 - 50 = -12Q^2 + 49Q - 50$$

and at the quantity-price combination of (2.04, 34.71) profit is zero.

6.6 Constraint Optimization

It was mentioned earlier that if the variable or total cost function of a firm is known then the two types of constraint optimization problems discussed at the beginning of this chapter would be very simple. Recall that the first type of optimization assumes that the firm's level of output is predetermined at Q_0 and the firm seeks to minimize its costs. This is an interesting and challenging problem within the context of a firm's least cost combination of recourses, given its production function (see Chap. 10). But in the current context where total cost function of a firm is given, it becomes a trivial question. If, for example, a competitive firm with the $TC(Q)$ function decides to produce Q_0 units of output its total cost would naturally amounts to $TC(Q_0)$. In this case the price established in the market must be greater than $\dfrac{TC(Q_0)}{Q_0}$ for the firm to make a profit:

$$TR(Q) > TC(Q) \quad \longrightarrow \quad P * Q_0 > TC(Q_0) \quad \text{leading to} \quad P > \frac{TC(Q_0)}{Q_0}$$

If, for example, a firm with the following TC function,

$$TC(Q) = Q^3 - 10Q^2 + 45Q + 100$$

decides to produce 20 units of output, its total cost would be

$$TC(20) = (20)^3 - 10(20)^2 + 45(20) + 100 = 5000$$

If price of a unit of this good in the market is less than \$250, this firm would obviously suffer losses.

The second type optimization problem is slightly more challenging. In this case the firm's total cost is predetermined, and it must find its level of output if operates in a competitive market, or its level of output and price if operates in a noncompetitive market. As an example assume that our competitive firm decides to set their total cost exogenously at 8000. What would be the firm's level of output?

$$TC_0 = 8000 = Q^3 - 10Q^2 + 45Q + 100 \quad \longrightarrow \quad Q^3 - 10Q^2 + 45Q - 7900 = 0$$

Here the problem is reduced to finding zeros of a cubic function. Using the numerical method for solving cubic equations described in the appendix to this chapter, $Q \approx 23$ is a zero of the function. Thus unless this firm could sell its product at approximately \$340 or more per unit, it would incur losses.

If this firm operates in a noncompetitive environment and the demand equation for its output is $P = 393 - Q$, then the firm can command $P = \$370$ in the market and earn \$510 profit.

6.7 Elasticity of Demand and Supply

6.7.1 Elasticity of a Function

Elasticity of a function $y = f(x)$ at a point x is defined as the rate of proportional change in y per unit proportional change in x. Elasticity measures sensitivity or responsiveness of y to small relative changes in x. If x changes by a small increment to $x + \Delta x$ the proportional change in x is $\dfrac{\Delta x}{x}$ (note that $\dfrac{\Delta x}{x}$ multiplied by 100 is proportional change stated as a percentages.) Similarly the proportional change in y (due to change in x) is $\dfrac{\Delta y}{y}$. Based on this definition, the elasticity of the function $y = f(x)$ at point x is expressed as

$$E = \frac{\dfrac{\Delta y}{y}}{\dfrac{\Delta x}{x}} = \frac{\Delta y}{\Delta x} \frac{x}{y} \tag{6.16}$$

We denote the limiting case of (6.16) when we let Δx become infinitesimally small or theoretically approaching zero by ϵ,

$$\epsilon = \lim_{\Delta x \to 0} E = \lim_{\Delta x \to 0} \frac{\Delta y}{\Delta x} \frac{x}{y}$$

But as $\Delta x \to 0$ $\dfrac{\Delta y}{\Delta x}$ becomes $\dfrac{dy}{dx}$, or the derivative of the function at point x. Therefore, the elasticity of a function at point x denoted by $\epsilon(x)$ is

$$\epsilon(x) = \frac{dy}{dx} \frac{x}{y} = f'(x) \frac{x}{f(x)}$$

Clearly $\epsilon(x)$ varies with different values of x. We refer to $\epsilon(x)$ as the *elasticity function* of $f(x)$. It should be abundantly clear that elasticity is a unit-less quantity.

Example

Let $y = f(x) = 3x^2 - 2$. What is the elasticity of the function at $x = 2$?
   We first find the derivate of the function

$$\frac{dy}{dx} = f'(x) = 6x$$

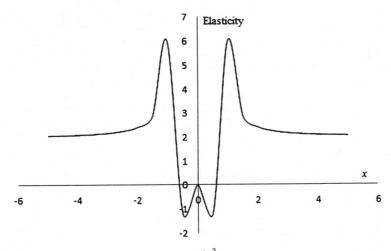

Fig. 6.15 Graph of elasticity function $\epsilon(x) = \dfrac{6x^2}{3x^2 - 2}$

Then the elasticity function of y is

$$\epsilon(x) = \frac{dy}{dx}\frac{x}{y} = 6x * \frac{x}{3x^2 - 2} = \frac{6x^2}{3x^2 - 2}$$

The graph of this function, Fig. 6.15, indicates changes in elasticity for various values of x. The elasticity function evaluated at $x = 2$ is

$$\epsilon(2) = \frac{6(2)^2}{3(2)^2 - 2} = 2.4$$

The elasticity of the function at $x = 2$ is 2.4. What is the meaning of this number? To express that at point 2 elasticity of this function is 2.4 means that a small proportional change in x when x is 2 will lead to 2.4 times proportional change in y. Let's verify this assertion.

Assume x changes from 2 to 2.001. This is a $\dfrac{(2.001 - 2)}{2} = 0.0005$ proportional change, or 0.05 % increase, in x. At 2 and 2.001 values of the function are

$$y = f(2) = 3(2)^2 - 2 = 10 \quad \text{and} \quad f(2.001) = 3(2.001)^2 - 2 = 10.0120$$

making the proportional change in y equal to $\dfrac{10.012 - 10}{10} = 0.0012$ or 0.12 %. This number is the same as $2.4 * 0.0005 = 0.0012$ or 0.12 %. For larger changes in x the approximation is still relatively good. If, for example, x changes from 2 to 2.1 (a 5 % increase), the value of the function increases by approximately $5 * 2.4 = 12$ %. To check this we find that the value of the function at $x = 2.1$ is

$f(2.1) = 3 * (2.1)^2 - 2 = 11.23$ and the percentage increase in the function is $\dfrac{11.23 - 10}{10} * 100 = 12.3$, or close to 12%. Generally elasticity is an accurate measure of proportional change in the function in a close neighborhood of x.

For another illustration, elasticity of the function at $x = 3$ is

$$\epsilon(x) = \frac{xf'(x)}{f(x)} = \frac{3f'(3)}{f(3)} = \frac{3 * 6(3)}{3 * (3)^2 - 2} = \frac{54}{25} = 2.16$$

which is different from function's elasticity at $x = 2$.

Students should be reminded again that elasticity is a unit-less quantity.

6.7.2 Elasticity of Demand and Supply

Consider the market demand function

$$Q_d = f(P)$$

Applying the concept of elasticity to this demand function, we can express the *price elasticity of demand* by

$$\epsilon(p) = \frac{dQ_d}{dP} \frac{P}{Q} = \frac{Pf'(P)}{f(P)}$$

Since price and quantity demanded move in opposite direction, that is, $\dfrac{dQ_d}{dP} = f'(P) < 0$, the price elasticity of demand is always negative. Note that price elasticity of demand is a function of price and varies as price changes. We call $\epsilon(p)$ the *demand elasticity function*. $\epsilon(p)$ evaluated at a certain price provides the price elasticity of demand at that price. Three categories of price elasticity are

1. When $|\epsilon(p)| > 1$ the demand is *price elastic*, implying that change in P leads to a proportionally larger change in Q.
2. When $|\epsilon(p)| < 1$ the demand is *price inelastic*, implying that change in P leads to a proportionally smaller change in Q.
3. Finally when $|\epsilon(p)| = 1$ the demand is *unit elastic* (or more formally *price unit elastic*), implying that a proportional change in P leads to the same proportional change in Q.

As we will see in the next section, whether demand for a product is elastic, inelastic, or unit elastic has significant implications for both consumers and producers.

Example

Consider the following linear demand function.

$$Q_d = f(P) = 20000 - 400P \qquad (6.17)$$

What is the price elasticity of demand if the price is $40?

Given that $\dfrac{dQ_d}{dP} = -400$ then

$$\epsilon(p) = \frac{dQ_d}{dP}\frac{P}{Q_d} = \frac{Pf'(P)}{f(P)} = \frac{-400P}{20000 - 400P} \qquad (6.18)$$

$\epsilon(p)$ evaluated at $P = 40$ leads to

$$\epsilon(p) = \frac{f'(40) * 40}{f(40)} = \frac{-400 * 40}{20000 - 400 * 40} = -\frac{16000}{4000} = -4$$

Since $|\epsilon(p)| = 4$, the demand at $40 is highly elastic and a 1% decline in price would generate 4% increase in quantity demanded. What is the elasticity of demand if price is $30?

$$\epsilon(p) = \frac{30 * f'(30)}{f(30)} = \frac{-400 * 30}{20000 - 400 * 30} = -\frac{12000}{8000} = -1.5$$

Note that at a price of $30 the demand function is significantly less elastic compared to a price of $40. This is a mathematical reiteration of general property of a downward sloping linear demand curve: demand over the segment of the curve corresponding to higher prices is more elastic compared to the segment corresponding to lower prices. Somewhere in between these two segments lies the unit elastic point.

At what price is the demand function unit elastic? We can find this price by setting $\epsilon(p)$ in (6.18) equal to -1 and solving for P

$$\epsilon(p) = \frac{-400P}{20000 - 400P} = -1$$

$$-400P = -20000 + 400P \quad \longrightarrow \quad 800P = 20000 \text{ and } P = 20000/800 = 25$$

Figure 6.16 graphs the elasticity function of the linear demand equation (6.17). This graph is prepared by ignoring the negative sign in the demand elasticity function (6.18). It is clear from the graph that the demand function is unit elastic at $P = 25$, i.e. $|\epsilon(25)| = 1$ At prices below 25, $|\epsilon(25)| < 1$ and demand is inelastic. At prices above 25, $|\epsilon(25)| > 1$ and demand is elastic.

In general, price elasticity of a linear demand function $Q = a - bP$ is

Fig. 6.16 Graph of the elasticity function
$$\epsilon(P) = \frac{-400P}{20000 - 400P}$$

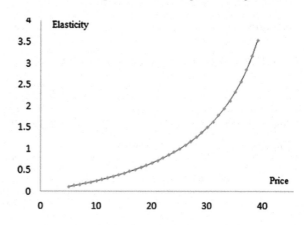

$$\epsilon(p) = -b \frac{P}{Q_d} = -\frac{bP}{Q_d}$$

and it is clear that larger the magnitude of b, greater would be the price elasticity. Since b is the slope of the demand curve, this means that steeper demand curves are more elastic. If, for example, we change the above demand function from $Q_d = 20000 - 400P$ to $Q_d = 20000 - 450P$ the price elasticity of demand at \$40 and \$30 will change from 4 and 1.5 to 9 and 2.1, respectively.

Now consider the following non-linear demand function.

$$Q_d = f(P) = 400 - 3P^2$$

What is price elasticity of demand when price is \$10?

$$f'(P) = -6P \quad \text{and} \quad f'(10) = -6 * 10 = -60$$
$$f(10) = 400 - 3(10)^2 = 100$$
$$\epsilon(10) = \frac{10f'(10)}{f(10)} = \frac{10(-60)}{100} = -6$$

A price elasticity of 6 at price 10 implies that a 1 % increase in price leads to 6 % decline in quantity demanded. At price 10 quantity demanded is 100. If price changes to 10.1 (a 1 % increase) the quantity demanded drops to $400 - 3(10.1)^2 = 93.97 \approx 94$, which is a 6 % decline. How elastic is demand at price 5?

$$f'(P) = -6P \quad \text{and} \quad f'(5) = -6 * 5 = -30$$
$$f(5) = 400 - 3(5)^2 = 325$$
$$\epsilon(5) = \frac{Pf'(p)}{f(P)} = -0.46$$

Comparing the coefficient of elasticity corresponding to prices 10 and 5 may tempt us to claim that our statement about magnitude of elasticity along a linear demand curve is also true for a nonlinear demand function. That is not, unfortunately, always the case. In the exercise section you will be asked to show that some demand functions have constant elasticity over the whole range of prices.

We can extend the concept of elasticity to measure sensitivity or responsiveness of demand to changes in consumers' income or prices of related goods. These elasticities are calculated as proportional change in quantity demanded to per unit proportional change in income or price of complement or substitute good. Consider the demand function expressed by

$$Q_d = f(P, I, P_c, P_s) \tag{6.19}$$

where P is the price the good itself, I is income of consumers, and P_c and P_s are the prices of complement and substitute goods. By partially differentiating expression (6.19) with respect to P, I, P_c, and P_s we can define the following elasticities:

$$(a) \quad \epsilon(p) = \frac{\partial Q_d}{\partial P} \frac{P}{Q_d}$$

$$(b) \quad \epsilon(i) = \frac{\partial Q_d}{\partial I} \frac{I}{Q_d}$$

$$(c) \quad \epsilon(p_c) = \frac{\partial Q_d}{\partial P_c} \frac{P_c}{Q_d}$$

$$(d) \quad \epsilon(p_s) = \frac{\partial Q_d}{\partial P_s} \frac{P_s}{Q_d}$$

(a) is, of course, the price elasticity of demand. But to avoid confusion when prices of other goods are included in the demand function, it is called *own price elasticity of demand*. (b) is *income elasticity of demand*. (c) and (d) are called *cross-price elasticity of demand*. They measure proportional change in quantity demanded per unit proportional change in price of a complement or substitute good.

Example

Assume the monthly demand function for a commodity is

$$Q_d = 2.5I - 12P_1 - 2P^2$$

where I is the household's average income per month, P_1 is the price of another good, and P is the price of the good itself. What is the relationship between these two goods? What are own price, cross price, and income elasticity of demand if household's average monthly income is $1,500, $P_1 = \$30$, and $P = \$20$?

Since P_1 is entered with a negative sign in the demand equation, these two goods must be complementary goods. To calculate own price, income, and cross price elasticities, we first find the value of the function at the given values for I, P_1, and P

$$Q_d = 2.5I - 12P_1 - 2P^2 = 2.5 * 1500 - 12 * 30 - 2(20)2 = 2590$$

Next we find

$$\frac{\partial Q_d}{\partial P} = -4P|_{P=20} = -80$$

and own price elasticity is

$$\epsilon(p) = \frac{\partial Q_d}{\partial P} \frac{P}{Q_d} = -80 \frac{20}{2590} = -0.62$$

Given $\dfrac{\partial Q_d}{\partial I} = 2.5$ and $\dfrac{\partial Q_d}{\partial P_1} = -12$, the income and cross-price elasticities are

$$\epsilon(i) = \frac{\partial Q_d}{\partial I} \frac{I}{Q_d} = 2.5 \frac{1500}{2590} = 1.45$$

$$\epsilon(p_1) = \frac{\partial Q_d}{\partial P_1} \frac{P_1}{Q_d} = -12 \frac{30}{2590} = -0.14$$

One can similarly use the concept of elasticity to define various *price elasticities of supply*. Consider the following supply function

$$Q_s = -50 + 5P \qquad P > 10$$

The supply elasticity function associated with this supply function is

$$\eta(p) = \frac{dQ_s}{dP} \frac{P}{Q_s} = 5 * \frac{P}{-50 + 5P} = \frac{5P}{-50 + 5P}$$

Figure 6.17 shows the graph of the supply elasticity function associated with the supply function $Q_s = -50 + 5P$.

Note that for $P > 10$ the supply elasticity function is always greater than 1, that is,

$$\eta(p) > 1 \quad \forall P > 10$$

As P increases, $\eta(p)$ asymptotically approaches 1. This means our supply function becomes unit elastic at very large values of P.

6.7.3 Price Elasticity of Demand and a Firm's Total Revenue

Assume a firm operating in a non-competitive market seeks a combination of price and quantity that maximizes its revenue. Suppose the firm faces a linear demand function

$$Q = a - bP$$

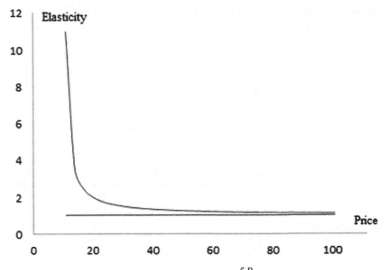

Fig. 6.17 Graph of supply elasticity function $\eta(P) = \dfrac{5P}{-50 + 5P}$

Solving for P, we have

$$P = \frac{a}{b} - \frac{1}{b}Q$$

The firm's total revenue is

$$TR = P * Q = (\frac{a}{b} - \frac{1}{b}Q) * Q = \frac{a}{b}Q - \frac{1}{b}Q^2$$

The FOC for maximizing TR is

$$\frac{dTR}{dQ} = MR = \frac{a}{b} - \frac{2}{b}Q = 0 \quad \longrightarrow \quad Q = \frac{a}{2}$$

From the demand function, we have

$$P = \frac{a}{b} - \frac{1}{b}Q = \frac{a}{b} - \frac{1}{b}\frac{a}{2} = \frac{a}{2b}$$

The revenue maximizing level of output occurs at $Q = a/2$ (where marginal revenue MR is zero, i.e. MR curve intersect the Q axis) and $P = a/2b$. This combination occurs at a point on the demand curve where the demand is exactly unit elastic (point A in Fig. 6.18). To show this we write the elasticity function and evaluate it at this point

Fig. 6.18 Demand elasticity
at various segments of demand
curve

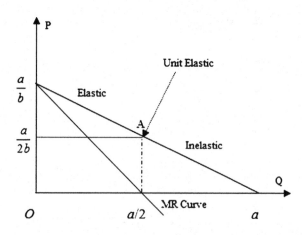

$$\epsilon(p) = \frac{dQ}{dP}\frac{P}{Q} = -b\frac{P}{a-bP} = -\frac{bP}{a-bP}$$

$$\epsilon\left(\frac{a}{2b}\right) = -\frac{-b\left(\dfrac{a}{2b}\right)}{a-b\left(\dfrac{a}{2b}\right)} = -1$$

This result is not limited to linear demand equations. It is true for any demand function. Next we provide a general proof of the assertion that at the profit maximizing price-output combination the price elasticity of demand is -1, or it is unit elastic.

Let $P = f(Q)$ be a demand function. The total revenue function is

$$TR = P * Q = Qf(Q)$$

and MR is

$$MR = \frac{dTR}{dQ} = f(Q) + f'(Q)Q = P + \frac{dP}{dQ}Q = P\left(1 + \frac{dP}{dQ}\frac{Q}{P}\right)$$

But $\dfrac{dP}{dQ}\dfrac{Q}{P}$ is the reciprocal of price elasticity of demand $\epsilon(p)$, therefore

$$MR = P\left(1 + \frac{1}{\epsilon(p)}\right)$$

It is clear that at the point on the demand curve where demand is unit elastic ($\epsilon(p) = -1$) MR is zero and the quantity/price combination at this point maximizes total revenue. If a firm operates on the inelastic part of the demand curve, its best strategy for enhancing revenue would be to increase price, moving upward along the demand curve. But as soon as the firm, in its journey along the demand curve, crosses point A and enters the "Elastic" territory, a price increase would lead to a decline in revenue. Point A is, therefore, the firm's optimal position. If this firm initially operates on the elastic segment of the demand curve, it can raise its revenue by reducing its price and moving downward along the demand curve. This process will continue to increase revenue as long the firm does not move to the "Inelastic" segment of the demand curve. Point A, again, is therefore the optimal position.

To illustrate this point, consider the linear demand function

$$Q_d = f(P) = 20000 - 400\,P$$

This demand function's elasticity for prices $40 and $30 are 4 and 1.5. The function's elasticity at price $25 is 1. Quantity demanded associated with these prices are 4000, 8000, and 10000 and corresponding total revenues are 16000, 24000, and 25000. Clearly the firm's revenue is the highest at price $25, where elasticity of demand is 1.

6.8 Exercises

1. For the following total cost functions, determine the total cost TC average total cost ATC, average variable cost AVC, average fixed cost AFC, and marginal cost MC for the specified level of output

   (a) $TC = 4Q^2 + 100$ $\hspace{5cm}$ $Q = 20$

   (b) $TC = \dfrac{1}{2}Q^2 + 150Q + 20000$ $\hspace{2.5cm}$ $Q = 500$

   (c) $TC = 0.01Q^2 + 140Q + 50000$ $\hspace{2.5cm}$ $Q = 1000$

   (d) $TC = 0.02Q^3 - 6Q^2 + 6800Q + 20000$ $\hspace{1cm}$ $Q = 100$

   (e) $TC = 0.03Q^3 - 0.6Q^2 + 5Q + 3700$ $\hspace{1.5cm}$ $Q = 10$

   (f) $TC = 0.08Q^3 - 1.8Q^2 + 50Q + 678$ $\hspace{1.5cm}$ $Q = 20$

2. For the cubic total cost functions given in problem (1), find

   (a) the minimum average total cost.
   (b) the minimum average variable cost.
   (c) the minimum marginal cost.

3. A firm operating in a competitive market has the following quadratic total cost function
$$TC = 10Q^2 + 5Q + 400$$

If the market price of the good is $165 per unit.

(a) What is the firm's profit maximizing level of output? What is its profit?

(b) What is the long run price established in this market?

(c) What is the output and profit of the firm at the long run price?

4. A competitive firm has the following quadratic TC function

$$TC = 5Q^2 + 200$$

Assume the market price P is $175 per unit.

(a) How much it costs the firm to produce the 11th unit of output?

(b) What is the marginal profit of the 11th unit of output?

(c) What is the firm's profit maximizing level of output? What is its profit?

(d) What is marginal profit of the last unit of output?

(e) What is the long-run price established in this market?

(f) What is the output and profit for the firm in the long-run?

Revenue Maximization

In some industries a firm's profit maximization objective is basically a revenue maximization problem. A good example is the operation of sport clubs, especially when the franchises do not own the arena or stadium they play in. With a stadium/arena lease and players' contracts in place, the cost of operation of a team for a season is very much known. Therefore, profit is largely determined by revenue from ticket sales and merchandising. Managing rental apartment complexes and hotels are other variations on this theme.

5. The New York-area MLS (Major League Soccer) team, Red Bulls (formerly MetroStar), plays in a newly built soccer-specific stadium located in Harrison, New Jersey. The new stadium has a 25000 seating capacity. If the average ticket price is $30, the average attendance per game is 15000. When the ticket price is $25, the average attendance is 20000.

(a) Find the demand function for the team's ticket, assuming it is linear.

(b) What price should the team charge for a ticket in order to maximize revenue?

(c) What would the attendance be at the revenue maximizing ticket price?

(d) Show that the price elasticity of demand at the revenue maximizing price and quantity is 1.

6. *Easy Living Corporation* manages a 1000-unit rental apartment complex. If rent is set at $2000 a month, the complex would have a 90 % occupancy rate. A 100 % occupancy rate is achieved by reducing the rent to $1800 a month. What are the

revenue maximizing rent and the occupancy rate, assuming a linear demand function?

7. *Grand Paradise* resort and casino in Las Vegas has 2500 rooms. The hotel management knows from past experience that a room rate of $100 a night results in full occupancy. It is also known that a 10 % increase in rate leads to a 1 % decline in the occupancy rate. If demand is linear, what rate should the hotel charge to maximize revenue? What would the occupancy rate be?

8. All firms operating in a perfectly competitive market have the variable cost function

$$VC = 0.2Q^2 + 2Q$$

and the firms' fixed cost is 120. The market demand and supply functions for this good, denoted by Q_{md} and Q_{ms} are given by

$$Q_{md} = 107 - 4P \qquad Q_{ms} = -37 + 2P$$

where Q_{md} and Q_{ms} are in units of 1000.

(a) What is a typical firm's profit maximizing level of output?

(b) What is a firm's maximum profit?

(c) How many firms are in the market?

(d) If similar firms enter the market, what would the long-run price be in this market?

(e) What is a firm's output in the long-run?

9. The variable cost function of a competitive firm is given by

$$VC = Q^3 - 40Q^2 + 1600Q$$

Assume the firm's fixed cost is $1500.

(a) If the market price for a unit of Q is $1400, determine the firm's profit maximizing level of out and its profit.

(b) What is the market long run price?

(c) What is the firm's shutdown price?

10. Using the rules developed in the text, evaluate the following costs functions

$$(a) \quad TC = \frac{1}{5}Q^3 + 25Q^2 + 1600Q$$

$$(b) \quad TC = \frac{1}{5}Q^3 - 25Q^2 + 1600Q$$

$$(c) \quad TC = 20Q^2 - 20Q - 100$$
$$(d) \quad TC = 2Q^2 - 10Q + 100$$

11. A competitive firm has a quadratic cost function $TC = aQ^2 + bQ + c$. Is it possible to determine this firm's shutdown price?

12. A monopolistically competitive firm has the following total cost function

$$TC = 0.03Q^2 + 12Q + 500$$

If the demand function for the firm's product is

$$P = 33 - 0.005Q$$

(a) Find the firm's profit maximizing level of output and its profit.

(b) What is the firm's long run price-quantity combination?

(c) What is the equation of the dislocated demand curve associated with the long run situation?

13. A monopolistically competitive firm has the following total cost function

$$TC = 4Q^2 + 30Q + 400$$

If the demand function for the firm's product is

$$P = 600 - 30Q \qquad \text{where} \quad Q < 20$$

(a) Find the firm's profit maximizing level of output and its profit.

(b) What is the firm's long-run price-quantity combination?

(c) What is the equation of the dislocated demand curve associated with the long run situation?

14. A monopolistically competitive firm has the following total cost function

$$TC = Q^3 - 10Q^2 + 45Q + 100$$

If the demand function for the firm's product is

$$P = 100 - 10Q$$

(a) Find the firm's profit maximizing level of output, price, and profit.

(b) What is the firm's long run price-quantity combination?

(c) What is the equation of the dislocated demand curve?

15. A monopolist has the following cubic cost function

$$TC = 0.08Q^3 - 8.2Q^2 + 2000Q + 32000$$

If the market demand function for the firm's product is

$$P = 3400 - 14Q$$

What is the monopolist's profit maximizing combination of price and quantity? What is the firm's profit?

16. A monopolist has the following cubic cost function

$$TC = 0.2Q^3 - 7Q^2 + 136Q + 100$$

If the market demand function for the firm's product is

$$P = 1000 - 7Q$$

What is the monopolist's profit maximizing combination of price and quantity? What is the firm's profit?

17. Due to adverse economic conditions, demand for goods produced by the monopolist in problem (16) declines, to the extent that the firm's above normal profit is eliminated. What is the monopolist's new optimal combination of price and quantity? What is the equation of the dislocated market demand curve?

18. A monopolist firm with the following cubic cost function

$$TC = 3Q^3 - 40Q^2 + 250Q + 900$$

faces the following market demand function for its product

$$P = 2000 - 10Q$$

The management of the firm decides to follow a three-stage market strategy

1. find the level of output that minimizes its average total cost
2. find the level of output that maximizes its total revenue
3. choose the average of the two levels for actual production

(a) What is the level of output in this strategy compared to the profit maximizing output?

(b) What is the price level in this case compared to the profit maximizing case?

Average Cost Pricing

Instead of pursuing profit maximization through marginal cost pricing, the regulated natural monopolists are required to follow the average cost pricing. In the average cost pricing, a natural monopolist firm sets its price at the level where the demand curve intersects the average cost ATC curve. This policy, compared to marginal cost pricing, leads to production of a larger volume of output at a relatively lower price.

19. What is the level of output and price if a natural monopolist with the cost function

$$TC = 0.3Q^3 - 8Q^2 + 120Q$$

and market demand function

$$P = 100 - Q$$

follows the average cost pricing? (From 2 possible solutions choose the larger one.)
Compare this solution to the profit maximizing solution.

20. A firm is competitive in both product and resource market. The firm's short run production function is given by

$$Q = 0.5L^{1/2}$$

where L is the labor input. Assume the wage rate is $10 and the firm's fixed cost is 100. If the price of the good in the market is $1200 per unit

(a) What is the firm's profit maximizing level of output and employment?
(b) What is the firm's total wage bill?
(c) What is its profit?

21. Continued from Problem (20): Assume new firms enter this market and due to additional supply price declines to $1000 per unit. Redo parts (a), (b), and (c).

22. Continued from Problem (21): Assume that due to strong demand for labor the wage rate in the labor market is increased by 10%. Redo parts (a), (b), and (c) under.

23. A monopolistically competitive firm has a short run production function given by

$$Q = 0.5L^{1/2}$$

Assume the wage rate is $5 and the firm's fixed cost is 828. If demand for the firm's output is

$$P = -3Q + 460$$

(a) What is the firm's profit maximizing level of output and employment?
(b) What is the firm's total wage bill?
(c) What is its profit?

24. Continued from Problem (23): If new firms enter the mark and as the result the firm's demand curve is dislocate

   (a) What is the firm's long-run level of output and employment?

   (b) What is the equation of the firm's dislocated demand curve?

25. A monopolistically competitive firm has a short run production function given by

$$Q = 2L^{0.5}$$

Assume the wage rate is $10, the firm's fixed cost is 500, and demand for the firm's output is

$$P = 600 - 3Q$$

Suppose the firm's goal is to maximize its sales revenue. What is the firm's revenue maximizing output, price, employment, and profit?
What is the firm's profit maximizing output, price, employment, and profit?

26. Write the average (A) and marginal (M) functions of the following functions. Show that the value of the average function A at its extrema (minimum or maximum) is the same as of the M function, that is, M intersects A at its extreme point.

   (1) $y = x^2 + 2x + 2$

   (2) $z = -2w^3 + 5w^2 + 10w$

   (3) $TC = 5Q^2 - 25Q + 240$

27. A firm operating in a competitive market has the following long run total cost function

$$TC = Q^3 - 24Q^2 + 200Q$$

Find the long run equilibrium price for this good.

28. A firm operating in a competitive market has the following production function

$$Q = 30L^2 - L^3$$

If this firm uses 18 units of labor and the price of the good in the market is $20, what wage must this firm pay its workers? If this firm's fixed cost is 30,000, what is its profit?

29. A firm in a non-competitive market has the total cost function

$$TC = \frac{1}{3}Q^3 - 6Q^2 + 50Q + 45$$

The demand function for this firm's product is

$$P = 35 - 2Q$$

(a) Write the firm's marginal revenue function.

(b) Find the firm's profit maximizing level of output and price.

30. A firm operating in a competitive market has the following production function

$$Q = 10L - L^2$$

If the firm's output sells for $10 per unit and it pays $40 per unit of labor, what is the firm's equilibrium level of labor utilization?

31. Suppose the market demand function is

$$Q_d = 200 - 5P$$

(a) Determine the price elasticity of demand at prices $10, $15, and $30.

(b) Determine the price at which the demand function is unit elastic.

(c) Determine the range of prices over which the demand function is (a) elastic and (b) inelastic.

32. Suppose the market supply function is

$$Q_s = -500 + 50P \qquad\qquad \text{where} \quad P > 10$$

(a) Determine the price elasticity of supply η at prices $15, $20, and $25.

(b) Is there a price at which this supply function is unit elastic?

33. Suppose the market supply function is given by

$$Q_s = -c + dP \qquad\qquad \text{where} \quad c, d > 0 \qquad\qquad \text{and} \quad P > \frac{c}{d}$$

Show that this supply function is elastic at all prices above $\frac{c}{d}$.

34. Suppose the market demand function is

$$Q_d = aP^\lambda \qquad\qquad \text{where} \quad a > 0 \quad \text{and} \quad \lambda < 0$$

Show that this demand function has constant elasticity equal to λ.

35. Determine the price elasticity of demand at prices $5, $6, and $8 for

$$Q_d = 600 - 2P^2$$

At what price is demand unit elastic?

36. Suppose the market demand function is

$$Q_d = 5I + 10P_1 - 0.4P^2$$

where I is income in \$1000 and P_1 is price of a related good.

(a) What is the relation between these two goods?
(b) Calculate the income, cross price, and own price elasticities of demand if $I = \$50,000$, $P_1 = \$10$ and $P = \$20$.
(c) At what price is demand unit elastic if $I = \$50,000$ and $P_1 = \$10$?

37. Suppose the market supply function is

$$Q_s = 1.2P^2 - 0.5P_E - 10P_L$$

where P_E and P_L are price of energy and labor, respectively.
Calculate the own price elasticity and cross price elasticities of energy and labor if $P = 20$, $P_E = 70$, and $P_L = 40$.

38. Assume the following nonlinear demand function.

$$Q_d = 100 - 2\sqrt{P}$$

(a) At what price is the price elasticity unitary?
(b) At what price is the price elasticity -1.5?

39. Suppose the market demand function is

$$Q_d = 8I - 9P_1 - 0.5P^2$$

where I is income in \$1000 and P_1 is price of a related good.

(a) What is the relation between these two goods?

(b) Calculate the income, cross price, and own price elasticities of demand if $I = \$40,000$, $P_1 = \$10$ and $P = \$15$.

(c) At what price is demand unit elastic if $I = \$40,000$ and $P_1 = \$10$?

40. Suppose the market demand function is

$$Q_d = 6I - 1.2P^2 - 0.4P$$

where I is income in \$1000. At what price is demand unit elastic if $I = \$57,000$?

Appendix A: Linear Demand Function and Marginal Revenue

The marginal revenue function associated with a linear demand function faced by a firm in a noncompetitive market is the first derivative of the firm's total revenue TR function.

If the demand equation is

$$P = a - bQ$$

then the total revenue function is

$$TR(Q) = PQ = (a - bQ)Q = aQ - bQ^2$$

By differentiating TR with respect to Q we get the marginal revenue function

$$MR = \frac{dTR(Q)}{dQ} = a - 2bQ$$

Compare MR with the demand function. They both have the same y-intercept, a, but the slope of the MR function is twice that of the demand function. The x-intercept of demand is $\frac{a}{b}$. The x-intercept of MR is $\frac{a}{2b}$; that is MR cuts the horizontal axis at a point half the distance of the point cut by the demand curve. This information is very helpful in graphing MR curves.

Relationship Between ATC, AVC, and MC Functions

Given $y = f(x)$, the average and marginal functions of y are $A = \frac{y}{x} = \frac{f(x)}{x}$ and $M = y' = f'(x)$. It is easy to show that functions A and M have the same value at the point where A reaches an extremum (maximum or minimum). To find the relative extremum of A we differentiate the function with respect to x and set the result equal to zero. Using the quotient rule, we have

$$\frac{dA}{dx} = \frac{xf'(x) - f(x)}{x^2} = 0 \quad \longrightarrow \quad xf'(x) - f(x) = 0 \quad \text{and}$$

$$f'(x) = \frac{f(x)}{x}$$

which means that at its extremum M has the same value as A.

Graphically, the marginal curve intersect the average curve at its maximum or minimum point. If the original function is a total cost function, then this result suggests that the marginal cost curve must cut the average variable and average total cost curves at their minimum points.

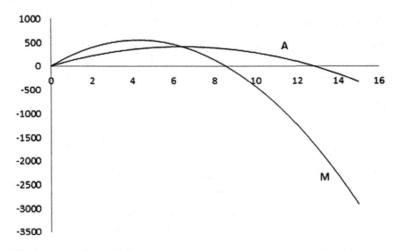

Fig. 6.19 Average and marginal curves

Example 1

$$\text{Let} \quad y = -10x^3 + 128x^2 + 3x$$
$$A = y/x = (-10x^3 + 128x^2 + 3x)/x = -10x^2 + 128x + 3$$
$$M = dy/dx = -30x^2 + 256x + 3$$

Figure 6.19 shows the graph of the average and marginal functions. Note that the marginal function intersects the average function at its maximum point.

Maximum of the average function A occurs at

$$dA/dx = -20x + 128 = 0 \quad \longrightarrow \quad x = 128/20 = 6.4$$

and the value of the average function at $x = 6.4$ is $A(6.4) = 412.6$. We can verify that function A's maximum point with coordinate $(6.4, 412.6)$ is also a point on the marginal function

$$M(6.4) = -30(6.4)^2 + 256(6.4) + 3 = 412.6$$

Example 2

Consider a cubic total cost function

$$TC = 0.5Q^3 - 3Q^2 + 12Q + 50$$

The average and marginal functions (which are, of course, the average and marginal cost functions) are

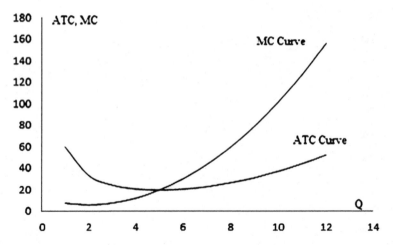

Fig. 6.20 *ATC* and MC curves

$$A = \frac{TC}{Q} = 0.5Q^2 - 3Q + 12 + \frac{50}{Q}$$

$$M = \frac{dTC}{dQ} = 1.5Q^2 - 6Q + 12$$

As Fig. 6.20 indicates, the curve of marginal function M intersects the curve of average function A at its lowest or minimum point $Q = 5$. At this point both functions have a value of 19.5.

Example 3

Consider a quadratic total cost function

$$TC = 10Q^2 - 35Q + 400$$

The associated average and marginal functions are

$$A = \frac{TC}{Q} = 10Q - 35 + \frac{400}{Q}$$

$$M = \frac{dTC}{dQ} = 20Q - 35$$

Here M is a linear function and cuts A at its minimum (6.33, 91.5).

Appendix B: Finding Roots of Third and Higher Degree Polynomials and Other Nonlinear Equations

Finding Roots of Third and Higher Degree Polynomials and Other Nonlinear Equations

In the present chapter we encountered the challenge of solving a cubic equation on several occasions. The *Quadratic formula* can be used for solving *quadratic equations*. There are also formulas for solving cubic and fourth-degree equations, but they are extremely complicated and time consuming. For polynomials of degree 5 or higher there is no such formula. In fact, the French mathematician Evariste Galois proved that it is impossible to find a general formula for the roots of an nth-degree equation if n is larger than 4. Our challenge is not limited to solving cubic or higher degree polynomial equations. As we will see in the following chapters, finding the equilibrium values of the endogenous variables in many non-linear economic models requires solving not only higher degree polynomials, but also exponential, logarithmic, and other nonlinear equations. These equations are not easy to solve, so we must turn to *numerical methods* for approximating root(s) of these equations.

Consider a function $y = f(x)$. The *zeros* of the function $f(x)$ are the values of x that make the function equal to zero. That is, if x^* is a zero of $f(x)$, then $f(x^*) = 0$.

The zeros of a function are points where the graph of the function crosses the x-axis. In other words, they are the coordinates of intersection between the graph of the function and the horizontal axis. If the graph cuts the x-axis from below, then values of the function must change from negative to positive. If it cuts the axis from above, then the change must be from positive to negative. The number of times the graph cuts the x-axis determines the number of zeros of the function. Obviously, if the graph of a function remains entirely above or below the x-axis then function does not have a zero.

The *roots* of an equation $f(x) = 0$ are simply the zeros of $f(x)$, therefore in finding the zeros of a function, we are also finding the roots of the related equation. This connection provides a simple numerical/graphical method for finding roots of cubic functions or solving polynomials of high degree like

$$x^4 - 3x^2 + 10x - 20 = 0$$

nonlinear equations like

$$P(3)^{0.02P} - \frac{57}{P} = 0$$

and even trigonometric equations similar to

$$sin(x^2) + 3cos(x) - 2x = 0$$

To find the roots of an equation we first graph it as a function and then examine the graph in the vicinity of its intersection with the x-axis. We use this method for finding zeros of several cubic functions discussed in this chapter and several nonlinear equations appearing in the next chapters.

We can, of course, graph a function on a Graphing Calculator and zoom in to read the roots. But achieving accuracy with this method is not possible if the root is an irrational number, or has a large number of decimal places. The newer generations of some calculators (HP, and TI) can solve quadratic and cubic functions, but they are not capable of solving complicated nonlinear equations. There is a number of mathematical software, like Maple, Matlab and Mathematica, which easily find roots of functions. But they are expensive and one has to invest time and effort to learn the software's language and its implementation. A small version of Mathematica called WolframAlpha is accessible on the Internet and we will introduce a couple of its commands for plotting and solving nonlinear equations at the end of this appendix. Since our goal here is pedagogical and not high speed computing, we will first discuss a simple method that could be implemented using more widely available Microsoft Excel.

Suppose we found ourselves in need of solving the following cubic equation related to a firm's cost of production

$$\frac{1}{3}Q^3 - 7Q^2 + 111Q - 2500 = 0$$

This is easily accomplished by using *Microsoft Excel*. To start *Excel* we double click on its icon.[8] A blank spreadsheet will appear with cell A1 selected. To generate the graph of a function, we must enter a range of values for the independent variable in this column. Luckily in this equation, and many others that we have to deal with, the variable Q is the level of output and cannot be negative. Therefore we limit ourselves to only nonnegative values for the function

$$f(Q) = 1/3Q^3 - 7Q^2 + 111Q - 2500$$

First we enter 0 in cell A1. In cell A2 we enter the text

$= A1 + 1$ which is recognized as a formula because of the "=" sign

After hitting the *Enter* key the number 1 should appear in cell A2. Next, we copy cell A2 and paste it to a number of successive cells below in column A. To do this, we follow these steps:

1. Select cell A2, which becomes an active cell. The active cell has a small black box in the lower right corner called *fill handle*.
2. Point the mouse at the fill handle; when the mouse pointer becomes a black cross, click and drag to fill cells with copies of the formula.

---

[8] My presentation here is based on the assumption (very likely unrealistic) that the reader is not familiar with Excel. For this reason the mechanics of implementing Excel is explained in an elementary fashion. Those who know Excel can use more advanced features of the software.

Table 6.1 Function $f(Q) = \frac{1}{3}Q^3 - 7Q^2 + 111Q - 2500$

First iteration		Second iteration		Third iteration		Fourth iteration	
Q	$f(Q)$	Q	$f(Q)$	Q	$f(Q)$	Q	$f(Q)$
0	−2500.00	21.1	−143.06	21.61	−6.33	21.631	−0.54
1	−2395.67	21.2	−116.84	21.62	−3.58	21.632	−0.27
2	−2303.33	21.3	−90.33	21.63	−0.82	21.633	0.01
3	−2221.00	21.4	−63.54	21.64	1.94	21.634	0.28
4	−2146.67	21.5	−36.46	21.65	4.71	21.635	0.56
5	−2078.33	21.6	−9.09	21.66	7.47	21.636	0.84
6	−2014.00	21.7	18.57	21.67	10.24	21.637	1.11
7	−1951.67	21.8	46.53	21.68	13.02	21.638	1.39
8	−1889.33	21.9	74.78	21.69	15.79	21.639	1.67
9	−1825.00	22.0	103.33	21.70	18.57	21.640	1.94
10	−1756.67						
11	−1682.33						
12	−1600.00						
13	−1507.67						
14	−1403.33						
15	−1285.00						
20	−413.33						
21	−169.00						
22	103.33						
23	405.67						
24	740.00						
25	1108.33						
26	1512.67						
27	1955.00						
28	2437.33						
29	2961.67						

We can now copy A2 and paste it to cells A3 to A30 (based on experience, we may not need to paste to cells beyond A30). After this step we have 30 values from 0 to 29 in column A. We next move to cell B1. In this cell we enter the formula

$$= (1/3) * (A1\^3) - 7 * (A1\^2) + 111 * A1 - 2500$$

and press the Enter key. The number -2500 should appear in cell B1. We copy cell B1 and paste it to cells B2 to B30. Numbers appearing in the first 30 cells of columns A and B are listed in Table 6.1 under "First Iteration". A quick inspection of values under $f(Q)$ shows that the function changes from negative to positive values somewhere between $Q = 21$ and $Q = 22$. At $Q = 21$ the value of function $f(21) = -169.00$. At $Q = 22$ the function is $f(22) = 103.33$. Clearly $f(Q)$ converges to zero at a Q between 21 and 22. The graph of the function, Fig. 6.21, visually verifies this assessment. The curve cuts the Q-axis right after point marked 20. This result helps

Fig. 6.21 Graph of $f(Q) = \frac{1}{3}Q^3 - 7Q^2 + 111Q - 2500$

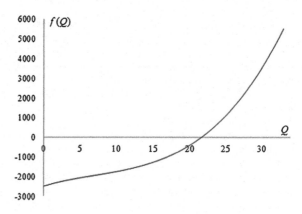

us to concentrate our search in a much narrower range between 21 and 22. In the second iteration we move from 21 towards 22 by small increments of 0.1.

We start the second iteration by entering 21.1 in cell C1. In cell C2 we enter the formula

$$= C1 + 0.1$$

After hitting the Enter key 21.2 should appear in this cell. Copy C2 and paste it to the next 8 cells in the C column. Next, we copy cell B1 to cell D1. This transfers the formula $= (1/3) * (C1^\wedge 3) - 7 * (C1^\wedge 2) + 111 * C1 - 2500$ over to D1. We then copy cell D1 to cells D2 to D10. Numbers appearing in the first 10 cells of columns C and D are listed in the table under "Second Iteration." A review of numbers in this column shows that the function approaches zero between 21.6 and 21.7. Value of the function changed from -9.09 to 18.57 when Q moves from 21.6 to 21.7. We have now narrowed the solution to a much shorter interval of 21.6 and 21.7.

The third iteration starts by entering 21.61 in cell E1. In cell E2 we enter the formula

$$= E1 + 0.01$$

After pressing the Enter key, number 21.62 should appear in cell E2. The rest of the steps in the third iteration from this point are exactly similar to those in the second iteration. Numbers appearing in the first 10 cells of columns E and F are given in the table under "Third Iteration." The third iteration brackets the answer in the range 21.63 and 21.64, which is a very narrow interval. The fourth iteration involves moving from 21.63 by increments of 0.001. After creating the 10 values from 21.631 to 21.640 in the first 10 cells of column G, we copy the formula from cell B1 to cells H1 and then from there to cells H2 to H10. The result is listed under "Fourth Iteration."

The figures for $f(Q)$ after the fourth iteration clearly indicates that value of the function converged almost to zero when $Q = 21.633$. The graph of the function in the range of 21.631 to 21.64 indicates that the curve crosses the Q-axis at 21.633

Fig. 6.22 Graph of $f(Q) = 1/3Q^3 - 7Q^2 + 111Q - 2500$ over the interval $21.631 < Q < 21.64$

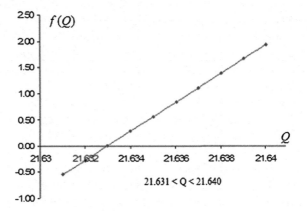

(see Fig. 6.22); that is, the root of the equation $\frac{1}{3}Q^3 - 7Q^2 + 111Q - 2500 = 0$ is, to 3 decimal places, 21.633.

Notice that in Fig. 6.22 over the narrow range of 21.631 to 21.64 the graph of the function is almost a straight line. This is the motivation behind the linear approximation of a nonlinear function over a relatively small interval. Most of the numerical root-finder algorithms are based on this very idea.

The next problem requires a slightly different approach to the first or initial iteration. Assume a firm's total cost function is given as

$$TC = 0.0003Q^3 - 0.04Q^2 + 5Q + 3700$$

and we are asked to determine the firm's minimum average total cost. Given that $ATC = \dfrac{TC}{Q}$, we have

$$ATC = \frac{0.0003Q^3 - 0.04Q^2 + 5Q + 3700}{Q}$$

$$ATC = 0.0003Q^2 - 0.04Q + 5 + \frac{3700}{Q}$$

The FOC for a minimum is

$$\frac{dATC}{dQ} = 0.0006Q - 0.04 - \frac{3700}{Q^2} = 0$$

After multiplying both sides by Q^2 we end up with the cubic equation

$$0.0006Q^3 - 0.04Q^2 - 3700 = 0$$

Due to the very small coefficient of the cubic term, we will run into trouble in the first iteration if we increment Q by one unit. The trick is to increment Q by

Table 6.2 Function $f(Q) = 0.0006Q^3 - 0.04Q^2 - 3700$

First iteration		Second iteration		Third iteration	
Q	$f(Q)$	Q	$f(Q)$	Q	$f(Q)$
1	−3700.04	202	−386.72	208.1	−25.09
11	−3704.04	203	−329.1	208.2	−18.95
21	−3712.08	204	−270.84	208.3	−12.81
31	−3720.57	205	−211.93	208.4	−6.67
41	−3725.89	206	−152.35	208.5	−0.51
51	−3724.45	207	−92.11	208.6	5.65
61	−3712.65	208	−31.21	208.7	11.82
71	−3686.89	209	30.36	208.8	17.99
81	−3643.58	210	92.6	208.9	24.17
91	−3579.1	211	155.52	209	30.36
101	−3489.86				
111	−3372.26				
121	−3222.7				
131	−3037.59				
141	−2813.31				
151	−2546.27				
161	−2232.87				
171	−1869.51				
181	−1452.6				
191	−978.52				
201	−443.68				
211	155.52				
221	822.68				
231	1561.39				
241	2375.27				

a larger unit, in this case by 10. The values of the function $f(Q) = 0.0006Q^3 - 0.04Q^2 - 3700$ after the first iteration shows that a root of this equation must be a number between 201 and 211 (see Table 6.2). The second iteration narrows the range to between 208 and 209. The third iteration narrows the range further between 208.5 and 208.6. It seems that 208.51 should be a very good approximation of the root of the function (the exact root is 208.5083) (Fig. 6.23).

At $Q = 208.51$ the firm's ATC reaches its minimum of 27.45. At this level of output The SOC for minimum is satisfied.

$$ATC'(Q) = 0.0006 + \frac{7400}{Q^3} > 0$$

This numerical method used for solving cubic equations can be easily adopted for solving polynomials of higher degree or other complicated or transcendental equations. For example, by applying this method to the following fourth degree equation

$$x^4 - 3x^2 + 10x - 20 = 0$$

Fig. 6.23 Graph of $f(Q) = 0.0006Q^3 - 0.04Q^2 - 3700$

Fig. 6.24 Graph of $f(x) = x^4 - 3x^2 + 10x - 20$

Fig. 6.25 Graph of $f(x) = sin(x^2) + 3cos(x) - 2x$

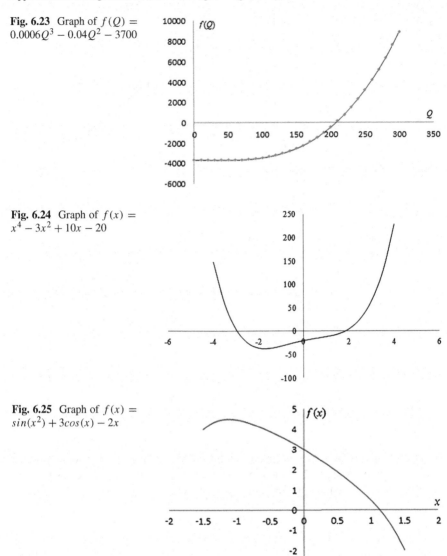

we find that it has two roots $x = -2.94833$ and $x = 1.85223$. Figure 6.24 depicts the graph of this function.

The trigonometric equation mentioned earlier $sin(x^2) + 3cos(x) - 2x = 0$ has a root at $x = 1.124$. Figure 6.25 is the graph of $f(x) = sin(x^2) + 3cos(x) - 2x$ drawn for values of $-1.5 \le x \le 1.5$ and Fig. 6.26 is the graph of the same function drawn over a much narrower interval for x between 1 and 1.2.

Fig. 6.26 Graph of $f(x) = sinx^2 + 3cosx - 2x$ over the interval $1 < x < 1.2$

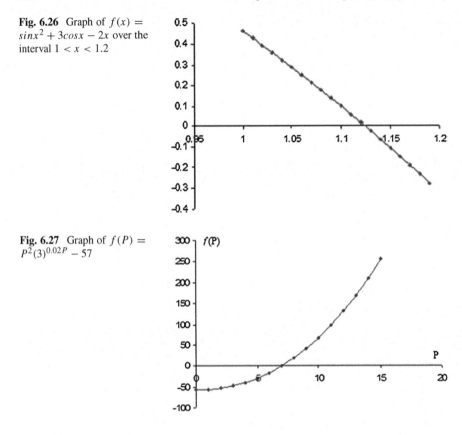

Fig. 6.27 Graph of $f(P) = P^2(3)^{0.02P} - 57$

We close this section by working through a problem that we will encounter in the next chapter. In Example 2 of the next chapter (Chap. 7) we find ourselves in need of solving the following equation for the equilibrium market price

$$P^2(3)^{0.02P} - 57 = 0$$

Figure 6.27 is the graph of $f(P) = P^2(3)^{0.02P} - 57 = 0$. By applying our method to the this equation we found that this function has a zero at 6.9916.

Using WolframAlpha, Maple or R for Solving Nonlinear Equations

To access WolframAlpha go to the site http://www.WolframAlpha.com. The software's opening page is similar to Google's opening page. It consists of the software's logo with a rectangular inquiry box with a little equal sign on the right hand side. The statement on top of the box reads "Enter what you want to **calculate** or **know about**". To find a solution to an equation, linear or nonlinear, we use the *solve* command, which in its simplest form appears as *solve(expr, var)*, where *expr* is *expression* of equation and *var* is the variable. A more useful version of the *solve* command is *solve(expr, var, dom)*, where *dom* specifies the domain of the function. Since most of the variables used in equations and models in economics are non-negative, the natural choice for *dom* is non-negative real numbers.

To solve the first equation we discussed earlier,

$$\frac{1}{3}Q^3 - 7Q^2 + 111Q - 25000 = 0$$

we simply type the following in the inquiry box

$$\text{solve}\left[(1/3)Q^{\wedge}3 - 7Q^{\wedge}2 + 111Q - 2500 = 0, \ Q > 0\right]$$

and then click the equal sign on the right hand side of the box. In a fraction of a second Wolframalpha produces the result $Q \approx 21.6330$. Note that this result is exactly the same as the one we found using Excel. Actually for solving equations we don't need to use the 'solve' command format. The same result is achieved by simply typing

$$1/3Q^3 - 7Q^2 + 111Q - 25000 = 0, \ Q > 0$$

in the inquiry box and clicking the equal sign. To solve

$$0.0006Q^3 - 0.04Q^2 - 3700 = 0$$

we enter

$$\text{solve}\left[0.0006Q^{\wedge}3 - 0.04Q^{\wedge}2 - 3700 = 0, \ Q > 0\right]$$

and get the answer as $Q \approx 208.508$. To solve

$$x^4 - 3x^2 + 10x - 20 = 0$$

we enter the following solve command

$$\text{solve}\left[x^{\wedge}4 - 3x^{\wedge}2 + 10x - 20 = 0, \ x\right]$$

or simply

$$x^{\wedge}4 - 3x^{\wedge}2 + 10x - 20 = 0$$

and the two real roots found are -2.9483 and 1.8522 (there are two complex results that we ignore). Readers are encouraged to verify that WolframAlpha easily finds roots of the exponential function

$$P(3)^{0.02P} - \frac{57}{P} = 0$$

and the trigonometric function

$$sin(x^2) + 3cos(x) - 2x = 0$$

Using Maple[9] to solve equations is equally easy. The simplest command for solving equations is

$$fsolve(equation)$$

For example, to solve

$$x^4 - 3x^2 + 10x - 20 = 0$$

we enter

$$fsolve\ (x^\wedge4 - 3x^\wedge2 + 10x - 20 = 0)$$

The third piece of software we can utilize is **R**, which is a powerful free statistical software developed collaboratively by a large number of contributors from all over the world. It is an integrated suite of software facilities for data manipulation, calculation and graphic display. You can download the software by entering http://www.r-project.org in the URL bar of your web browser. That will take you to the website "The R Project for Statistical Computing". Inside the rectangle "Getting Started:" click on "downloadR" and follow the instructions. **R** runs on variety of Windows, Mac, and UNIX platforms. **R** has powerful facilities for matrix and linear algebra operations that we will discuss in more details in Chap. 13 "Matrices and Their Applications." Students of economics and related fields who must apply a variety of statistical and econometric tools in their future work related projects will greatly benefit from learning **R**.

The *polyroot* function of **R** can be used for solving polynomials. The format is

$$polyroot(c(a_0, a_1, a_2, \ldots a_n))$$

where $a_0, a_1, a_2, \ldots a_n$ are the coefficients of the nth degree equation

$$a_0 + a_1x + a_2x^2 + a_x^3 + \cdots + a_nx^n = 0$$

To solve, for example,

$$x^4 - 3x^2 + 10x - 20 = 0$$

we write it as

$$-20 + 10x - 3x^2 + x^4 = 0$$

---

[9] Most colleges and universities have license for a mathematical software like Maple.

and use

$$\text{polyroot}(c(-20, 10, -3, 0, 1))$$

Note that since there is no cubic term in this equation its coefficient is entered as 0. To solve

$$\frac{1}{3}Q^3 - 7Q^2 + 111Q - 25000 = 0$$

we use

$$\text{polyroot}(c(-25000, 111, -7, 1/3))$$

R has facilities for solving both nonlinear equations and systems of nonlinear equations.

Last, and currently most accessible mathematical software, is Microsoft Mathematics which is part of Windows 7 and higher.

Basic Plotting

A graph of a function provides a good visual assessment of its behavior over certain intervals and specially facilitates locating zero(s) of the function. Two helpful plot commands in WolframAlpha are

1. plot [f, {x, x_{min}, x_{max}}]
2. plot [{$f_1, f_2, ...$}, {x, x_{min}, x_{max}}

The first command plots f as a function of x over the interval defined by the minimum and maximum values of x. For the graph of the function

$$y = x^4 - 3x^2 + 10x - 20$$

over the interval $[-4, 4]$ we enter

$$\text{plot } [y = x^4 - 3x^2 + 10x - 20, \ \{x, -4, 4\}]$$

The second command plots several functions together. For example, if demand and supply equations are given as

$$Q_d = \frac{100P + 1500}{P^2 + 10} \quad \text{and}$$
$$Q_s = 10P^{0.7}$$

we can plot (Fig. 6.28) the functions together by entering

$$\text{plot } [\{(100P + 1500)/(P^2 + 10), 10P^{0.7}\}, \ \{P, 0, 20\}]$$

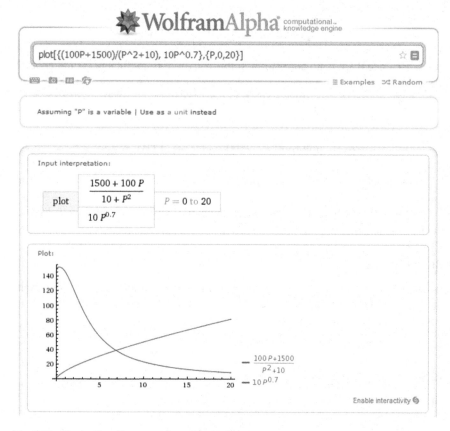

Fig. 6.28 Graph of nonlinear supply and demand

Maple's plot (f, x) plots the function $f(x)$ over the interval -10 to 10. plot $(f, x = x_0..x_1)$ plots $f(x)$ over the interval x_0 to x_1. plot $(\{f_1, f_2, \ldots\}, h, v)$ plots several function over the horizontal range h and vertical range v.

Both WolframAlpha and Maple have powerful three dimensional plotting capability. We will see a sample of them in Chap. 10. To learn more about variations and options of plot facility in WolframAlpha and see a sample of plots, go to http://reference.wolfram.com/mathematica/ref/Plot.html.

Exercises

Find the zero(s) of the following cubic functions

1. $Y = (1/5)x^3 + 25x^2 + 1600$
2. $TC = (1/5)Q^3 - 25Q^2 + 1600$
3. $TC = 20Q^3 - 20Q^2 - 100$
4. $Y = 2x^3 - 10x^2 + 100$

5. $TC = 2Q^3 - 20Q^2 + 100$

6. $Y = (1/2)x^3 - 15x^2 - 100$

7. $TC = (1/2)Q^3 - 10Q^2 + 100$

8. $TC = (1/2)Q^3 + 15Q^2 100$

9. $Y = x^3 - 19x^2 + 118x - 240$

Chapter 7
Nonlinear Models

There is a host of economic models—related to firms, markets, sectors, and the whole economy—that are extensively, but only graphically, presented in various levels of micro-macro-economics studies and mathematical economics. The main reason for adopting a graphical, rather than mathematical, approach is that in most cases the solution to the mathematical model requires solving polynomials of a third or higher degree or other complicated nonlinear equations.

One way to avoid complications is to work with linear functions by assuming a linear world. Despite strong evidence that the world is a nonlinear system, it is a common practice in mathematical economic textbooks to express supply, demand, consumption, investment, and many other functions as linear.

In a passage in his widely praised book *The Black Swan*, Nassim Taleb (philosopher, mathematician, statistician, historian, Wharton School graduate, and former Credit Swiss First Boston fund manager) argues that in spite of undeniable evidence that the world is a nonlinear system, people prefer linearity because

> With linearities, relationships between variables are clear, crisp, and constant ...easy to grasp in a single sentence .... Nonlinear relationships can vary; perhaps the best way to describe them is to say that they cannot be expressed verbally in a way that does justice to them.
>
> ...nonlinear relationships are ubiquitous in life. Linear relationships are truly the exception; we only focus on them in classrooms and textbooks because they are easier to understand. Yesterday afternoon I tried to take a fresh look around me to catalog what I could see during my day that was linear. I could not find anything.[1]

Real life practitioners—from econometricians to mathematical programmers to industrial engineers—specify and estimate a variety of nonlinear economic models. The two functional forms most frequently used for demand, for instance, are semi-log and log-linear functions. Another simple example from macroeconomics is the widely used functional form for the real money demand expressed as a double logarithmic function[2]

---

[1] Nassim Nicholas Taleb, *The Black Swan, pp 88–89*.

[2] Logarithmic and Exponential functions will be discussed in detail in Chap. 9.

S. Vali, *Principles of Mathematical Economics,*   193
Mathematics Textbooks for Science and Engineering 3,
DOI: 10.2991/978-94-6239-036-2_7, © Atlantis Press and the authors 2014

$$ln(M^d/P)_t = \beta_0 + \beta_1 ln(Y_t) + \beta_2 ln(r_t) + \epsilon_t$$

One objective of this chapter is to introduce a number of examples related to various nonlinear economic models and to show that with a certain amount of care an inherently non-linear system can be transformed into a linear or quadratic model and subsequently solved, in many cases with small or tolerable errors. The other, and maybe more important, objective is to show that we have reached a level of computational sophistication that many of these nonlinear models can be fairly easily handled. As was discussed earlier, the currently available tools, some like freely available WolframAlpha , remove the barrier to the introduction of these models in microeconomics and mathematical economics textbooks. In what follows we offer a number of examples related to a variety of nonlinear economic models and their corresponding solutions or approximations. More nonlinear models will be introduced in subsequent chapters.

7.1 Nonlinear Supply and Demand Models

Example 1

As a simple example of a nonlinear model consider the following market demand and supply function:

$$Q_d = \sqrt{520 - 8P} \qquad\qquad\qquad Q_s = 4\sqrt{3P - 4}$$

The market equilibrium price is determined by setting quantity supplied equal to quantity demanded

$$\sqrt{520 - 8P} = 4\sqrt{3P - 4}$$

and then solving for P. After squaring both side of the equation, we have

$$520 - 8P = 16(3P - 4) = 48P - 64 \quad\longrightarrow\quad 56P = 584 \text{ and } P^* = 10.43$$

and subsequently $Q^* = 20.89$.

Note that if we write the excess demand function as

$$ED(Q) = Q_d - Q_s = \sqrt{520 - 8P} - 4\sqrt{3P - 4}$$

then finding the equilibrium price is equivalent to finding the zero of the excess demand function or root of the excess demand equation

$$ED(Q) = Q_d - Q_s = \sqrt{520 - 8P} - 4\sqrt{3P - 4} = 0$$

Table 7.1 Demand, supply and excess demand

P	Q_d	Q_s	$ED(Q)$
6	21.7256	14.9666	6.7589
7	21.5407	16.4924	5.0482
8	21.3542	17.8885	3.4656
9	21.1660	19.1833	1.9827
10	20.9762	20.3961	**0.5801**
11	20.7846	21.5407	**−0.7560**
12	20.5913	22.6274	−2.0362
13	20.3961	23.6643	−3.2682
14	20.1990	24.6577	−4.4586
15	20.0000	25.6125	−5.6125
16	19.7990	26.5330	−6.7340
17	19.5959	27.4226	−7.8267

Fig. 7.1 Graph of excess demand function $\sqrt{520 - 8P} - 4\sqrt{3P - 4}$

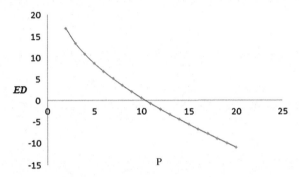

Alternatively, we can linearly approximate both the supply and demand functions and then solve the linear system. A visual inspection of Table 7.1 and the graph of the excess demand functions (Fig. 7.1) indicates that the equilibrium price is between 10 and 11. At $P = 10$ the excess demand $ED = 0.5801$. At $P = 11$ $ED = -0.7650$, indicating that the ED curve crosses the horizontal axis (i.e. the P axis) at a point between 10 and 11 (see the last column of Table 7.1).

At $P = 10$, the quantity demanded $Q_d = 20.9762$ and at $P = 11$ $Q_d = 20.7846$. Using these two points, we can write the linear demand function as

$$Q_d - 20.9762 = \frac{20.9762 - 20.7846}{10 - 11} (P - 10)$$

which upon simplification gives us

$$Q_d = 22.8922 - 0.1916P \tag{I}$$

Similarly, at $P = 10$, $Q_s = 20.3961$ and at $P = 11$, $Q_s = 21.5407$. Using these two points we write the linear supply function as

$$Q_s - 20.3961 = \frac{20.3961 - 21.5407}{10 - 11}(P - 10) \quad \text{and}$$

$$Q_s = 8.9501 + 1.1446P \tag{II}$$

Solving (I) and (II) together as a linear system yields $P^* = 10.434$ and $Q^* = 20.89$ which are extremely close to the original nonlinear solution. This shows that one can linearly approximate nonlinear functions with a high degree of accuracy over narrow intervals.

Example 2

The above example is a rather simple and benign nonlinear system that could be easily solved. Now let's consider a little more complicated system consisting of the following demand and supply functions:

$$Q_d = \frac{100P + 1500}{P^2 + 10}$$

$$Q_s = 10P^{0.7}$$

Here demand is a rational function and supply is a polynomial of degree 0.7.

To find the equilibrium price algebraically, we follow the familiar routine

$$Q_d = Q_s \quad \longrightarrow \quad \frac{100P + 1500}{P^2 + 10} = 10P^{0.7}$$

$$10P^{0.7}(P^2 + 10) = 100P + 1500$$

leading to

$$P^{2.7} + 10P^{0.7} - 10P - 150 = 0 \tag{7.1}$$

Equation (7.1) is not a run-of-the-mill familiar equation, like linear or quadratic, and does not yield an analytical solution. Thus we have to turn to approximation methods for finding the root(s) of the equation.

Solving (7.1) is equivalent to finding the zero(s) of the function

$$f(P) = P^{2.7} + 10P^{0.7} - 10P - 150 \tag{7.2}$$

or the zeros of the excess demand function $ED(P)$

$$ED(P) = \frac{100P + 1500}{P^2 + 10} - 10P^{0.7}$$

Table 7.2 provides values of Q_d, Q_s, and $f(P)$ for selected values of P. Figure 7.2 shows the graph of $f(P)$ in (7.2).

As was discussed in the second Appendix of Chap. 6, there are a number of ways that one can solve Eq. (7.1); from use of a graphic calculator, mathematical

Table 7.2 Values of Q_d, Q_s and $f(P)$ for different values of P

P	Q_d	Q_s	$f(P)$
1	145.45	10.00	−149.000
2	121.43	16.25	−147.257
3	94.74	21.58	−139.004
4	73.08	26.39	−121.386
5	57.14	30.85	−92.019
6	**45.65**	**35.05**	**−48.763**
7	**37.29**	**39.05**	**10.367**
8	31.08	42.87	87.245
9	26.37	46.56	183.654
10	22.73	50.12	301.306
11	19.85	53.58	441.853
12	17.53	56.94	606.895
13	15.64	60.22	797.987

Fig. 7.2 Graph of $f(P) = P^{2.7} + 10P^{0.7} - 10P - 150$

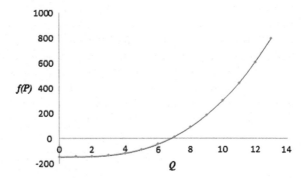

software, to high level computer programming. But as it was mentioned, each of these methods may have their own drawbacks.

Dating back to ancient civilizations, from the Sumerian to the Greeks, solving equations or Root-finding has been among the most fascinating mathematical undertakings. In the middle ages Persian and Arab mathematicians and astronomers expanded the field and made important contribution. But it was Isaac Newton's global root-finding algorithm which opened a new era of numerical analysis based on iterative methods. The Newton (sometimes called Newton-Raphson) *approximation method* is the grand daddy of all root-finding techniques. Most of these methods are in fact a variation of the Newton-Raphson approximation algorithm.

A requirement of the Newton-Raphson algorithm is that we have to evaluate the derivative of the function and go through a number of iterations. This can be a computationally involved and time consuming process, and sometimes even intractable. There is also no guarantee that after a number of iterations the estimated solution converges to the actual solution.

I will neither discuss the Newton algorithm nor pursue the question of convergence here but instead offer two alternative methods; *linearization* and the *Taylor expansion* methods for solving a nonlinear equation such as (1). Interestingly, the Newton approximation will turn out to be a special case of the Taylor expansion method so widely used in mathematical economics textbooks for deriving the first- and second-order conditions for a maximum or minimum but never used for solving models. The Taylor expansion method will be presented in Sect. 7.3.

7.2 Linear Approximation of the Model (Linearization)

As it was demonstrated in Example 1, a fairly straight forward strategy for approximating the solution to nonlinear supply and demand models is to *linearize* the supply and demand functions over a narrow interval containing the equilibrium price and then solve the resultant linear system for the equilibrium price and quantity. A visual inspection of Table 7.2, graphs of the supply and demand functions, and the graph of $f(P)$ indicates that in this example the equilibrium price falls in the range of 6–7. Note that the value of the $f(P)$ function changes from negative to positive when P moves from 6 to 7, indicating that the $f(P)$ curve cuts the P axis in the interval.

Note from Table 7.2 that at $P = 6$, values of the demand, supply, and excess demand functions are 45.65, 35.05, and -48.763 respectively. At $P = 7$ the corresponding values are 37.29, 39.05, and 10.36.

P	Q_d	Q_s	$f(x)$
6	45.65	35.05	-48.763
7	37.29	39.05	10.367

By using this information, we write the linear demand, supply, and excess demand functions over the interval [6, 7] as

$$Q_d = 95.81 - 8.36P$$
$$Q_s = 11.05 + 4P$$
$$f(P) = -403.543 + 59.139P$$

By solving the supply and demand equations simultaneously we find the equilibrium price $P = 6.858$, which in turn yields the equilibrium quantity $Q = 38.48$ units. The solution to $f(P) = 0$ generates a slightly different P due to round off errors incurred in the process of linearizing the function $f(P)$.

7.3 Linear and Quadratic Approximation Using Taylor Series

Another alternative approximation method is to express a nonlinear function as poly-nomials using Taylor *power series expansion*. A function $f(x)$ can be expanded *at, about, centered at* or in the *neighborhood* of x_0, a specific value of x, and thus expressed as a power series

$$f(x) = c_0 + c_1(x-x_0) + c_2(x-x_0)^2 + c_3(x-x_0)^3 + \cdots = \sum_{i=0}^{\infty} c_i(x-x_0)^i \qquad (7.3)$$

Our task is to determine the coefficients $c_0, c_1, c_2, \ldots$.
   If in (7.3) $x = x_0$ then $c_0 = f(x_0)$.
   By differentiating the function $f(x)$, we have

$$f'(x) = c_1 + 2c_2(x-x_0) + 3c_3(x-x_0)^2 + 4c_4(x-x_0)^3 + 5c_5(x-x_0)^4 + \cdots \qquad (7.4)$$

If we evaluate $f'(x)$ at x_0, then we have $c_1 = f'(x_0)$.
   By taking the derivative of (7.4), we have

$$f^{(2)}(x) = f''(x) = 2c_2 + 2 * 3c_3(x-x_0) + 3 * 4c_4(x-x_0)^2 + 4 * 5c_5(x-x_0)^3 + \cdots \qquad (7.5)$$

and by letting $x = x_0$ in (7.5), we get

$$2c_2 = f''(x_0) \quad \longrightarrow \quad c_2 = \frac{f''(x_0)}{2} = \frac{f^{(2)}(x_0)}{2!}$$

[we adopt the convention that $f^{(n)}(x)$ represents the *n*th derivative of $f(x)$].
   By repeating the same procedure and differentiating (7.5), we have

$$f^{(3)}(x) = 2 * 3c_3 + 2 * 3 * 4c_4(x-x_0) + 3 * 4 * 5c_5(x-x_0)^2 + 4 * 5(x-x_0)^3 + \cdots \qquad (7.6)$$

and by letting $x = x_0$ in (7.6) we can find c_3 as

$$2 * 3\, c_3 = f^{(3)}(x_0) \quad \longrightarrow \quad c_3 = \frac{f^{(3)}(x_0)}{2 * 3} = \frac{f^{(3)}(x_0)}{3!}$$

We repeat this procedure one more time by differentiating (7.6)

$$f^{(4)}(x) = 2 * 3 * 4c_4 + 2 * 3 * 4 * 5c_5(x - x_0) + 3 * 4 * 5(x - x_0)^2 + \cdots$$

Evaluating $f^{(4)}(x)$ at x_0 results in

$$f^{(4)}(x_0) = 2 * 3 * 4c_4 \quad \longrightarrow \quad c_4 = \frac{f^{(4)}(x_0)}{2 * 3 * 4} = \frac{f^{(4)}(x_0)}{4!}$$

By closely examining the expressions for c_1, c_2, c_3 and c_4, we can discern the pattern and write the general expression for the nth coefficient c_n as

$$c_n = \frac{f^{(n)}(x_0)}{n!}$$

By replacing c_i by their values, we can write the original power series in (7.3) as the Taylor series expansion of the function $f(x)$ at, about, centered at or in the neighborhood of x_0

$$f(x) = f(x_0) + \frac{f^{(1)}(x_0)}{1!}(x - x_0) + \frac{f^{(2)}(x_0)}{2!}(x - x_0)^2$$
$$+ \frac{f^{(3)}(x_0)}{3!}(x - x_0)^3 + \cdots$$
$$= \sum_{n=0}^{\infty} \frac{f^{(n)}(x_0)}{n!}(x - x_0)^n \qquad (7.7)$$

where by convention $f^{(1)}, f^{(2)}, f^{(3)}, \ldots, f^{(n)}$ denote the first, second, third, ..., nth derivative of $f(x)$.[3]

We can write (7.7) as

$$f(x) = f(x_0) + \frac{f^{(1)}(x_0)}{1!}(x - x_0) + \frac{f^{(2)}(x_0)}{2!}(x - x_0)^2 + R \qquad (7.8)$$

where R, called the remainder, represents the sum of all the remaining terms beyond the quadratic term.

If we pick only the first two terms of the series in (7.8), we get a *first degree* (sometimes called *first order*) or *linear approximation* to the function. The first three terms of (7.8) provides a *second degree* (also known as *second order*) or *quadratic approximation* to the function.

$$f(x) = f(x_0) + \frac{f^{(1)}(x_0)}{1!}(x - x_0) \qquad \text{First order or linear approximation}$$

$$f(x) = f(x_0) + \frac{f^{(1)}(x_0)}{1!}(x - x_0) + \frac{f^{(2)}(x_0)}{2!}(x - x_0)^2 \qquad \text{Second order}$$

$$\text{or quadratic approximation}$$

Our main interest in the Taylor expansion is to approximate the zero(s) of the functions. To this end we try to use the graph or the table of values for the function in order to select the initial estimate of x_0 as close as possible to the zero of the function. In the case of linear approximation, the zero of the function is

---

[3] For the special case $x_0 = 0$ the Taylor series is known as the **Maclaurin Series**.

$$f(x) \approx f(x_0) + f'(x_0)(x - x_0) = 0 \quad \longrightarrow \quad f(x_0) + f'(x_0)x - f'(x_0)x_0 = 0$$
$$f'(x_0)x = f'(x_0)x_0 - f(x_0)$$

If $f'(x_0) \neq 0$, we can solve this equation for x

$$x = \frac{f'(x_0)x_0 - f(x_0)}{f'(x_0)} \quad \text{which can be simplified as}$$
$$x = x_0 - \frac{f(x_0)}{f'(x_0)}$$

Here x is the new approximation of the zero of the function, which can be subsequently used for another round of more accurate approximation. If we denote this value of x by x_1, then the new, and hopefully more accurate, zero of the function can be approximated by

$$x = x_1 - \frac{f(x_1)}{f'(x_1)}$$

By going through this process or iterations several times, we can improve the accuracy of our estimate. This iterative process is the Newton approximation method. Newton method is in fact a special case of Taylor expansion, the linear expansion.

Next we apply the Taylor expansion method to solve the nonlinear function $f(P)$ in (7.2). We first expand $f(P)$ around $P = 6$, where the function is closest to zero, and then find both the first and second degree approximations of the zero of the function.

$$f(P) = P^{2.7} + 10P^{0.7} - 10P - 150$$
$$f(6) = -48.7634$$
$$f'(P) = 2.7P^{1.7} + 7P^{-0.3} - 10$$
$$f'(6) = 50.87267$$

Then the linear approximation to $f(P)$ about 6, denoted as $f_l(P)$, is

$$f_l(P) = f(6) + f'(6)(P - 6)$$

leading to

$$f_l(P) = -48.7634 + 50.87267(P - 6) = -353.9994 + 50.87267P \qquad (7.9)$$

The zero of $f_l(P)$ in (7.9) is $-353.9994 + 50.87267P = 0$ leading to the solution $P = 6.96$. By substituting this value for P in the demand and supply functions and averaging the two values,[4] we arrived at $Q = 38.23$ units.

Next we derive the quadratic approximation to the $f(P)$ function

---

[4] Due to roundoff and approximation errors we get slightly different answers for output from the supply and demand equations. Averaging the two values is the best way to reconcile the difference.

$$f''(P) = 4.59P^{0.7} - 2.1P^{-1.3}$$
$$f''(6) = 15.8841$$

Denoting the quadratic approximation of $f(P)$ by $f_q(P)$, we have

$$\begin{aligned} f_q(P) &= f(6) + f'(6)(P - 6) + \frac{f''(6)}{2}(P - 6)^2 \\ &= -353.9994 + 50.87267P + 7.9421(P - 6)^2 \\ &= 7.9421P^2 - 44.4322P - 68.0848 \end{aligned}$$

Zero(s) of this quadratic function is

$$7.9421P^2 - 44.4322P - 68.0848 = 0$$

The only acceptable solution is $P = 6.8467$ yielding $Q \approx 38.5$ units. Given that by using WolframAlpha or Maple we find the actual solution to four decimal places to be $P = 6.8437$, the quadratic approximation seems to be fairly accurate. The *APE* (Absolute Percentage Error) of the linear and quadratic estimations are 1.7, and 0.044, respectively.

Using WolframAlpha for solving equation (7.2) [see the Chap. 6 Appendix] we simply invoke the *solve[expr, vars]* command which solves the system *expr* of equations or inequalities for the variables *vars*. In our case all we need to do is to enter

$$\text{solve } [P^{\wedge}2.7 + 10P^{\wedge}0.7 - 10P - 150 = 0, \ P > 0]$$

in the box and click on the equal sign located at the right side of the box. The software then gives you the solution $P = 6.84368$. As it was mentioned in Chap. 6, we don't need to use the 'solve' command format. The same result is achieved by simply typing $P^{\wedge}2.7 + 10P^{\wedge}0.7 - 10P - 150 = 0, P > 0$ in the inquiry box and clicking the equal sign.

Using Maple to solve the equation is equally easy. The appropriate command in Maple is

$$\text{fsolve } [P^{\wedge}2.7 + 10P^{\wedge}0.7 - 10P - 150 = 0, \ P = 0..20]$$

Examples of Nonlinear Models

7.3.1 Long Run Competitive Market Equilibrium Price

Example 3

Assume all firms in a competitive industry have the same *TC* function

$$TC = Q^3 - 24Q^2 + 200Q + 500$$

What is the long run market equilibrium price?

As it was discussed in an earlier chapter, the long run market price is the minimum of the ATC function (or alternatively at the point of intersection between the MC and ATC curves).

$$ATC(Q) = Q^2 - 24Q + 200 + \frac{500}{Q}$$

The FOC for a stationary point is

$$ATC'(Q) = 2Q - 24 - \frac{500}{Q^2} = 0$$

which upon simplification gives us

$$2Q^3 - 24Q^2 - 500 = 0$$

which is equivalent to finding the zero(s) of the function

$$f(Q) = 2Q^3 - 24Q^2 - 500$$

Next we find the first and the second derivatives of $f(Q)$ for the Taylor series expansion of the function

$$f'(Q) = 6Q^2 - 48Q$$
$$f''(Q) = 12Q - 48$$

Table 7.3 provides values for the function $f(Q)$ and its first and second derivatives evaluated at some selected values for Q.

The second column of the table and the graph of $f(Q)$ (Fig. 7.3) clearly indicate that the zero of the function is between 13 and 14. We utilize this information and write the first three terms of the Taylor series centered at 13

$$f(13) = -162$$
$$f'(13) = 390$$
$$f''(13) = 108$$

The first three terms of the Taylor approximation to $f(Q)$ is

$$f(Q) = -162 + 390(Q - 13) + \frac{108}{2}(Q - 13)^2 \qquad (7.10)$$

If we choose only the first two terms of (7.10), then we have the first degree or linear approximation of $f(Q)$, denote by $f_l(Q)$

Table 7.3 Values of $f(Q)$, $f'(Q)$, $f''(Q)$ for selected values of P

Q	$f(Q)$	$f'(Q)$	$f''(Q)$
6	−932	−72	24
7	−990	−42	36
8	−1012	0	48
9	−986	54	60
10	−900	120	72
11	−742	198	84
12	−500	288	96
13	**−162**	**390**	**108**
14	**284**	**504**	**120**
15	850	630	132
16	1548	768	144
17	2390	918	156

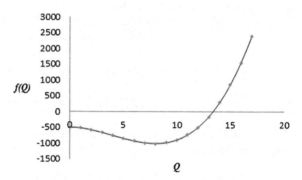

Fig. 7.3 Graph of $f(Q) = 2Q^3 - 24Q^2 - 500$

$$f_l(Q) = -162 + 390(Q - 13) = -5232 + 390Q$$

The zero of the this linear function is

$$f_l(Q) = 0 \quad \longrightarrow \quad -5232 + 390Q = 0 \quad \text{leading to} \quad Q = 13.42$$

By choosing all three terms in (7.10), we will have the second degree or quadratic approximation to $f(Q)$, denoted by $f_q(Q)$

$$f_q(Q) = 54Q^2 - 1014Q + 3894$$

The zero(s) of this quadratic function is

$$54Q^2 - 1014Q + 3894 = 0$$

The above quadratic equation has two roots, 13.394 and 5.384. Since we look for solution in the neighborhood of $Q = 13$, the acceptable answer is 13.394 (the actual solution to four decimal places is 13.3939). By substituting $Q = 13.394$ in the ACT function, we find the long run market price

$$P_{LR} = ATC(13.394) = 13.394^2 - 24 * 13.394 + 200 + \frac{500}{13.394} = 95.27$$

As it was noted in previous examples, as another alternative we can linearize the function $f(Q)$ over the interval $(13, 14)$ and then determine the zero of the converted linear function. Given that $f(13) = -162$ and $f(14) = 284$, the equation of the line joining points $(13, -162)$ and $(14, 284)$ is then

$$f(Q) + 162 = \frac{284 + 162}{14 - 13}(Q - 13)$$
$$f(Q) = -5960 + 446Q \quad \text{and}$$
$$f(Q) = 0 \quad \longrightarrow \quad -5960 + 446Q = 0 \quad \text{leading to} \quad Q = 13.36$$

which is an accurate approximation.

7.3.2 Terms of Trade

Example 4

Consider two countries A and B. Assume that country A has comparative advantage in production of y and country B in production of x. The amount of y (in million units) country A is willing to offer in exchange for x (in million units) is given by the following *offer functions*

$$y = \frac{0.5x(55 - x)^2}{2400} \qquad (7.11)$$

The amount of x country B is willing to offer for a given amount of y is

$$x = \frac{25y - y^2}{5} \qquad (7.12)$$

We want to determine the amount of x and y traded at equilibrium.

We substitute for x from (7.12) in (7.11) and after a series of algebraic manipulations we find ourselves in possession of the following sixth degree equation

$$(25y - y^2)(550 - 50y + 2y^2)^2 - 2400000y = 0$$

Solving this equation is equivalent to finding the zero(s) of the function

$$f(y) = (25y - y^2)(550 - 50y + 2y^2)^2 - 2400000y$$

Table 7.4 Values of $f(y)$ for selected values of y

y	3	4	5	6	7	8
$f(y)$	4331784	2657616	**250000**	**−2580024**	−5610696	−8689376

Fig. 7.4 Graph of $f(y) = (25y - y^2)(550 - 50y + 2y^2)^2 - 2400000y$

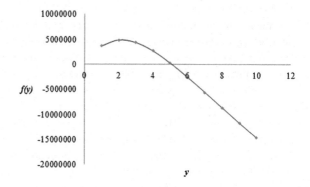

The values of the function $f(y)$ for selected values of y are given in Table 7.4. Figure 7.4 shows the graph of this function. It is clear from the table and the graph that the equation has a non-zero real solution between $y = 5$ and $y = 6$. From WolframAlpha this solution is $y = 5.0932$. By substituting this value for y in (12) we get $x = 20.278$. This means that country A is willing to trade 5.0932 million units of good y for 20.278 million units of country B's good x, i.e. the physical terms of trade is 3.98 unit of x per 1 unit of y.

7.3.3 Joint Production

Example 5

A firm with a certain amount of resources (raw materials, labor, and machinery) produces two goods. If the firm produces Q_1 units of good 1, then with the rest of the resources it could produce

$$Q_2 = 100 - \frac{700}{841 - Q_1^2} \quad Q_1 < 29 \qquad (7.13)$$

units of good 2. If the price of good 1 P_1 is $50 per unit and that of good 2 P_2 is $80 per unit, what combination of good 1 and 2 maximizes the firm's revenue?

We can formulate the problem as:

$$\textit{Maximize} \quad TR(Q_1, Q_2) = P_1 Q_1 + P_2 Q_2 = 50 Q_1 + 80 Q_2$$

$$\textit{Subject to} \quad Q_2 = 100 - \frac{700}{841 - Q_1^2} \tag{7.14}$$

After substituting for Q_2 from (7.14) into the revenue function, we get

$$TR(Q_1) = 50 Q_1 + 80 \left(100 - \frac{700}{841 - Q_1^2} \right)$$

$$TR(Q_1) = 50 Q_1 + 8000 - \frac{56000}{841 - Q_1^2}$$

The *FOC* for maximizing *TR* is

$$TR'(Q_1) = \frac{dTR}{dQ_1} = 50 - \frac{112000 Q_1}{(841 - Q_1^2)^2} = 0 \tag{7.15}$$

By re-labeling $TR'(Q_1)$ as $f(Q_1)$ and after simplification, we write (7.15)

$$f(Q_1) = Q_1^4 - 1682 Q_1^2 - 2240 Q_1 + 707281 = 0$$

To have some idea about the behavior of this function, we graph it for selected values of Q_1. A close inspection of Table 7.5 or the graph of $f(Q_1)$, Fig. 7.5 indicates that our fourth degree equation has a real root between 24 and 25 and another between 33 and 34. Since the second root violates the condition $Q_1 < 29$ in (7.13), we concentrate on determining the root in the interval 24–25.

The next step is to expand this function about 25 using Taylor expansion.

$$f(25) = -9344$$
$$f'(Q_1) = 4 Q_1^3 - 3364 Q_1 - 2240$$
$$f'(25) = -23840$$
$$f''(Q_1) = 12 Q_1^2 - 3364$$
$$f''(25) = 4136$$

Table 7.5 Values of $f(Q_1)$ for certain values of Q_1

Q_l	23	24	25	30	33	34	35
$f(Q_l)$	45824	16465	−9344	−63719	−12416	23065	69056

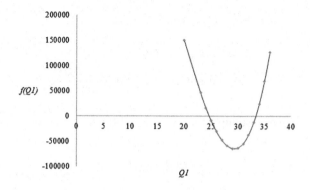

Fig. 7.5 Graph of $f(Q_1) = Q_1^4 - 1682Q_1^2 - 2240Q_1 + 707281$

Then the first degree or linear Taylor approximation of function $f(Q_1)$ about 25 is

$$f_l(Q_1) = f(25) + f'(25)(Q_1 - 25) = -9344 - 23840(Q_1 - 25)$$

which after simplification becomes

$$f_l(Q_1) = 586656 - 23840Q_1$$

The quadratic approximation is

$$
\begin{aligned}
f_q(Q_1) &= f(25) + f'(25)(Q_1 - 25) + \frac{f''(Q_1)}{2}(Q_1 - 25)^2 \\
&= -9344 - 23840(Q_1 - 25) + \frac{4136}{2}(Q_1 - 25)^2 \\
&= 586656 - 23840Q_1 + 2068(Q_1^2 + 625 - 50Q_1) \\
&= 2068Q_1^2 - 127240Q_1 + 1879156
\end{aligned}
$$

Using the linear form, we approximate the zero of the function as

$$f_l(Q_1) = 586656 - 23840Q_1 = 0 \quad \longrightarrow \quad Q_1 = \frac{586656}{23840} = 24.61 \text{ units}$$

By substituting this value for Q_1 in (7.13), we have

$$Q_2 = 100 - \frac{700}{841 - Q_1^2} = 100 - \frac{700}{841 - (24.61)^2} = 97.03 \text{ units}$$

Therefore, the linear approximation to the combination of goods which maximizes the firm's revenue is $(24.61, 97.03)$.

Using the quadratic form, we approximate the zero of the function by solving the quadratic equation

$$f(Q_1) = 2068Q_1^2 - 127240Q_1 + 1879156 = 0$$

There are two solutions, 24.62 and 36.9. The second solution violates the condition $Q_1 < 29$ in (7.13), hence we discard it. From (7.13) we find $Q_2 = 97.02$.

We can, of course, solve the original 4th degree polynomial

$$Q_1^4 - 1682Q_1^2 - 2240Q_1 + 707281 = 0$$

using Maple, WoframAlpha, or the polyroot function from **R**. Recall from the Chap. 6 Appendix that to use the polyroot function in **R** we must enter the following command

$$\text{polyroot}(c(707281, -2240, -1682, 0, 1))$$

R gives two exact real answers, 24.62033 and 33.38349.

7.3.4 Nonlinear Demand and Optimal Price and Output for a Noncompetitive Firm

Example 6

A construction company is planning to build small low-cost housing units in a low-income neighborhood. Assume that the market demand for housing in this community is

$$Q_d = 1000 - 10P^{0.4}$$

and the firm's cost function is

$$TC(Q) = Q^3 - 20Q^2 + 950Q + 650000$$

What is the developer's profit maximizing combination of housing price and number of housing units? What is the firm's profit?

To write the profit function $\Pi = TR - TC$ we first rewrite the market demand function as

$$P = (100 - 0.1Q)^{2.5}$$

and the total revenue function as

$$TR(Q) = PQ = (100 - 0.1Q)^{2.5}Q$$

leading to

$$\Pi(Q) = TR(Q) - TC(Q) = (100 - 0.1Q)^{2.5}Q - Q^3 + 20Q^2 - 950Q - 650000$$

The first order condition for profit maximization is

$$\Pi'(Q) = \frac{dTR}{dQ} - \frac{dTC}{dQ} = MR - MC = 0$$

after renaming $\Pi'(Q)$ to $\varphi(Q)$, we have

$$\varphi(Q) = -0.25Q(100 - 0.1Q)^{1.5} + (100 - 0.1Q)^{2.5} - 3Q^2 + 40Q - 950 = 0$$

Table 7.6 indicates that function $\varphi(Q)$ has a zero in the range of $125 < Q < 130$, where the value of the function changes from positive to negative (Fig. 7.6).

To find the linear or first order approximation of the zero of $\varphi(Q)$, we expand the function about $Q = 125$.

$$\varphi(125) = -0.25 * 125 * (100 - 0.1 * 125)^{1.5} + (100 - 0.1 * 125)^{2.5}$$
$$- 3(125)^2 + 40 * 125 - 950$$
$$\varphi(125) = 3214.925$$
$$\varphi'(Q) = -0.50(100 - 0.1Q)^{1.5} + 0.0375Q(100 - 0.1Q)^{0.5} - 6Q + 40$$
$$\varphi(125) = -1075.396$$

Thus the linear approximation is

$$\varphi(Q) = \varphi(125) + \varphi'(125)(Q - 125)$$
$$= 3214.925 - 1075.396(Q - 125)$$
$$\varphi(Q) = 137639.4 - 1075.396Q$$

Table 7.6 Values of $\varphi(Q)$ for selected values of Q

Q	$\varphi(Q)$
20	88874
50	69940.5
110	18786.90
115	13720.39
120	8529.76
125	**3214.92**
130	**−2224.23**
135	−7787.81
140	−13475.91
150	−25226.09

Fig. 7.6 Graph of $\varphi(Q) =$ $-0.25Q(100 - 0.1Q)^{1.5} +$ $(100 - 0.1Q)^{2.5} - 3Q^2 +$ $40Q - 950$

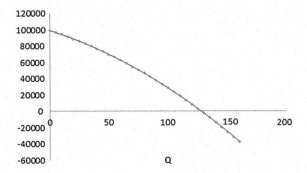

The zero of this function is $\varphi(Q) = 137639.4 - 1075.396Q = 0$ leading to a solution $Q = 128$ units with the price of each unit $P = \$71,000$. With this combination, the developer realizes a \$6,546,928 profit.

Alternatively, we can linearize the marginal revenue function $MR(Q)$ over the interval $125 < Q < 130$ and then solve the profit maximizing condition $MR(Q) - MC(Q) = 0$ for the optimal solution. From the total revenue function

$$TR(Q) = PQ = (100 - 0.1Q)^{2.5}Q$$

we derive the marginal revenue function as

$$MR(Q) = (100 - 0.1Q)^{2.5} - 0.25Q(100 - 0.1Q)^{1.5}$$

Values for marginal revenues at $Q = 125$ and $Q = 130$ are 46039.92 and 44225.77, respectively. Therefore, the equation of the line joining the points $(125, 46039.92)$ and $(130, 44225.77)$ on the MR curve is

$$MR = 91393.67 - 362.83Q$$

which is the linear approximation of the nonlinear MR function over the 125–130 interval. The next step is to solve $MR - MC = 0$ for the profit maximizing level of output Q, that is

$$91393.67 - 362.83Q - 3Q^2 + 40Q - 950 = 0 \longrightarrow 3Q^2 + 322.83 - 90443.67 = 0$$

This quadratic equation has an acceptable solution at $Q = 128$ units.

Dislocated Demand and Normal Profit Combination of Price and Output for a Noncompetitive Firm

Example 6 Continued

Now let's assume that as soon as this developer embarks on building 128 housing units, a massive crisis in the financial market leads to a collapse of the housing market and a dramatic decline in the demand for new housing. Assume the market forces push the developer to settle for a break-even combination of units and price, i.e. a normal profit. What is this combination? What is the equation of the *new* or *dislocated* demand function?

As it was discussed in Chap. 6, the break-even combination occurs at the point where the dislocated demand curve becomes tangent to the average total cost curve. At the point of tangency, the slope of the dislocated demand function is the same as that of the *ATC* function. Since the dislocated demand curve is obtained by a downward parallel shift of the original demand curve, its equation can be expressed as

$$P = (100 - 0.1Q)^{2.5} - C$$

where C is a constant that will be determined later.

The slope of the dislocated demand function

$$P = (100 - 0.1Q)^{2.5} - C \quad \text{is}$$
$$\frac{dP}{dQ} = -0.25(100 - 0.1Q)^{1.5}$$

The slope of the *ATC* function is

$$\frac{dATC}{dQ} = \frac{2Q^3 - 20Q^2 - 650000}{Q^2}$$

Thus at the point of tangency, we have

$$-0.25(100 - 0.1Q)^{1.5} = \frac{2Q^3 - 20Q^2 - 650000}{Q^2}$$

leading to the equation

$$2Q^3 + 0.25Q^2(100 - 0.1Q)^{1.5} - 20Q^2 - 650000 = 0$$

Based on the numbers in Table 7.7 and the graph of $\psi(Q)$ (Fig. 7.7), where $\psi(Q)$ is

$$\psi(Q) = 2Q^3 + 0.25Q^2(100 - 0.1Q)^{1.5} - 20Q^2 - 650000$$

Table 7.7 Values of $\psi(Q)$ for certain values of Q

Q	$\psi(Q)$
15	−592760.9
20	−544984.9
25	−480822.6
30	−399048.7
35	−298436.6
40	−177758.4
45	**−35784.5**
50	**128715.9**
55	316975.3
60	530227.4
65	769707.7

Fig. 7.7 Graph of $\psi(Q) = 2Q^3 + 0.25Q^2(100 - 0.1Q)^{1.5} - 20Q^2 - 650000$

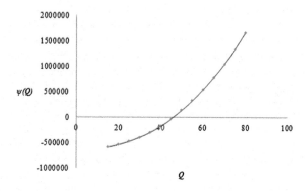

this equation has a solution between 45 and 50.

The first order Taylor expansion of $\psi(Q)$ about 45 is

$$\psi(Q) = -35784.51 + 30606(Q - 45)$$

leading to a solution $Q = 46.16 \approx 46$ units and $P = ATC(46) = \$16,241$.

We can now determine the value of C and the equation of the dislocated demand function

$$P = (100 - 0.1Q)^{2.5} - C$$
$$16241 = (100 - 0.1 * 46)^{2.5} - C \longrightarrow C = 72617.8$$

The equation of the *new* or *dislocated* demand is

$$P_{new} = (100 - 0.1Q)^{2.5} - 72617.8$$

7.3.5 *Optimal Price and Output for a Multi-Plant Firm*

Example 7

A firm operates two plants. Assume the first and second plant total cost functions, denoted as $TC(Q_1)$ and $TC(Q_2)$, are

$$TC(Q_1) = \frac{1}{3}Q_1^3 - 6Q_1^2 + 50Q_1 + 45$$
$$TC(Q_2) = 2Q_2^3 - 60Q_2^2 + 950Q_2 + 600$$

where Q_1 and Q_2 are the output of plant number 1 and 2, respectively. Assume that the market demand for the firm's product is

$$P = 900 - 3Q$$

where $Q = Q_1 + Q_2$ represents firm's total output. Our objective is to determine the firm profit-maximizing level of output and price, and to find the optimal division of output between its two plants.

At the optimal level of output, it must be true that

$$MC(Q_1) \ = \ MC(Q_2) \ = \ MR(Q)$$

Given

$$MC(Q_1) = Q_1^2 - 12Q_1 + 50$$
$$MC(Q_2) = 6Q_2^2 - 120Q_2 + 950$$

and

$$MR(Q) = 900 - 6Q = 900 - 6(Q_1 + Q_2)$$

we have

$$Q_1^2 - 12Q_1 + 50 = 900 - 6(Q_1 + Q_2)$$
$$6Q_2^2 - 120Q_2 + 950 = 900 - 6(Q_1 + Q_2)$$

which leads to the following nonlinear system of two equations with two unknowns

$$\begin{cases} Q_1^2 - 6Q_1 + 6Q_2 = 850 \\ 6Q_2^2 - 114Q_2 + 6Q_1 = -50 \end{cases} \tag{7.16}$$

We can express the first equation in (7.16) as

$$Q_1^2 - 6Q_1 + (6Q_2 - 850) = 0$$

and subsequently solve for Q_1 in terms of Q_2

$$Q_1 = \frac{6 \pm \sqrt{3436 - 24Q_2}}{2} = 3 \pm \sqrt{859 - 6Q_2} \qquad (7.17)$$

From the two possible answers in (7.17), we select $Q_1 = 3 + \sqrt{859 - 6Q_2}$. Students should be able to verify that $Q_1 = 3 - \sqrt{859 - 6Q_2}$ leads to an unacceptable negative value for the output of plant 1.

After substituting $Q_1 = 3 + \sqrt{859 - 6Q_2}$ for Q_1 in the second equation of (7.16), we have

$$6Q_2^2 - 114Q_2 + 18 + 6\sqrt{859 - 6Q_2} = -50 \quad \text{or}$$
$$6Q_2^2 - 114Q_2 + 6\sqrt{859 - 6Q_2} + 68 = 0$$

The solution(s) to this nonlinear equation is (are) the zero(s) of the function

$$f(Q_2) = 6Q_2^2 - 114Q_2 + 6\sqrt{859 - 6Q_2} + 68$$

As Table 7.8 and the graph of $f(Q_2)$ in Fig. 7.8 indicate this function has two zeros, one in the range of 2–2.5 and the other in the range of 16.5–17.

We expand $f(Q_2)$ around 2.5 and 16.5 and determine the first order approximations.

$$f'(Q_2) = 12Q_2 - 114 - \frac{18}{\sqrt{859 - 6Q_2}}$$

Table 7.8 Values of $f(Q_2)$ for certain values of Q_2

Q_2	1	2	2.5	6	16.5	17	17.5	18
$f(Q_2)$	135.24	38.62	−5.19	−227.87	−14.09	29.08	75.25	124.43

Fig. 7.8 Graph of $f(Q_2) = 6Q_2^2 - 114Q_2 + 6\sqrt{859 - 6Q_2} + 68$

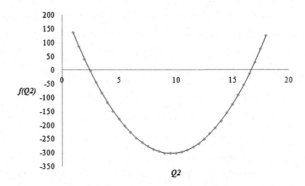

With the following information

$$f(2.5) = -5.19$$
$$f'(2.5) = -84.62$$
$$f(16.5) = -14.09$$
$$f'(16.5) = 83.35$$

we can write two linear approximations to $f(Q_2)$ as

$$f_{l1}(Q_2) = -5.19 - 84.62(Q_2 - 2.5) = 206.36 - 84.62Q_2$$
$$f_{l2}(Q_2) = -14.09 + 83.35(Q_2 - 16.5) = -1389.32 + 83.35Q_2$$

Setting the first function equal to zero leads to a solution $Q_2 = 2.439$. Similarly, by setting the second function equal to zero, we find the second solution as $Q_2 = 16.667$. But since the marginal cost of the second plant $MC(Q_2)$ for $Q_2 = 2.439$ is 693.01 and for $Q_2 = 16.667$ is 616.7, we discard 2.439 in favor of 16.667.

By substituting 16.667 for Q_2 in either equations in (7.16), we determine the optimal value of Q_1 as 30.55 units. With the optimal combination of output for each plant (30.55, 16.667), the firm's total output is $Q = Q_1 + Q_2 = 47.217$ units, with price $P = 900 - 3 * 47.217 = 758.35$ per unit. This combination generates 21,304.02 of profit. With this combination the marginal costs of both plants and their marginal revenues are all equal to 616.7 ($Q_2 = 2.347$ leads to $Q_1 = 32.06$ and this combination generates 18,424.25 of profit which is less than profit of 21,304.02 generated by the combination 30.55, 16.667).

7.3.6 Nonlinear National Income Model

Example 8

We close this series of examples by examining a nonlinear macroeconomic model. In the Keynesian national income determination model, consumption, investment, and other functions are conveniently assumed to be linear. Here we depart from the linearity assumption and construct a nonlinear model around a nonlinear consumption function. Assume the following model for a closed economy[5]

$$Y = C + I + G_0 \tag{7.18}$$
$$C = 1291 + 0.538Y - 0.00000578Y^2$$
$$I = 20 + 0.2Y$$

---

[5] The consumption function in this model is a twisted version of Ralph D. Husby's estimated function, appeared in his article in 1985 *Atlantic Economic Journal*.

and exogenously given government expenditure $G_0 = 500$.

After substituting for $C, I,$ and G_0 in (7.18), we have

$$Y = 1291 + 0.538Y - 0.00000578Y^2 + 20 + 0.2Y + 500$$
$$Y = 1811 + 0.738Y - 0.00000578Y^2$$

and finally
$$0.00000578Y^2 + 0.262Y - 1811 = 0$$

Solution(s) to this quadratic equation are

$$Y = \frac{-0.262 \pm \sqrt{0.262^2 - 4(0.00000578)(-1811)}}{2(0.00000578)} = \frac{-0.262 \pm 0.3324}{0.00001156}$$

Since Y cannot be negative, we must pick $+0.3323$, which leads to

$$Y^* = \frac{-0.262 + 0.3324}{0.00001156} = 6093.16 \quad \text{and}$$
$$C^* = 1291 + 0.538(6093.16) - 0.00000578(6093.16)^2 = 4354.53$$
$$I^* = 20 + 0.2(6093.16) = 1238.63$$

For our nonlinear consumption function, the marginal propensity to consume MPC is

$$MPC = \frac{dC}{dY} = 0.538 - 2*0.00000578Y = 0.538 - 0.00001156Y \qquad (7.19)$$

At the equilibrium national income the MPC is

$$\frac{dC}{dY}(6093.16) = 0.538 - 0.00001156(6093.16) = 0.468$$

Note that in (7.19) the MPC is inversely related to the level of income. In the national income models with a linear consumption function, the marginal propensity to consume is the same or constant at all income levels. But the result of our nonlinear model is more in tune with the findings of many empirical studies that show the households' MPC decline as they move to higher income bracket.

But we should be aware of the *fallacy of composition* here. What is true at the microeconomic level may not necessarily be true at the macroeconomic level. If anything, the dramatic rise in the share of consumption in the *GDP* of the United States in the last 15–20 years—due partly to the lax credit market and the stock and housing bubbles—might be cited as a counter example. Readers should also be reminded that the marginal propensity to consume in the national income determination model is the national or aggregated average of *MPC*s of various income groups.

We next derive the general form of the income multiplier for a nonlinear model. If we present the nonlinear consumption and investment functions as

$$C = C_0 + f(Y)$$
$$I = I_0 + g(Y)$$

then the national income identity will be

$$Y = C + I + G_0 = C_0 + f(Y) + I_0 + g(Y) + G_0$$
$$Y - f(Y) - g(Y) = C_0 + I_0 + G_0 = AE_0 \tag{7.20}$$

where AE_0 represents *aggregate exogenous (autonomous) expenditure.*

By defining the multiplier as the rate of change in income (output) in response to a change in autonomous expenditure, we determine it by differentiating both sides of (7.20) with respect to AE_0

$$\frac{dY}{dAE_0} - f'(Y)\frac{d(Y)}{dAE_0} - g'(Y)\frac{d(Y)}{dAE_0} = 1$$
$$\frac{dY}{dAE_0}[1 - f'(Y) - g'(Y)] = 1$$

The multiplier m, subsequently, is

$$m = \frac{dY}{dAE_0} = \frac{1}{1 - f'(Y) - g'(Y)} \tag{7.21}$$

It can be seen that m in (7.21) is a function of Y and thus must be evaluated at the equilibrium level of income. It should be clear that in (7.19) $f'(Y)$ is the *MPC* and $g'(Y)$ is the *marginal propensity to invest MPI*. An interesting feature of this nonlinear national income model is that as income declines substantially the marginal propensity to consume and invest both decline, which leads to a larger multiplier. Thus a relatively large does of government expenditure can get the economy out a of a sever recession. This is a much more realistic macroeconomic picture than the linear model with constant multiplier.

To determine our model's multiplier, we first find $f'(Y)$ and $g'(Y)$. From the consumption and investment functions we have

$$f(Y) = 0.538Y - 0.00000578Y^2$$
$$g(Y) = 0.2Y$$

and

$$f'(Y) = 0.538 - 0.00001156Y$$
$$f'(6093.16) = 0.4675$$
$$g'(Y) = 0.2$$

Therefore

$$m = \frac{1}{1 - 0.4675 - 0.2} = 3.0$$

7.4 Exercises

1. Assume the cubic function $f(x) = x^3 - 15x^2 - 600$

   (a) Create a table for the values of x from 6 to 20 and their corresponding values for $f(x)$.

   (b) Graph the $f(x)$ function.

   (c) Use the information from parts (a) and (b) and find the first and second degree approximations to the cubic function using the Taylor expansion series.

2. Write the linear expansion of the following demand and supply functions about $P = 6$ separately and then solve the linear system for the market equilibrium price and quantity. Compare your results with the results from Example 2.

$$Q_d = \frac{100P + 1500}{P^2 + 10} \qquad\qquad Q_s = 10P^{0.7}$$

3. Assume all firms in a perfectly competitive market have the following total cost function

$$TC(Q) = \frac{1}{12}Q^3 - 2.5Q^2 + 30Q + 100$$

   Follow steps discussed in Example 3 and find the linear approximation of the market long run equilibrium price.

4. Find linear and quadratic approximations for the market long run price if all firms in a competitive market have the following TC function

$$TC = Q^3 - 15Q^2 + 120Q + 600$$

5. The market demand function for product of a firm with the total cost function

$$TC(Q) = 2Q^3 - 10Q^2 + 30Q + 100$$

   is given by

$$P = 65000 - 8\sqrt{Q}$$

   (a) Find the firm's profit-maximizing level of output and price.

   (b) Assume the economy experiences a severe recession and the demand for the firm's product declines to the level that the firm can only earn a normal profit. Determine the normal profit price and quantity and the equation of the dislocated demand function.

6. Assume the following market demand and supply functions

$$Q_d = 100 - 5P^{0.7}$$
$$Q_s = (2P - 10)^{1.1}$$

   (a) Graph the functions.
   (b) Graph the excess demand function.
   (c) Use the Taylor expansion and find the first degree or linear approximation of market equilibrium price and quantity.

7. Consider the following market demand and supply functions

$$Q_d = \frac{700}{2P + 3} - 10$$
$$Q_s = (2P - 15)^{0.7}$$

   (a) Graph the excess demand function.
   (b) Use the Taylor expansion of the excess demand function and find the first degree approximation for the market equilibrium price and quantity.

8. Assume the following market demand and supply functions

$$Q_d = 1000 \left(\frac{P + 25}{P^2 + 5} \right)$$
$$Q_s = 100P^{0.4}$$

   (a) Graph the excess supply function.
   (b) Use the Taylor expansion of the excess supply function and find the first degree approximation for the market equilibrium price and quantity.
   (c) Write the linear expansion of the demand and supply functions separately and then solve the linear system for the market equilibrium price and quantity. Compare your results with those from part (b).

9. A firm with a certain amount of resources produces two goods. If the firm produces Q_1 units of good one, then it could produce

$$Q_2 = 85 - \frac{700}{900 - Q_1^2} \qquad\qquad Q_1 < 30$$

   units of good two. Assume price of good one P_1 is \$60 per unit and that of good two P_2 is \$90 per unit.

   (a) Use the first degree Taylor approximation and determine the combination of good one and two that maximizes the firm's revenue.

(b) Use the quadratic Taylor approximation and determine the combination of good one and two that maximizes the firm's revenue.

10. Consider the following market supply and demand functions

$$Q_s = (2P - 10)^{0.6}$$
$$Q_d = \frac{700}{2P + 3} - 10$$

(a) Graph the excess supply function.

(b) Find the linear approximation of excess supply function and determine the equilibrium price and quantity.

11. A firm operates two plants. Total cost functions for the two plants, denoted as $TC(Q_1)$ and $TC(Q_2)$, are

$$TC(Q_1) = 3Q_1^3 - 5Q_1^2 + 50Q_1 + 400$$
$$TC(Q_2) = Q_2^3 - 5Q_2^2 + 20Q_2 + 100$$

Assume that the market demand for the firm's product is

$$P = 400 - 5Q$$

Determine the firm's profit maximizing level of output and price, and find the optimal division of output between its two plants.

12. A firm operates two plants. Assume that the first and second plant total cost functions, denoted as $TC(Q_1)$ and $TC(Q_2)$, are

$$TC(Q_1) = 0.6Q_1^3 - 3Q_1^2 + 6Q_1 + 50$$
$$TC(Q_2) = Q_2^3 - 10Q_2^2 + 45Q_2 + 100$$

Assume that the market demand for the firm's product is

$$Q = 250 - 2P$$

Determine the firm's profit maximizing level of output and price, and find the optimal division of output between its two plants.

13. A firm operates two plants. Assume that the first and second plant total cost functions, denoted as $TC(Q_1)$ and $TC(Q_2)$, are

$$TC(Q_1) = 0.2Q_1^3 - 7Q_1^2 + 136Q_1 + 100$$
$$TC(Q_2) = 3Q_2^3 - 4Q_2^2 + 25Q_2 + 900$$

Assume that the market demand for the firm's product is

$$P = 1000 - 5Q$$

Determine the monopolist's profit maximizing level of output and price, and find the optimal division of output between its two plants.

14. In Example 6, linear approximate the MR function over the interval $125 < Q < 130$, and then find the profit-maximizing number of housing units along with their price.

15. In Example 6, linear approximate the demand function over the interval $125 < Q < 140$ and then find the profit-maximizing number of housing units along with their price.

16. Assume the following market supply and demand functions

$$Q_s = 10 \, e^{0.1P} - 20$$
$$Q_d = 150 \, e^{-0.2P}$$

Determine the equilibrium price and quantity.

17. A firm operating in a noncompetitive market has the following cost function

$$TC = Q^3 - 20Q^2 + 150Q + 1000$$

The market demand function for the firm's product is

$$Q = 3000 - 200P^{0.8}$$

Find the firm's profit-maximizing combination of price and quantity.

18. Consider a market for a commodity with the following demand and supply functions:

$$Q_d = \frac{100p + 1500}{p^2 + 10}$$
$$Q_s = p(3)^{0.02p} - 10$$

Use the Taylor expansion and find the first degree or linear approximation of the market equilibrium price and quantity.

19. Find the equilibrium of the following market

$$Q_d = 100 - 10P^{0.3}$$
$$Q_s = -10 + 5P^{1.1}$$

20. Assume a Keynesian model for a closed economy with

$$C = 506 + 0.9177Y - 0.0001355Y^2$$
$$I = 35 + 0.15Y$$

If $G_0 = 545$, determine the equilibrium values of the endogenous variables.

21. Assume a Keynesian model for a closed economy with

$$C = 506 + 0.9177Y - 0.0001355Y^2$$
$$I = 30 + 0.15Y + 4.3\sqrt{Y}$$

Assume $G_0 = 545$ and write the reduced form of the model. Linearly approximate Y by expanding the reduced form around 3390.

22. Using time series data for the United States from 1990 to 2006,[6] the following Consumption and Investment functions are estimated

$$C = -238.64 + 0.683Y + 0.0001355Y^2$$
$$I = -219.0 + 0.18Y$$

Assume a government expenditure of $2,216.8 billion and a net export of $-615.4 billion,

(a) determine the equilibrium national income, consumption and investment.

(b) What is the model multiplier?

23. By considering consumption C a function of disposable income Y_d and using the same data set as in problem (22), the following consumption function is estimated

$$C = -1016.48 + 0.966Y_d - 0.0000087Y_d^2$$

An investment and tax function is also estimated

$$I = -219.0 + 0.18Y$$
$$T = 0.1043Y$$

Assume a government expenditure of $2,216.8 billion and a net export of $-615.4 billion,

(a) determine the equilibrium national income, consumption and investment.

(b) find the multiplier.

---

[6] *Economic Report of the President* 2008. Statistical Tables.

Chapter 8
Additional Topics in Perfect and Imperfect Competition

8.1 Competitive Firms and Market Supply Functions

A major step in completing a market model is deriving the supply function of a typical firm and the whole market supply function. The supply function of a competitive firm is the *relevant* part of the firm's marginal cost MC function, which is the segment of the MC curve above the shutdown price, or the minimum average variable cost. If we have a well-behaved cubic total cost function, then the marginal cost curve conveniently intersects both the average total and average variable cost curves at their minimum points, providing a guide for drawing the curve.

Deriving the supply function is equally easy. In order to determine the boundary of values that the price can take, we first find the shutdown price by following the familiar steps:

$$\text{Total Cost} = \text{Total Variable Cost} + \text{Fixed Cost}$$
$$TC = TVC + FC$$

Assuming

$$TVC = f(Q), \qquad \text{then}$$
$$TC = f(Q) + FC$$
$$AVC = \text{Average Variable Cost} = \frac{TVC}{Q} = \frac{f(Q)}{Q}$$

The first order condition for minimum of *AVC* is

$$\frac{dAVC}{dQ} = \frac{f'(Q)Q - f(Q)}{Q^2} = 0 \ \longrightarrow \ f'(Q)Q - f(Q) = 0$$

S. Vali, *Principles of Mathematical Economics*,
Mathematics Textbooks for Science and Engineering 3,
DOI: 10.2991/978-94-6239-036-2_8, © Atlantis Press and the authors 2014

Assuming $Q° > 0$ is a solution to the above equation, then the shutdown price P_{sd} is

$$P_{sd} = AVC(Q°)$$

Thus the supply equation of a competitive firm is

$$P = MC(Q) = f'(Q) \qquad P \geq P_{sd}$$

But this formulation of the supply function does not lend itself to aggregation. If the competitive market is populated by n similar firms, then deriving the market supply function through the aggregation of all n firms' supply functions (horizontal summation) is not possible unless we convert $P = f'(Q)$ to the more familiar expression $Q = g(P)$. This requires that we solve $P - f'(Q) = 0$ for Q to arrive at $Q = g(P)$. In that case the supply function of the ith firm ($i = 1, 2, 3, \ldots, n$) can be expressed in "aggregation-friendly" form as

$$Q_i = \begin{cases} 0 & \text{for } P < P_{sd} \\ g(P) & \text{for } P \geq P_{sd} \end{cases}$$

If all firms in the market have the same cost structure, then the market supply function is

$$Q_{ms} = \sum_{i=1}^{n} Q_i = n\, g(P)$$

The advantage of this formulation, as we will see in the next section, is that if the market demand function is known, we can determine the market equilibrium price and quantity. This turns the price P in the competitive model from an exogenous variable to an endogenous variable, making the task of comparative static analysis easier.

Example 1

To derive the supply equation of the firm with the total cost function

$$TC = \frac{1}{12}Q^3 - 2.5Q^2 + 30Q + 100$$

we first write the AVC function as

$$AVC(Q) = \frac{VC(Q)}{Q} = \frac{\frac{1}{12}Q^3 - 2.5Q^2 + 30Q}{Q} = \frac{1}{12}Q^2 - 2.5Q + 30$$

Next we find the minimum of this function:

$$\frac{dAVC}{dQ} = \frac{1}{6}Q - 2.5 = 0 \longrightarrow Q = 15$$

$$\frac{d^2AVC}{dQ^2} = 1/6 > 0, \quad \text{satisfying the } SOC \text{ for a minimum.}$$

At $Q = 15$ $AVC(15) = \frac{1}{12}(15)^2 - 2.5(15) + 30 = 18.75 - 37.5 + 30 = 11.25$

Therefore, this firm's shutdown price is $P = \$11.25$.
Since a competitive firm's supply equation is $P = MC$ (as long as the price is at or above the shutdown price) we write

$$P = MC = \frac{1}{4}Q^2 - 5Q + 30$$
$$4P = Q^2 - 20Q + 120 \longrightarrow Q^2 - 20Q + 120 - 4P = 0$$

Solving this quadratic equation for Q, we have

$$Q = \frac{20 \pm \sqrt{400 - 4(120 - 4P)}}{2} = \frac{20 \pm \sqrt{16P - 80}}{2} = 10 \pm 2(P - 5)^{1/2}$$

Since a supply function must be positively sloped and noting that at $P = 11.25$ the output level must be 15, then only

$$Q = 10 + 2(P - 5)^{1/2}$$

is an acceptable solution.
The firm's supply equation can now be expressed as

$$Q = \begin{cases} 0 & \text{for} \quad P < 11.25 \\ 10 + 2(P - 5)^{1/2} & \text{for} \quad P \geq 11.25 \end{cases}$$

8.2 From Firm to Industry: A Comprehensive Firm and Market Equilibrium Model for Competitive Market

Example 2

Consider a competitive firm with the following total cost function

$$TC = 0.2Q^3 - 6Q^2 + 80Q + 100$$

Let P denote the market price of the good. We know that the supply function of this firm is the firm's marginal cost MC function with prices equal and above the firm's shutdown price, i.e. the minimum average variable cost AVC. To derive the supply function we must first determine the shutdown price.

$$AVC(Q) = \frac{TVC(Q)}{Q} = \frac{0.2Q^3 - 6Q^2 + 80Q}{Q} = 0.2Q^2 - 6Q + 80$$

$$\frac{dAVC}{dQ} = 0.4Q - 6 = 0 \longrightarrow Q = 15$$

Value of AVC at $Q = 15$ is

$$AVC(15) = 0.2(15)^2 - 6(15) + 80 = 35$$

Thus the firm's shutdown price is \$35.

Using the equilibrium condition for a profit-maximizing firm, $MC = P$, we have

$$MC = \frac{dTC}{dQ} = 0.6Q^2 - 12Q + 80 = P$$

$0.6Q^2 - 12Q + 80 - P = 0$ simplified to $Q^2 - 20Q + 1.667(80 - P) = 0$

Solving this quadratic equation for Q, we have

$$Q = \frac{20 \pm \sqrt{400 - 6.668(80 - P)}}{2} = 10 \pm (1.667P - 33.33)^{1/2}$$

To pick the right solution, we use the fact that at the shutdown price $P = 35$ the level of output $Q = 15$. The acceptable solution, therefore, is $Q = 10 + (1.667P - 33.33)^{1/2}$ and the firm's supply function is expressed as

$$Q = \begin{cases} 0 & \text{for } P < 35 \\ 10 + (1.66P - 33.33)^{1/2} & \text{for } P \geq 35 \end{cases}$$

Now assume that there are 100 firms with identical total cost functions operating in this market. The market supply function Q_{ms} is the aggregate (horizontal summation) of the 100 individual firms' supply functions, that is,

$$Q_{ms} = \sum_{i=1}^{100} \left[10 + (1.667P - 33.33)^{1/2} \right] = 1000 + 100 \, (1.667P - 33.33)^{1/2} \tag{8.1}$$

Further assume that the market demand function is given as

$$Q_{md} = 2000 - 5.85P \tag{8.2}$$

By solving (8.1) and (8.2) simultaneously,[1] we find the market equilibrium price and quantity are $P^* \approx 50$ and $Q^* = 1707.4$ units. Each firm's level of output is $1707.4/100 = 17.074$ units with total revenue $TR = 50 * 17.074 = 853.70$, total costs $TC = 712.28$, and profit $\pi = 141.42$.

Firms in this industry are making above-normal profits, and this is a powerful incentive for new firms to enter the market. This process drives the market price down to the level of minimum average total cost. At this price the above-normal profit is eliminated and the market reaches the long run equilibrium. As usual, to determine the long run price we must find the minimum average total cost ATC:

$$ATC(Q) = \frac{TC(Q)}{Q} = \frac{0.2Q^3 - 6Q^2 + 80Q + 100}{Q} = 0.2Q^2 - 6Q + 80 + \frac{100}{Q}$$

$$\frac{dATC}{dQ} = 0.4Q - 6 - \frac{100}{Q^2} = 0 \longrightarrow 0.4Q^3 - 6Q^2 - 100 = 0$$

The solution to the above cubic equation, using the numerical method described in the Appendix B of Chap. 6, is $Q = 15.98$ units and the long run price P^* is equal to $ATC(15.98) = 41.45$.

Ignoring the possibility of additional output by the existing firms, we can show that the market reaches this equilibrium after 10 new firms, each having the same cost structure, enter the market. With an additional 10 firms, the market supply function (8.1) changes to

$$Q_{ms} = 1100 + 110(1.667P - 33.33)^{1/2}$$

By solving the new supply equation with the market demand function (8.2), we find the new market equilibrium price and quantity as $P^* = 41.45$ and $Q^* = 1757.5$ units. Note that the market equilibrium price is exactly equal to the minimum average cost of each firm. At this price firms earn normal profits. Let's check this claim. An individual firm's output is $1757.5/110 = 15.98$ units and each firm earns $41.45 * 15.98 = 662.3$. Each firm's total cost is $TC = 0.2(15.98)^3 - 6(15.98)^2 + 80(15.98) + 100 = 662.3$, and $\pi = 0$.

Exactly how the number of new entries into the market (10 firms) was determined is illustrated in the following example.

Example 3

Assume that a competitive industry consists of 100 identical firms with the following total cost function

$$TC = 0.4Q^3 - 12Q^2 + 150Q + 240$$

---

[1] If you consider solving a non-linear system of equations a little tedious and messy, you can find the linear approximation to the non-linear supply function (8.1) and then solve the resulting linear market model. You would be surprised by the accuracy of the linear solution.

If the market demand function is given as

$$Q_{md} = 3645 - 25P$$

how can we determine the market equilibrium price and a firm's profit-maximizing output for both the short run and the long run? To reach the long run market equilibrium, how many new firms must enter or leave the market? Our first task is to derive a typical firm's supply equation. Next, by aggregating over all firms' supply equations we arrive at the market supply function. We then find the market short run equilibrium price and quantity by solving the market supply and demand equations simultaneously. If at this price firms realize above (or below) normal profit, new firms enter (or some of the existing firms leave) the market. This entry into or exit from the market pushes the market price toward its long run equilibrium, which is the minimum average cost or average unit cost. At this price the market reaches its long run equilibrium, with all firms' profit or loss eliminated.

$$AVC = 0.4Q^2 - 12Q + 150$$

$$\frac{dAVC}{dQ} = 0.8Q - 12 = 0 \longrightarrow Q = 15 \text{ units}$$

$$AVC(15) = 0.4(15)^2 - 12 * 15 + 150 = 60$$

Thus the shutdown price is $60
From the profit-maximizing condition $MC = P$, we have

$$1.2Q^2 - 24Q + 150 = P \longrightarrow 1.2Q^2 - 24Q + (150 - P) = 0$$

Solutions to the above quadratic equation are

$$Q = 10 \pm 0.417(4.8P - 144)^{1/2}$$

Since at price 60 the firm's level of output is 15, the right answer for a firm supply equation is

$$Q_s = 10 + 0.417(4.8P - 144)^{1/2}$$

By aggregating over 100 firms, we find the market supply function Q_{ms} as

$$Q_{ms} = 1000 + 41.7(4.8P - 144)^{1/2}$$

By solving the market supply and demand simultaneously, we find the market equilibrium price and quantity as $P^* \approx 80$ and $Q^* = 1645$ respectively, with the individual firm's profit maximizing level of output equal to $1645/100 = 16.45$ units.

Firms' positive economic profit of $TR - TC = 1316 - 1240.83 = 75.17$ lures new firms into the industry. After a sufficient number of entries, the market price drops to the lowest ATC and process reaches its conclusion. To reach the long run equilibrium, how many new firms must enter the market? To answer this question, we must first determine the long run price or the minimum ATC.

$$ATC = 0.4Q^2 - 12Q + 150 + \frac{240}{Q}$$

$$dATC/dQ = 0.8Q - 12 - \frac{240}{Q^2} = 0$$

$$0.8Q^3 - 12Q^2 - 240 = 0$$

The acceptable solution is $Q = 16.15$ units leading to $ATC(16.15) = 75.4$, which is the long run equilibrium price. We determine the market equilibrium quantity from the market demand function by plugging in 75.4 for the price

$$Q_{md} = 3645 - 25P = 3645 - 25 * 75.4 = 1760 \text{ units}$$

This means that at equilibrium the quantity supplied to the market by producers must be 1760 units. Let the number of firms in the market be denoted by n. With n producers the new market supply function is

$$Q_{ms} = n[10 + 0.417(4.8P - 144)^{1/2}]$$

Since the equilibrium quantity demanded is 1760, we must have

$$1760 = n[10 + 0.417(4.8 * 75.4 - 144)^{1/2}]$$

Solving for n we get $n \approx 109$, which means that the long run equilibrium is reached after 9 additional firms enter the market. Alternatively, if the individual firms each produce 16.15 units and the total supply to the market is 1760 units, then there must be $1760/16.15 \approx 109$ firms in the market, an increase of 9 new firms.

8.2.1 Aggregation and Related Problems

In Examples 2 and 3 we derived the aggregated supply function for 100 firms in the industry. It is very tempting to think that we can arrive at this aggregate function by first aggregating the total cost function over 100 firms and then deriving the market supply as the relevant marginal cost function. This temptation must be strongly resisted because this procedure leads to an incorrect supply function. This is another example of the aggregation problem mentioned in earlier chapters.

In the above numerical example we assumed that all 100 firms in the industry have exactly the same total cost function. However, in the textbook presentation of perfect competition, *product homogeneity*, not *firm homogeneity*, is one of the several assumptions specified to ensure that the market for a good is perfectly competitive.

Interestingly, firm homogeneity is a long run by-product of existence of a competitive market, which manifests itself in homogeneity of cost structure.

In the short run, however, firms with different cost structures can operate in a competitive market, and this makes aggregation over firms and determining market equilibrium complicated. To illustrate, we offer another example, and in the process of working out answers to the questions raised in this example, we discuss issues related to the aggregation problem.

Example 4

Assume 300 identical firms operate in a competitive market. If a typical firm has the following total cost function

$$TC = Q^3 - 10Q^2 + 45Q + 100$$

and the market demand function is

$$Q_{md} = 2500 - 20P$$

We want to determine

(a) the short run market equilibrium price and quantity.

(b) the profit maximizing level of output for each firm.

(c) the normal profit market equilibrium price and quantity.

(d) the number of firms entering into (or exiting from) the market in the long run.

(a) To determine the market equilibrium we must first find the market supply function. The market supply function is the aggregate of the supply functions of all 300 firms. Therefore, we must first find an individual firm's supply equation. To do this we need the shutdown price P_{sd}, which is the minimum of the AVC.

$$AVC = \frac{VC}{Q} = \frac{Q^3 - 10Q^2 + 45Q}{Q} = Q^2 - 10Q + 45$$

The *FOC* for a minimum

$$\frac{dAVC}{dQ} = 2Q - 10 = 0 \longrightarrow Q = 5$$

And $P_{sd} = AVC(5) = 25 - 50 + 45 = 20$. Utilizing the profit maximizing condition for a competitive industry, we write

$$MC = P \text{ or } MC - P = 0 \longrightarrow 3Q^2 - 20Q + 45 - P = 0$$

Solving this equation for Q, we have

$$Q = \frac{10}{3} \pm \frac{1}{3}\sqrt{3P - 35}$$

To check whether $+$ or $-$ is the right answer we use the shutdown price and quantity combination $(20, 5)$ as a test. It is clear that when the price is 20 we get $Q = 5$ if the $+$ is used. Also, note that the minus sign implies an inverse relationship between Q and P which cannot be a supply relationship. Hence the resulting supply equation of a typical firm is

$$Q_i = \begin{cases} 0 & \text{for } P < 20 \\ \frac{10}{3} + \frac{1}{3}\sqrt{3P - 35} & \text{for } P \geq 20 \end{cases} \qquad i = 1, 2, ..., 300 \qquad (8.3)$$

The market supply function is then

$$Q_{ms} = \sum_{i=1}^{300} Q_i = 300\left(\frac{10}{3} + \frac{1}{3}\sqrt{3P - 35}\right) = 1000 + 100\sqrt{3P - 35}$$

The market equilibrium is determined by

$$Q_{md} = Q_{ms} \longrightarrow 2500 - 20P = 1000 + 100\sqrt{3P - 35}$$

Or

$$15 - 0.2P = \sqrt{3P - 35} \longrightarrow (15 - 0.2P)^2 = 3P - 35$$

$$225 + 0.04P^2 - 6P = 3P - 35 \longrightarrow 0.04P^2 - 9P + 260 = 0$$

There are two solutions to the above equation, 34.0382 and 190.962. The acceptable solution is $P^* = 34.0382$ with the market equilibrium quantity of $Q^* = 2500 - 20(34.0382) = 1819.236$. The second solution must be discarded because it leads to a negative quantity.

(b) The profit-maximizing (loss-minimizing) level of output for each firm is $Q = 1819.236/300 = 6.064$ units. With the price/output combination of $34.0382/6.064$ each firm loss is

$$TR - TC = 34.0382 * 6.064 - (6.064)^3 + 10 * (6.064)^2 - 45 * 6.064 - 100 = -21.74$$

(c) The long run, break-even, or normal profit market equilibrium price is the minimum of the average total cost ATC :

$$ATC = \frac{TC}{Q} = \frac{(Q^3 - 10Q^2 + 45Q + 100)}{Q} = Q^2 - 10Q + 45 + \frac{100}{Q}$$

The FOC

$$\frac{dATC}{dQ} = 2Q - 10 - 100/Q^2 = 0 \longrightarrow 2Q^3 - 10Q^2 - 100 = 0$$

A real solution to the above cubic function is $Q = 6.27131$ units. Then the break-even or normal profit price is $P_{np} = ATC(6.27131) = 37.562$.

In the absence of an increase in demand or any change in the cost structure of the firm and industry, a move from a loss of 21.74 to the break-even point could only occur if some firms exited from the market, which will be discussed next.

(d) Using the market demand function and the long run price, we find the long run market quantity as

$$Q = 2500 - 20 * 37.562 = 1748.76$$

And since each firm produces 6.27131 units, then the number of firms must be

$$\frac{1748.76}{6.27131} = 278.9 \approx 279 \text{ firms}$$

Then in order for the market to reach the normal profit equilibrium, 21 firms of the current 300 firms must exit the market. With only 279 firms in the market the new market supply function, using (8.3), will be

$$Q'_{ms} = 279 \left(\frac{10}{3} + \frac{1}{3}\sqrt{3P - 35} \right) = 930 + 93\sqrt{3P - 35}$$

which indicates a leftward shift in the supply curve. If we solve this supply function with the demand function, we arrive at the break-even or normal profit equilibrium $P = 37.536$ and $Q = 1749.8$ (The small discrepancy between prices and quantities are due to the round off error caused by rounding the number of remaining firms from 278.8 to 279.)

There are two ways that this situation, i.e. firms forced to leave the market, can be avoided:

(1) If there is an increase in demand for this product.
(2) If a firm can change the cost structure of its operation by introducing new cost-saving technology or another innovative production arrangement.

An increase in demand manifests itself by a rightward shift in the demand curve. For the market to the reach break-even or normal profit equilibrium without firms being forced out of business, the demand curve must move outward to a new location, with a new Y-intercept. Let's denote the new demand function by

$$Q'_{md} = \beta - 20P$$

From part (c) we know that the break-even or normal profit (price, output) combination for a single firm is (37.562, 6.27131). If a change in the demand condition allows all 300 firms to stay in business then the aggregated market output would be 1881.393 units. Therefore, the demand curve must shift to the new position with new intercept

$$\beta = 1881.393 + 20 * 37.562 = 2632.63$$

Subsequently the new demand equation must be

$$Q'_{md} = 2632.63 - 20P$$

Alternatively, this situation can be avoided by a change in the cost structure of a firm in danger of elimination. If by introduction of new cost-saving technology, or other innovative production arrangement, this firm lowers its production cost by a certain percentage, it could cut its losses and even earn a short run above normal profits. To illustrate, assume a firm (or a few firms) acquires a new technology that reduces the cost of production by 10%. This changes the firm total cost function from

$$TC = Q^3 - 10Q^2 + 45Q + 100$$

to

$$TC_1 = (1 - 0.1)TC = 0.9TC = 0.9Q^3 - 9Q^2 + 40.5Q + 90$$

If the market price is the same as the price established in part (a), the profit-optimizing level of output for the firm would be a solution to

$$MC - P = 0 \longrightarrow 2.7Q^2 - 18Q + 40.5 - 34.0382 = 0$$

which is $Q = 6.2859$ units. With a (price, quantity) combination of (34.0382, 6.2859) the firm not only cuts its losses of 21.74 but makes 1.46 units of profit.

A Short Digression

It is actually possible to determine the percentage of the cut (if feasible) a firm must make in order to have a certain amount of profit. To illustrate, assume this firm's goal is to make 10 units of profit when the price of the good in the market is 34.0382. What proportion of cuts in cost is needed to achieve this goal? Let's denote this proportion by ρ. The new total cost function TC_2 is then

$$TC_2 = (1 - \rho)TC = (1 - \rho)(Q^3 - 10Q^2 + 45Q + 100)$$

From the profit-maximizing condition we have

$$MC_2 = P \text{ or } (1 - \rho)MC = P$$

$$(1 - \rho)(3Q^2 - 20Q + 45) = P \tag{8.4}$$

And the desire to make 10 units of profits translates to

$$TR - TC_2 = 10 \longrightarrow (1 - \rho)TC = TR - 10 = PQ - 10$$

$$(1 - \rho)(Q^3 - 10Q^2 + 45Q + 100) = PQ - 10 \tag{8.5}$$

Solving (8.4) and (8.5) simultaneously leads to the level of output Q and proportion of cuts ρ needed for a firm to achieve its objective. By dividing (8.5) by (8.4) we eliminate ρ

$$\frac{(1 - \rho)(Q^3 - 10Q^2 + 45Q + 100)}{(1 - \rho)(3Q^2 - 20Q + 45)} = \frac{(PQ - 10)}{P} = Q - \frac{10}{34.0382} = Q - 0.2938$$

After simplifying this equation, we have

$$2Q^3 - 10.8814Q^2 + 5.876Q - 113.22 = 0$$

A real solution to this cubic equation is $Q = 6.3734$. Now by using (8.4) we solve for ρ.

$$(1 - \rho)\left[3(6.3734)^2 - 20 * 6.3734 + 45\right] = 34.0382$$

$$39.389(1 - \rho) = 34.0382 \longrightarrow \rho = 0.1359$$

This means that this firm must cut its costs by 13.59 % and must produce 6.3734 units of output in order to make about 10 units of profit.

> ### End of Digression

Let's pick up the thread of our discussion before the short digression. It is possible for some firm to introduce new technology and cost containing strategies and thus avoid losses when the market is not in good shape. But if the entire industry adopts the same technology and strategies, which is almost certainly the case in a competitive industry and in the long run, then all firms' losses will not be eliminated, necessitating the inevitable departure of a number of financially weaker firms from the market.

To demonstrate, assume all 300 firms in the industry adopt the same technology and have the same $TC = 0.9Q^3 - 9Q^2 + 40.5Q + 90$. We must first derive a typical firm's supply function and then through aggregation derive the market supply function:

$$\frac{dAVC}{dQ} = 1.8Q - 9 = 0 \longrightarrow Q = 5$$

$$P_{sd} = AVC(5) = 18$$

$$MC - P = 0 \longrightarrow 2.7Q^2 - 18Q - 40.5 - P = 0$$

$$Q_i = \frac{1}{9}(30 + \sqrt{30P - 315}) \quad P \geq 18 \quad i = 1, 2, \ldots, 300$$

$$Q_{ms} = \frac{300}{9}(30 + \sqrt{30P - 315}) = 1000 + \frac{100}{3}\sqrt{30P - 315}$$

Now with the aid of the demand function we can determine the market equilibrium price and quantity.

$$1000 + \frac{100}{3}\sqrt{30P - 315} = 2500 - 20P$$

$$\sqrt{30P - 315} = \frac{3}{100}(1500 - 20P)$$

$$30P - 315 = (45 - 0.6P)^2$$

$$0.36P^2 - 84P + 2340 = 0$$

$$P = 32.3393 \longrightarrow Q = 1853.214$$

Each firm's share of the market is $\frac{1853.214}{300} = 6.1774$ units. A typical firm's total revenue and total cost are 199.772 and 208.8997, with a loss of 9.13, proving the assertion that what might be good for a few may not be good for all; the good old *fallacy of composition.*

Now assume that the industry consists of two groups of 150 firms. The firms in the first group all have the total cost function

$$TC_1 = Q^3 - 10Q^2 + 45Q + 100$$

And the other 150 in the second group have the total cost function

$$TC_2 = 0.2Q^3 - 3Q^2 + 35Q + 50$$

We already know that the supply function of individual firms in the first group is (8.3). By aggregating over the 150 firms, we have

$$Q_1 = \begin{cases} 0 & \text{for} \quad P < 20 \\ 500 + 50\sqrt{3P - 35} & \text{for} \quad P \geq 20 \end{cases} \qquad (8.6)$$

The individual firm's supply function in the second group is

$$Q_i = \begin{cases} 0 & \text{for} \quad P < 23.75 \\ 5 + 1.291\sqrt{P - 20} & \text{for} \quad P \geq 23.75 \quad i = 1, 2, \ldots, 150 \end{cases}$$

And the subsequent aggregated supply function for these 150 firms is

$$Q_2 = \begin{cases} 0 & \text{for} \quad P < 23.75 \\ 750 + 193.65\sqrt{P-20} & \text{for} \quad P \geq 23.75 \end{cases} \tag{8.7}$$

The whole market supply function is then the sum of Q_1 and Q_2, i.e. the aggregation of (8.6) and (8.7). This leads to the market supply function

$$Q_{ms} = \begin{cases} 0 & \text{for} \quad P < 20 \\ 500 + 50\sqrt{3P-35} & \text{for} \quad 20 \leq P < 23.75 \\ 1250 + 50\sqrt{3P-35} + 193.65\sqrt{P-20} & \text{for} \quad P \geq 23.75 \end{cases} \tag{8.8}$$

Next we face the complication of solving this supply function with the demand function $Q_{md} = 2500 - 20P$ for the market equilibrium price and quantity. Since we don't know what market price equilibrates supply and demand, we can't say what segment of the market supply function (8.8) must enter in the equilibrium condition. One feasible approach is to graph the supply and demand functions and by visual inspection determine the right segment of the supply curve that intersects the demand curve. A graph of the supply and demand functions would indicate that for market equilibrium we must solve the demand function with the third segment of the supply function.

$$1250 + 50\sqrt{3P-35} + 193.65\sqrt{P-20} = 2500 - 20P$$

$$20P + 50\sqrt{3P-35} + 193.65\sqrt{P-20} - 1250 = 0$$

The solution to this equation is $P = 25.01 \approx 25$. Subsequently the market equilibrium quantity is $Q = 1999.8 \approx 2000$ units.

Before moving to the next layer of complication, let's pause for a moment and try to imagine what the supply function would look like if the market is populated by hundreds of producers with different cost structures. No doubt that the supply function would have a large number of segments, but its graph would be much smoother.

The next problem is to determine the share of each producer in each group from the equilibrium quantity of 2000 units. To answer this question we must recall the narrative of the perfectly competitive market mechanism: the combined forces of supply and demand determine the market price, which each producer takes and uses in the search for its profit-maximizing level of output. Given that the market price is 25, all 150 firms in the first group find their profit-maximizing output by solving

$$MC_1 - P = 0 \longrightarrow 3Q^2 - 20Q + 45 - 25 = 0 \longrightarrow 3Q^2 - 20Q + 25 = 0$$

The solution to this equation is $Q = 5.442$ units. The profit-maximizing level of output for the second group is determined by

$$MC_2 - P = 0 \quad \longrightarrow \quad 0.6Q^2 - 6Q + 35 - 25 = 0 \quad \longrightarrow \quad 0.6Q^2 - 6Q + 10 = 0$$

leading to $Q = 7.887$ units. The first group's contribution to the market is $150 * 5.442 = 816.22$ units and the second group's is $150 * 7.887 = 1183.05$ units, thereby adding up to approximately 2000 units of total supply. Readers are encouraged to verify that the presence of the second group with a relatively better cost structure creates a more unpleasant business environment for the first group and hence deepens their losses.

8.3 Urge to Merge

Monopolizing a Competitive Market

The discussion of a *monopoly* in Microeconomics textbooks invariably leads to the conclusion that if it is not possible to realize a large economy of scale, then through consolidation of a competitively organized industry emerges a monopoly that produces less output and charges a higher price. But the Microeconomics textbooks' treatment of the monopolization of a competitive market and its aftermath are only diagrammatic. The objective of this section is to provide a mathematic treatment of this topic and highlight some of the aspects of modeling the monopolization of a competitive market which are not fully explored in the literature. This will be done by first introducing a theorem and its economics application, followed by a simple numerical example to demonstrate the delicate process of modeling monopolization of a competitive market.

In its most general form the theorem can be presented as follows:

Assume two functions $f(x)$ and $g(x)$ such that $\dfrac{d}{dx} f(x) < 0 \ \forall x \in R$ and $\dfrac{d}{dx} g(x) > 0 \ \forall x \in R$

We form the total function $T(x)$

$$T(x) = f(x)x$$

and the marginal function $M(x)$

$$M(x) = T'(x) = \frac{d}{dx} T(x) = f'(x)x + f(x)$$

If $x_c > 0$ is such that

$$f(x_c) = g(x_c) \quad \text{or} \quad g(x_c) - f(x_c) = 0 \tag{8.9}$$

and $x_m > 0$ is such that

$$f'(x_m)x_m + f(x_m) = g(x_m) \quad \text{or} \quad f'(x_m)x_m + f(x_m) - g(x_m) = 0 \qquad (8.10)$$

then $x_m < x_c$ and $f(x_m) > f(x_c)$.

We use 'proof by contradiction' to establish the validity of the assertion. Assume $x_m > x_c$. Since $g(x)$ is upward sloping then

$$g(x_m) > g(x_c) \qquad (8.11)$$

Given that $f(x)$ is downward sloping, then it must be the case that

$$\begin{aligned} f(x_m) &< f(x_c) \quad \text{or} \\ -f(x_m) &> -f(x_c) \end{aligned} \qquad (8.12)$$

By adding both sides of (8.11) and (8.12), we have

$$g(x_m) - f(x_m) > g(x_c) - f(x_c)$$

But from (8.9) the right hand side of this inequality is zero, leading to

$$g(x_m) - f(x_m) > 0 \quad \text{or} \quad g(x_m) > f(x_m)$$

This result contradicts the implication of (8.10). Since $f'(x_m) < 0$, for (8.10) to hold requires that $f(x_m) > g(x_m)$.

The application of this theorem to the monopolization of a competitive market is not that straightforward and requires some ground work.

Assume there are n firms in a competitive industry. Let P denote the market price and $MC_i(Q)$ the marginal cost function of the ith firm. If $\tilde{P}_i = min\{AVC_i(Q)\}$ denotes the ith firm's shutdown price, then

$$P = MC_i(Q) \qquad P \geq \tilde{P}_i \qquad i = 1, 2, ..., n \qquad (8.13)$$

is a typical firm's supply function. However, this formulation of the firms' supply function does not lend itself to aggregation. We can solve for Q in terms of P in (8.13) and arrive at

$$Q_i = g_i(P) \qquad \forall i$$

The supply function of the ith firm in an aggregation friendly form can now be expressed as

$$Q_i = \begin{cases} 0 & \text{for} \quad P < \tilde{P} \\ g_i(P) & \text{for} \quad P \geq \tilde{P} \end{cases}$$

and the market supply function as

$$Q = \sum_{i=1}^{n} Q_i = \sum_{i=1}^{n} g_i(P)$$

If all firms in the market have the same cost structure,[2] then the market supply equation can be expressed as

$$Q = ng(P) \tag{8.14}$$

If in (8.14) we solve for P in terms of Q the resulting $P = S(Q)$ would be the Marshallian market supply function. Further assume that $P = D(Q)$ is the market demand function.

With these elements in place, it is relatively easy to see that if $f(x)$ and $g(x)$ are the market demand and supply functions, respectively, then x_c is the competitive market equilibrium level of output Q_C, determined by solving

$$D(Q) - S(Q) = 0$$

If this market is monopolized, then the *single period* profit maximizing level of output of the consolidated firm Q_M, determined by

$$D'(Q)Q + D(Q) - S(Q) = 0$$

is the same as x_m, where $Q_M < Q_C$. Similarly, the perfectly competitive market equilibrium price P_C is equivalent to $f(x_c)$ and the monopolist single period profit maximizing price P_M is equivalent to $f(x_m)$, with $P_M > P_C$.

Example 5

Recall from Example 4 that due to weakness in consumers demand all 300 identical firms in the market were incurring losses. Also recall that in the absence of any improvement in the demand side of the market and without the possibility of implementing cost-saving technology, the normal-profit situation required that 21 firms leave the market. Under this circumstances, the managements of the firms facing financial crisis and the prospect of being forced out of the market (and out of their jobs) may inspired to merge with each other. Assume that through an intense merger/acquisition process the 300 firms consolidate into a big monopoly. Our objective is to show that even without any change in the economic environment this newly formed monopoly will make above normal profits, and subsequently demonstrate that under unfavorable business climates firms have strong incentives to merge with their rivals.

---

[2] As it was mentioned earlier, a perfectly competitive market requires *product homogeneity* and not *firm homogeneity*, but we make this assumption here to avoid additional "aggregation" complication which is not the core issue of this section.

After this market is monopolized, the market supply function becomes the supply function of the consolidated monopolist. This monopolist maximizes its *single period* profit by choosing the output level where its marginal cost is equal to the marginal revenue.

To find Q_M and P_M we must first solve the market supply function

$$Q = 1000 + 100\sqrt{3P - 35}$$

in terms of P, leading to

$$P = \frac{1}{30000}Q^2 - 0.0667Q + 45 \qquad P \geq 20$$

Next we derive the marginal revenue function as

$$Q = 2500 - 20P \longrightarrow P = 125 - 0.05Q \qquad (8.15)$$
$$TR = PQ = 125Q - 0.05Q^2$$
$$MR = \frac{dTR}{dQ} = 125 - 0.1Q$$

The profit-maximizing level of output Q_M is then determined by

$$\frac{1}{30000}Q^2 - 0.0667Q + 45 = 125 - 0.1Q \qquad (8.16)$$

After some simple manipulation of (8.16), we arrive at the quadratic equation

$$Q^2 + 1000Q - 2400000 = 0$$

The acceptable solution for the profit-maximizing level of output is 1127.88. By plugging in this value for Q in the demand function (8.15), we have

$$P = 125 - 0.05Q = 125 - 0.05 * 1127.88 = 68.61$$

Thus, the profit-maximizing (output, price) combination for the monopolist is (1127.88, 68.61). By comparing this combination with the profit maximizing combination under perfect competition ($Q^* = 1819.236$, $P^* = 34.0382$) it is, as expected, clear that the monopolist restricts the output and charges a higher price.

Figure 8.1 depicts the situation. In this graph Q_C and P_C are the output and price for competitive market and Q_M and P_M are output and price when the market is monopolized. It is clear from the graph that compared to the competitive optimal solution the monopolist profit maximization leads to lower output and higher price.

Before consolidation and under competition, an individual firm's profit-maximizing (loss-minimizing) level of output is 6.064 units. With the price 34.0382 in the market, each firm loses about 21.74 and all firms collectively lose $300 * 21.74 =$

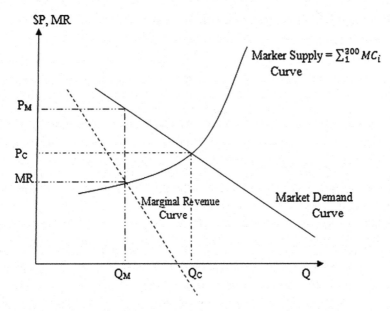

Fig. 8.1 Monopolizing a competitive market

6522. It was due to this loss that 21 firms were forced to leave the market, and subsequently the long run normal profit was reached at a price of 37.562 and a market output of 1748.76 units. But after the monopolization of the market, the consolidated firm makes an above normal profit. The monopolist's total revenue is

$$TR = P * Q = 68.61 * 1127.88 = 77383.85$$

From a total of 1127.88 units supplied to the market, each firm's share is $1127.88/300 = 3.76$ units and the total cost of production for each firm is

$$TC_i = (3.76)^3 - 10(3.76)^2 + 45 * 3.76 + 100 = 180.98 \quad i = 1, 2, ..., 300$$

Then the total cost of production for the monopolist is

$$TC = \sum_{i=1}^{300} TC_i = 300 * 180.98 = 54294$$

So, the monopolist's profit is

$$\Pi = TR - TC = 77383.85 - 54294 = 23089.85$$

8.4 Deriving the Trace of Optimal Output
for a Noncompetitive Firm

The main objective of this section[3] is to formulate a model for deriving the *trace of optimal output* (*trace* for short) of a monopoly or monopolistic competitive firm. I am using *trace* instead of *supply curve* to avoid the controversy surrounding the existence of supply curve for noncompetitive firms. Microeconomics textbooks tell us that a monopolist or monopolistic competitive firm has no supply curve independent of the demand curve for its product. Our intention here is to use this dependency and show that under a widely accepted and practiced assumption about changes in the market demand for product of a noncompetitive firm, it is possible to derive its locus of supply points, or its trace of optimal output.

True to the common definition of trace—"to follow the course or trail of something"—the trace function of a noncompetitive firm tracks the trail of any of the firm's optimal (price, quantity) combination to a specific and unique location of the demand curve. As demand changes over time, the locus of optimal (price, quantity) combinations or supply points forms the firm's trace.

To formalize the model, let's assume a noncompetitive firm with the total variable cost function $f(Q)$ and total cost function $TC = f(Q) + FC$. Further assume that the original demand function facing the firm is

$$P = \alpha + g(Q)$$

where α is its intercept and $\dfrac{d(P)}{dg(Q)} < 0$

The first step in the process of deriving a firm's trace function is to determine its shutdown price. The shutdown price occurs at the point of tangency between the average variable cost (AVC) curve and the *Dislocated Demand Curve* (DDC). For a monopolist the original demand curve and its associated marginal revenue (MR) curve would be dislocated due to the deterioration of economic conditions that adversely impact the demand for the good or service produced by the monopolist or an improvement in the business environment. Besides changes in the economic environment, the demand dislocation for a monopolistically competitive firm could also be due to firms entering or exiting the market. In either case the assumption is that in the short run changes in demand are exclusively due to exogenous variables, and the impact of these changes are transmitted only to parameter α. In other words, a change in demand manifests itself as a downward or upward *parallel* slide in the demand curve. This assumption implies that elasticities of non-price variables included in the demand equation are not constant. This prevents the demand curve from rotating when any of these variables, like income, change.[4] If "income" is a

---

[3] This section is reprinted from Vali, Shapoor 2012, "Deriving the Trace of Optimal Output," *Atlantic Economic Journal*, Vol 40, 49–60, 2012; with kind permission of the journal's editorial board.

[4] See Philip P. Graves and Robert L. Sexton, "Demand and Supply Curves: Rotation versus Shifts", *Atlantic Economic Journal*, 34(3), 361–364, 2006.

variable included in the equation and the constancy of income elasticity is a desired feature of the model, then the demand function should be of *log-linear* form (more about this in Chap. 9).

Now let's assume that the demand curve is dislocated and shifts to the point where it becomes tangent to the firm's AVC curve. Since these shifts are parallel then the demand equation at the point of tangency with the AVC curve will have the same slope, but a different intercept. Let's write this dislocated demand equation as

$$P = \beta + g(Q) \tag{8.17}$$

where β is the new intercept. At the point of tangency between the *AVC* and *DDC* curves we must have

$$\frac{dAVC}{dQ} = \frac{dP}{dQ} \tag{8.18}$$

Given that *AVC* is

$$AVC = \frac{TVC}{Q} = \frac{f(Q)}{Q}$$

then Eq. (8.18) can be expressed as

$$\frac{f'(Q)Q - f(Q)}{Q^2} = g'(Q)$$

or

$$f'(Q)Q - f(Q) - g'(Q)Q^2 = 0 \tag{8.19}$$

If $\tilde{Q}$ is the solution to equation (8.19), then

$$\tilde{P} = AVC(\tilde{Q})$$

is the shutdown price. Figure 8.2 depicts the situation. In this graph the original demand curve D_0 shifts to D_1 where it is tangent to the *AVC* curve.

Alternatively, the shutdown price is reached when the following two conditions are simultaneously satisfied

$$\begin{cases} AVC = P \\ MC = MR \end{cases}$$

Graphically, this occurs when the point of intersection of the MR and MC curves align with the point of tangency between the dislocated demand curve and the AVC curve. Using $TC = f(Q) + FC$ and $P = \beta + g(Q)$, we have

$$\begin{cases} \dfrac{f(Q)}{Q} = \beta + g(Q) \\ f'(Q) = \beta + g(Q) + Qg'(Q) \end{cases} \tag{8.20}$$

Fig. 8.2 Tangency of demand and ATC *curves* and shutdown price $\tilde{P}$

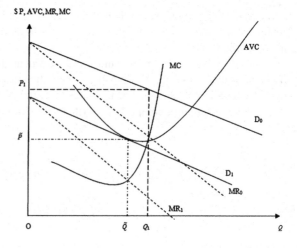

Equation (8.20) is a system of two equations with two unknowns β and Q. We can eliminate β by multiplying both sides of the second equation in (8.20) by -1 and adding it to the first equation. The result will be

$$\frac{f(Q)}{Q} - f'(Q) = -Qg'(Q)$$

which can be simplified into the same equation as (8.19).

The firm's marginal revenue equation is

$$MR = \frac{dTR}{dQ} = \frac{d}{dQ}\,[\beta Q + g(Q)Q]$$

$$MR = \beta + g'(Q)Q + g(Q)$$

Replacing β by $P - g(Q)$ from the dislocated demand Eq. (8.17), we have

$$MR = P + g'(Q)Q$$

Using the firm's profit-maximizing (or loss-minimizing) condition $MC = MR$, we have

$$f'(Q) = P + g'(Q)Q \tag{8.21}$$

which can be rewritten as

$$P = f'(Q) - g'(Q)Q \qquad\qquad P \geq \tilde{P} \tag{8.22}$$

For a more familiar expression, we can write (8.22) as

$$f'(Q) - g'(Q)Q - P = 0 \qquad (8.23)$$

and solve for Q in terms of P. If $Q = h(P)$ is the solution to (8.23) then

$$Q = \begin{cases} 0 & \text{for} \quad P < \tilde{P} \\ h(P) & \text{for} \quad P \geq \tilde{P} \end{cases}$$

is the *trace function* of the firm.

Note that for any arbitrary price P' the optimal level of output is $h(P')$ and the value of β can be determined as

$$\beta = P' - g[h(P')] \qquad (8.24)$$

Consequently the demand function corresponding to the optimal combination $[P', h(P')]$ would be

$$P = P' - g[h(P')] + g(Q)$$

Next we offer several numerical examples to demonstrate the working of the model.

Example 6

Assume a noncompetitive firm with the total cost function

$$TC = Q^3 - 10Q^2 + 45Q + 100$$

faces the following linear demand function for its product in the market

$$P = 250 - 2Q$$

We derive the trace function of the firm by first determining the shut down price. By noting that

$$f(Q) = Q^3 - 10Q^2 + 45Q \qquad \text{and} \qquad g(Q) = -2Q$$

we can utilize Eq. (8.19) and write

$$(3Q^2 - 20Q + 45)Q - (Q^3 - 10Q^2 + 45Q) - (-2)Q^2 = 0$$

leading to

$$2Q^3 - 8Q^2 = 0$$

with the solution $Q = 4$. Thus the shut down price is

$$\tilde{P} = AVC(4) = 21$$

Using Eq.(8.23), we have

$$3Q^2 - 20Q + 45 - (-2)Q - P = 0 \longrightarrow 3Q^2 - 18Q + (45 - P) = 0$$

The above equation has two solutions

$$Q = 3 \pm \frac{1}{3}\sqrt{3P - 54}$$

but since at the shutdown price $\tilde{P} = 21$, Q must be 4, then the only acceptable solution is

$$Q = 3 + \frac{1}{3}\sqrt{3P - 54}$$

leading to the firm's trace function

$$Q = \begin{cases} 0 & \text{for} \quad P < 21 \\ 3 + \frac{1}{3}\sqrt{3P - 54} & \text{for} \quad P \geq 21 \end{cases}$$

When the price is 227.29 the quantity supplied is

$$Q = 3 + \frac{1}{3}\sqrt{3 * 227.29 - 54} = 11.353 \quad \text{units}$$

By using (8.24) we can determine that this (price, quantity) combination is the profit-maximizing combination consistent with the original demand equation $P = 250 - 2Q$.

$$\beta = P' - g[h(P')] = 227.29 - [-2(11.353)] = 250$$

$$P = 250 - 2Q$$

With a price/quantity combination of (227.29, 11.353) this firm earns 1795.15 in profits. If the price drops to 38.23, then the quantity supplied will be reduced to 5.596 and the firm earns a normal profit. By using (8.24) we can trace this (price, quantity) combination to the dislocation of demand to $P = 49.42 - 2Q$. This demand curve is tangent to the average total cost (ATC) curve. We can also trace the (price, quantity) combination of (21, 4), which signifies the shutdown situation, to the demand falling to $P = 29 - 2Q$. 29 is the lower bound of β in the demand Eq. (8.18). As another example, the optimal level of output at $P = 45$ is 6 units, which is associated with the demand shifting to $P = 57 - 2Q$.

Example 7

In this example the total cost and demand functions are

$$TC = 0.2Q^3 - 7Q^2 + 136Q + 100$$
$$P = 100 - 1.5Q$$

By following the steps outlined in Example 6, we can derive the firm's trace function as

$$Q = \begin{cases} 0 & \text{for} \quad P < 77.56 \\ 10.417 + 0.833\sqrt{2.4P - 170.15} & \text{for} \quad P \geq 77.56 \end{cases}$$

In the prevailing economic environment captured by the demand equation, the best this firm can do is to minimize its losses by picking the (price, quantity) combination of $P = 78.90$ and $Q = 14.07$ units. The firm's loss with this combination is 74.79. If the situation improves the firm can eliminate its loss and earn a normal profit by reaching the combination $Q = 14.88$ units and $P = 82.84$. This requires that the demand curve shift upward, thereby reaching a new location with the equation $P = 105.16 - 1.5Q$. The demand equation $P = 98.187 - 1.5Q$ corresponds to the shutdown situation where the shutdown price is 77.56.

Example 8

Assume that our model consisting of the total cost function of a noncompetitive firm and the demand function for its product are

$$TC = 2Q^3 - 30Q^2 + 180Q + 400$$
$$P = 400 - 15Q + 0.4Q^2$$

Here our firm faces a non-linear demand function.

By noting that $f(Q) = 2Q^3 - 30Q^2 + 180Q$ and $g(Q) = -15Q + 0.4Q^2$ we can use (8.19) and determine the shut down price and quantity as $P = 83.32$ and $Q = 4.688$. Next we use (8.23) and write

$$6Q^2 - 60Q + 180 - (-15 + 0.5Q)Q - P = 0$$

which after simplification is

$$5.2Q^2 - 45Q + (180 - P) = 0$$

The solution to this quadratic equation is

$$Q = \frac{45 \pm \sqrt{2025 - 20.8(180 - P)}}{10.4}$$

$$Q = \frac{45 \pm \sqrt{20.8P - 1719}}{10.4}$$

Given that at $P = 83.32$ Q must be 4.688,

$$Q = \frac{45 + \sqrt{20.8P - 1719}}{10.4}$$

is the right solution. The trace function, therefore, must be expressed as

$$Q = \begin{cases} 0 & \text{for} \quad P < 83.32 \\ \frac{1}{10.4}(45 + \sqrt{20.8P - 1719}) & \text{for} \quad P \geq 83.32 \end{cases}$$

With the original demand function, this firm's maximum profit combination is $Q = 10.58$ and $P = 286.06$. This combination generates 1711.789 in profits. If for any reason the demand declines, business deteriorates and the profit is eliminated, then the firm normal profit combination will be $Q = 7.14$ and $P = 123.78$. The dislocated demand equation for normal profit is

$$P = 210.49 - 15Q + 0.4Q^2$$

To see how we arrived at this (price, quantity) combination and the associated demand equation, remember that the normal profit (price, quantity) combination occurs at the point of tangency between the demand curve and the ATC curve; that is, where

$$\frac{dATC}{dQ} = \frac{dP}{dQ}$$

Given that

$$ATC = \frac{TC}{Q} = \frac{2Q^3 - 30Q^2 + 180Q + 400}{Q} = 2Q^2 - 30Q + 180 + \frac{400}{Q}$$

then

$$4Q - 30 - \frac{400}{Q^2} = -15 + 0.8Q$$

leading to

$$3.2Q^3 - 15Q^2 - 400 = 0$$

The real solution to the above cubic function is $Q = 7.14$ and the corresponding price is

$$P = ATC(7.14) = 123.78$$

Using (8.24), we have

$$\beta = P' - g[h(P')] = 123.78 - [-15(7.14) + 0.4(7.14^2)] = 210.49$$

leading to the demand equation associated with the normal profit shown above. If, as another example, $P' = 200$ then $Q = 9.08$ and by using (8.25) β is

$$\beta = P' - g[h(P')] = 200 - [-15(9.08) + 0.4(9.08^2)] = 303.20$$

And the demand equation associated with price 200 would be

$$P = 303.20 - 15Q + 0.4Q^2$$

8.5 Nonlinear Demand Function and the Shutdown Price

If the demand function (8.17) is nonlinear then $\tilde{P} = AVC(\tilde{Q})$, where $\tilde{Q}$ is the solution to (8.19), may or may not be the shutdown price. To determine the shutdown price we first find the minimum of the following function,

$$P = f'(Q) - g'(Q)Q \qquad (8.25)$$

The first order condition for a minimum requires that

$$\frac{dP}{dQ} = f''(Q) - g'(Q) - Qg''(Q) = 0 \qquad (8.26)$$

Assume $\ddot{Q}$ is the real solution to (8.26), then the minimum price is

$$\ddot{P} = f'(\ddot{Q}) - \ddot{Q}\, g'(\ddot{Q}) \qquad (8.27)$$

If $TR(\ddot{Q}) - TVC(\ddot{Q}) > 0$ and $\ddot{P} < \tilde{P}$ then $\ddot{P}$ is the shutdown price, otherwise $\tilde{P}$ is the shutdown price. Thus, the "shutdown price" for a noncompetitive firm may not conform to the definition under perfect competition.

Example 9

Assume that our model consisting of the total cost function of a noncompetitive firm and the demand function for its product respectively are

$$TC = 2Q^3 - 30Q^2 + 180Q + 400$$
$$P = 400 - 25Q + 0.4Q^2$$

Here we use exactly the same total cost function as Example 8 but assume that our firm faces a different quadratic demand function. To find the locus of optimal output as the demand shifts, we write the demand function as

$$P = \beta - 25Q + 0.4Q^2 \tag{8.28}$$

with the intercept specified as a parameter. To maximize profit, we form the profit function

$$\pi(Q) = TR - TC = \beta Q - 25Q^2 + 0.4Q^3 - 2Q^3 + 30Q^2 - 180Q - 400$$

$$\pi(Q) = -1.6Q^3 + 5Q^2 + (\beta - 180)Q - 400$$

The first order condition for a maximum is

$$\frac{d\pi}{dQ} = -4.8Q^2 + 10Q + \beta - 180 = 0$$

The solution to the above quadratic equation is

$$Q = \frac{(10 \pm \sqrt{100 - 19.2(180 - \beta)})}{9.6} = \frac{10}{9.6} \pm \frac{\sqrt{100 - 19.2(180 - \beta)}}{9.6} \tag{8.29}$$

The second order condition for a maximum requires that

$$\frac{d^2\pi}{dQ^2} = -9.6Q + 10 < 0 \longrightarrow Q > \frac{10}{9.6}$$

Therefore in (8.29) the right answer must be

$$Q = \frac{10}{9.6} + \frac{\sqrt{100 - 19.2(180 - \beta)}}{9.6} = \frac{10}{9.6} + \frac{\sqrt{19.2\beta - 3356}}{9.6} \tag{8.30}$$

It is clear from (8.30) that for the optimization to have a real solution β must be equal to or greater than 174.792.

To calculate the price, we substitute for Q from (8.30) in (8.28)

$$P = \beta - 25\left(\frac{10}{9.6} + \frac{\sqrt{19.2\beta - 3356}}{9.6}\right) + 0.4\left(\frac{10}{9.6} + \frac{\sqrt{19.2\beta - 3356}}{9.6}\right)^2 \tag{8.31}$$

As demand for the good produced by this firm increases, that is, as the value of β increases and the demand curve shifts up, we *expect* to get higher quantity and prices from the profit optimization routine; that is, $\frac{dP}{d\beta} > 0$. Differentiating (8.31), we have

$$\frac{dP}{d\beta} = 1 - \frac{25}{\sqrt{19.2\beta - 3356}} + 0.083333 + \frac{8}{9.6\sqrt{19.2\beta - 3356}} > 0$$

which can be simplified to

$$\sqrt{19.2\beta - 3356} > \frac{232}{10.4} \longrightarrow 19.2\beta - 3356 > 497.63314$$

And finally

$$\beta > 200.7101 \tag{8.32}$$

Further, given that $f(Q) = 2Q^3 - 30Q^2 + 180Q$ and $g(Q) = -25Q + 0.4Q^2$, Eq. (8.19) can be expressed as

$$(6Q^2 - 60Q + 180)Q - (2Q^3 - 30Q^2 + 180Q) - (-25 + 0.8Q)Q^2 = 0$$

After simplification this equation is reduces to

$$3.2Q^3 - 5Q^2 = 0$$

which has a nonzero solution $\tilde{Q} = 1.5625$, resulting in the price

$$\tilde{P} = AVC(\tilde{Q}) = AVC(1.5625) = 2(1.5625)^2 - 30(1.5625) + 180 \approx 138$$

We determine the intercept of the demand equation corresponding to this (price, output) combination by using (8.31)

$$\beta = 138 + 25(1.5625) - 0.4(1.5625)^2 \longrightarrow \beta = 176.086$$

Thus the demand equation associated with (price, output)of (138, 1.5625) is

$$P = 176.086 - 25Q + 0.4Q^2 \tag{8.33}$$

Implication of (8.30), (8.32) and (8.33) is that if β is below 174.792 there is no real solution to the profit maximization. If β is larger than 174.792 but below 176.098, there is a real solution to the model, but the combination of price and output is such that the total revenue would be less than the total variable cost, so our firm has no incentive to operate. For values of β in the range of 176.098 and 200.714, the optimization results in higher outputs but lower prices, and only for values of β greater than 200.714 the optimization leads to higher outputs and higher prices.

Since the (price, output) combination (138, 1.5625) is such that the firm's total revenue of 215.63 is equal to its total variable cost, we call the demand function (8.33) the shutdown demand and 176.086 the shutdown β. If demand condition deteriorates such that the demand falls below (8.33), i.e. its intercept becomes smaller than 176.086, the firm's loss is more than the fixed cost and there is no reason for the firm to continue operation. However, interestingly, a price of 138 associated with this shutdown demand is not the shutdown price.

To determine the shutdown price and its associated demand function we use (8.22) and write

$$P = 5.2Q^2 - 35Q + 180 \tag{8.34}$$

The minimum of this function is

$$\frac{dP}{dQ} = 10.4Q - 35 = 0 \longrightarrow Q = \frac{35}{10.4} \approx 3.3654$$
$$P = 5.2(3.3654)^2 - 35(3.3654) + 180 \approx 121.1058$$

Since $TR(3.3654) - TVC(3.3654) = 407.57 - 342.23 > 0$ and $121.1058 < 138$ then the shutdown price is 121.1058. It should be clear from this example that, as it was mentioned earlier, the shutdown price for a noncompetitive firm facing a nonlinear demand function may not conform to the definition of shutdown price under perfect competition.

By solving (8.34) for Q, we have

$$Q = \frac{35 \pm \sqrt{(20.8P - 2519)}}{10.4}$$

Given the boundaries of β and their associated prices, the trace function must be expressed as

$$Q = \begin{cases} 0 & \text{for} \quad P < 121.1058 \\ \frac{1}{10.4}(35 \pm \sqrt{20.8P - 2519}) & \text{for} \quad 121.1058 \le P \le 138 \\ \frac{1}{10.4}(35 + \sqrt{20.8P - 2519}) & \text{for} \quad P > 138 \end{cases} \tag{8.35}$$

Figure 8.3 shows the graph of this trace function.

Fig. 8.3 Graph of trace function (8.35)

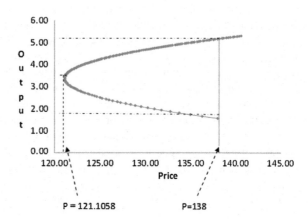

As the trace function (8.35) and graph of the trace function (Fig. 8.3) indicate, in the price range 121.1058–138 there are two possible level of optimal output, each corresponding to a specific demand function. A price of 135, for example, corresponds to $Q = 5$ and $Q = 1.731$. The (price, quantity) combination of (135, 5) is associated with the demand function $P = 250 - 25Q + 0.4Q^2$ and (135, 1.731) with $P = 177.1 - 25Q + 0.4Q^2$.

With the original demand function $P = 400 - 25Q + 0.4Q^2$ this firm's maximum profit combination is $Q = 7.891$ units and $P = 227.625$. This combination generates 861.19 of profit for the firm. If for any reason demand declines, business deteriorates and profit is eliminated, then the firm's normal profit combination will be $Q = 5.579$ units and $P = 146.58$. The dislocated demand equation for normal profit is

$$P = 273.60 - 25Q + 0.4Q^2$$

The graph of this function will be tangent to the ATC curve.

In another example, if $P = 200$ then $Q = 7.26$ and by using (8.18), we have

$$\beta = P' - g[h(P')] = 200 - [-25(7.26) + 0.4(7.26^2)] = 360.43$$

Thus the demand equation associated with a price of 200 would be

$$P = 360.43 - 25Q + 0.4Q^2$$

If we apply the same analysis for determining the shutdown price to Examples 6 and 7—where the demand functions are linear—we see that in both cases the combination of $\ddot{Q}$ and $\ddot{P}$ lead to $TR(\ddot{Q}) - TVC(\ddot{Q}) < 0$, suggesting that $\tilde{P}$ is in fact the correct shutdown price. One can analytically prove that it is indeed always the case for linear demand equations. It is also clear from Example 8, where our firm faced the demand function $P = \beta - 15Q + 0.4Q^2$ instead of $P = \beta - 25Q + 0.4Q^2$, $\tilde{P}$ and not $\ddot{P}$ is the shutdown price.

Deriving the trace of optimal output when a firm faces *semi-log* or *log linear* demand functions in the market is presented in Chap. 9, where logarithms and their applications in economics are discussed. It should be noted that the introduction of nonlinear demand functions into the model may lead to the situation that an analytical derivation of a firm's trace function may not be possible.

An Important Conclusion

The method for deriving the supply function of a firm under perfect competition discussed in the earlier part of this chapter (Sect. 8.1) is a special case of the method of deriving the trace under imperfect competition. For firms under perfect competition price is a given constant. This means that in Eq. (8.17), $g(Q) = 0$ and $P = \beta$.

Therefore the shutdown situation expressed in (8.19) simply leads to the minimum of the AVC function and Eq. (8.23) becomes $f'(Q) - P = 0$, which is the familiar $MC - P = 0$.

8.6 Exercises

1. Assume a competitive industry consists of 300 firms each with the following cost function

$$TC = Q^3 - 10Q^2 + 45Q + 100$$

   If the market demand function is

$$Q_d = 4000 - 20P$$

   what is the market short run equilibrium price and quantity?

2. Determine the long run market equilibrium price and quantity in Problem 1. How many new firms enter the market?

3. Assume a noncompetitive firm with a total cost function

$$TC = Q^3 - 20Q^2 + 150Q + 1000$$

   facing the following linear demand function for its product in the market

$$P = 250 - 2Q$$

   Derive the trace function of the firm.

4. All 150 firms in a competitive industry have the same cost structure captured by the following total cost function

$$TC = 2Q^3 - 60Q^2 + 950Q + 600$$

   If the market demand function is

$$Q_d = 4568.20 - 3.5P$$

   determine the market equilibrium price and quantity.

5. If one of the firms in Problem 4 acquires the other 149 firms in the industry and forms a giant monopoly, what would the market equilibrium price and quantity be?

• **Average Cost Pricing**

Instead of pursuing profit maximization through marginal cost pricing, the regulated natural monopolists are required to follow the average cost pricing. In the average cost pricing, a natural monopolist firm sets its price at the level where the demand curve intersects the average cost ATC curve. This policy, compared to marginal cost pricing, leads to the production of a larger volume of output at a relatively lower price.

6. If the newly formed monopolist in Problem 5 is regulated and required to follow the average cost pricing, what would the market equilibrium price and quantity be?

7. A monopolistically competitive firm has the following total cost function

$$TC = \frac{1}{3}Q^3 - 20Q^2 + 350Q + 100$$

If the demand function for the firm's product is

$$P = 350 - 10Q$$

(a) Derive the firm's trace function.

(b) What is the equation of the dislocated demand curve associated with normal profits?

(c) What is the equation of the dislocated demand curve associated with a price of $206?

8. A monopolist has the following cubic cost function

$$TC = 0.8Q^3 - 32Q^2 + 1000Q + 32000$$

If the market demand function for the firm's product is

$$P = 3400 - 14Q$$

(a) What is the monopolist's profit-maximizing combination of price and quantity?

(b) What is the firm's profit?

(c) Derive the firm's trace function.

(d) What is the equation of the dislocated demand curve associated with normal profits?

(e) What is the equation of the dislocated demand curve associated with a price of $3200?

9. A monopolist has the following cubic cost function

$$TC = 0.2Q^3 - 10Q^2 + 136Q + 100$$

If the market demand function for the firm's product is

$$P = 1000 - 7Q$$

(a) What is the monopolist's profit-maximizing combination of price and quantity?
(b) What is the firm's profit?
(c) Derive the firm's trace function.
(d) What is the equation of the dislocated demand curve associated with normal profits?
(e) What is the equation of the dislocated demand curve associated with a price of $3200?

10. A monopolist has the following cubic cost function

$$TC = 0.08Q^3 - 8.2Q^2 + 2000Q + 32000$$

If the market demand function for the firm's product is

$$P = 3400 - 4Q$$

(a) Derive the firm's trace function.
(b) What is the equation of the dislocated demand curve associated with normal profits?
(c) What is the equation of the dislocated demand curve associated with a price of $3200?

11. A monopolist has the following cubic cost function

$$TC = 0.4Q^3 - 7Q^2 + 70Q + 100$$

If the market demand function for the firm's product is

$$P = 2000 - 10Q$$

What is the monopolist's profit maximizing combination of price and quantity? What is the firm's profit?

12. Due to adverse economic conditions the demand for the good produced by the monopolist in Problem 11 declines to the extent that the firm's above normal profit is eliminated. What is the monopolist's new optimal combination of price and quantity? What is the equation of dislocated market demand curve?

13. A monopolist firm with the following cubic cost function

$$TC = 3Q^3 - 40Q^2 + 250Q + 900$$

faces the following market demand function for its product

$$P = 2000 - 10Q$$

The management of the firm decides to follow a three-stage market strategy,
    1. find the level of output that minimizes its average total cost

    2. find the level of output that maximizes its total revenue

    3. choose the average of the two levels for actual production.

(a) What is the level of output in this strategy compared to the profit-maximizing output?

(b) What is the price level in this case compared to the profit-maximizing case?

14. What is the level of output and price if a natural monopolist with the cost function

$$TC = 0.3Q^3 - 8Q^2 + 120Q$$

and market demand function

$$P = 100 - Q$$

follows the average cost pricing? (From two possible solutions choose the larger one). Compare this solution to the profit-maximizing solution.

15. If one of the firms in Problem 1 acquires the remaining 299 firms and creates a big monopoly, what would happen to the market equilibrium quantity and price?

16. A monopolistically competitive firm has the following total cost function

$$TC = 4Q^2 + 30Q + 400$$

If the demand function for the firm's product is

$$P = 600 - 30Q \qquad Q < 20$$

(a) Find the firm's profit-maximizing level of output, and its profit.

(b) What is the firm's long run (price, quantity) combination?

(c) What is the equation of the dislocated demand curve associated with the long run situation?

17. A monopolistically competitive firm has the following total cost function

$$TC = Q^3 - 10Q^2 + 45Q + 100$$

If the demand function for the firm's product is

$$P = 100 - 10Q$$

(a) Find the firm's profit maximizing level of output, price, and profit.

(b) What is the firm's long run (price, quantity) combination?

(c) What is the equation of dislocated demand curve associated with the long run situation?

18. A competitive industry consists of 500 identical firms with the following total cost function

$$TC = 0.6Q^3 - 3Q^2 + 6Q + 50$$

Assume the market demand function is

$$Q_{md} = 6000 - 10P$$

(a) What is the market supply function?

(b) Determine the short run market equilibrium price and quantity.

(c) What is the profit-maximizing level of output for each firm?

(d) What is the long run market equilibrium price and quantity?

(e) In the long run how many firms enter into (or exit from) the market?

19. Assume all n firms in a competitive industry have the following quadratic total cost function

$$TC_i = aQ_i^2 + bQ_i + c \qquad\qquad i = 1, 2, ..., n$$

Assume the market demand function is

$$Q = \alpha - \beta P$$

(a) Find the market equilibrium price P_C and quantity Q_C.

(b) Now assume that all n firms consolidate into a giant monopoly. Find the market equilibrium price P_M and quantity Q_M.

(c) Show that $Q_M < Q_C$ and $P_M > P_C$.

20. Assume the market demand and supply functions are

$$P = D(Q) \qquad \frac{dP}{dQ} < 0$$

$$P = S(Q) \qquad \frac{dP}{dQ} > 0$$

where $Q > 0$. We form the total revenue function TR

$$TR = PQ = D(Q)Q$$

and the marginal revenue function MR

$$MR(Q) = TR' = \frac{dT}{dQ} = D'(Q)Q + D(Q)$$

Assume $Q_C > 0$ is such that

$$D(Q_C) = S(Q_C) \quad \text{or} \quad D(Q_C) - S(Q_C) = 0$$

(this is the perfectly competitive market equilibrium). Assume $Q_M > 0$ is such that

$$D'(Q_M)Q_M + D(Q_M) = S(Q_M) \quad \text{or} \quad D'(Q_M)Q_M + D(Q_M) - S(Q_M) = 0$$

(this is the monopolized market equilibrium)
Show that $Q_M < Q_C$ and $P_M = D(Q_M) > P_C = D(Q_C)$.

A Sample of Maple Snippets for solving several types of problems

introduced in the last three Chapters.

Maple Snippet 1

(Texts in black are entered in Maple and texts in Blue are Maple's answer. Maple treats text after # sign as comment)

\# *This problem asks for finding the market equilibrium price and*

\# *quantity, given the demand Qd and supply Q_s functions.*

\# *Enter Qd and Q_s*

$$Qd := \frac{700}{(2\,P + 3)} - 10$$

$$\frac{700}{2\,P + 3} - 10$$

$$Qs := (2\,P + 15)^{0.7}$$

$$(2\,P + 15)^{0.7}$$

\# *Graph of the supply and demand functions*
$plot([\,Qs, Qd\,], P = 3 ..20, color = black\,)$

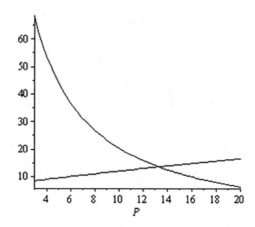

\# *Finding the equilibrium price*

$P := fsolve\,(Qs = Qd, P)$

$$13.3263503$$

Determining the equilibrium quantity
$Qe := Qs = Qd$

$$13.6066187 = 13.6066187$$

Maple Snippet 2

#Maple snippet for solving another market equilibrium model

$$Qd := 1000\left(\frac{(P + 25)}{(P^2 + 5)} \right)$$

$$\frac{1000\,(P + 25)}{P^2 + 5}$$

$Qs := 100\,P^{0.4}$

$$100\,P^{0.4}$$

#Plot the supply and demand function
$plot([Qd, Qs], P = 1..20, color = black)$

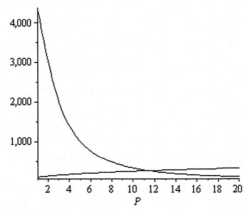

Form the excess demand function
$Exd := Qd - Qs$

$$\frac{1000\,(P + 25)}{P^2 + 5} - 100\,P^{0.4}$$

Graph of the excess demand function

$plot(Exd, P = 1..20)$

plot (Exd, P = 1..20) means graph Exd equation for values of P
from 1 to 20 with increment of 1

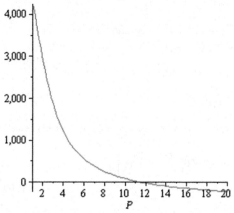

 # As the graph indicates, the market equilibrium is reached when excess
 # demand is eliminated. Set the excess demand function equal to zero and
 # solve for P
fsolve $(Exd = 0, P)$

$$11.5066002$$

 # We get exactly the same answer by
$P := fsolve\,(Qd = Qs, P)$

$$11.5066002$$

#Maple Snippet 3
#How to find maximum or minimum of a function

#Enter the function and select "function definition"

$f(x) := -3\,x^4 + 5\,x^3 + 2\,x^2 + 10\,x + 10$
$$x \rightarrow -3\,x^4 + 5\,x^3 + 2\,x^2 + 10\,x + 10$$

 # Find the first derivate of the function
$df(x) := diff\,(f(x), x)$

$$x \rightarrow \frac{d}{dx}\,f(x)$$

#Set the fist derivate function df(x) equal to zero and solve for x
$fsolve\,(df(x) = 0, x)$

$$1.72381008$$

#*This function has a stationay point at x=1.7238.*
#*We need the SOC. Take the second derivative of the function*
ddf (x) := diff (f(x), x$2)

$$x \rightarrow \frac{d^2}{dx^2} f(x)$$

#*Evaluate the second derivative at 1.7238*
$$ddf(x)\Big|_{x = 1.7238}$$

$$-51.259511$$

#*Since the second derivative evaluated at the stationary point is*
negative then the function has a maximum at this point.

#*Find value of the function at this pont   f(1.7238)*

$$32.3030195$$

#*plot the function*
plot(f(x), x = -2..3, color = black)

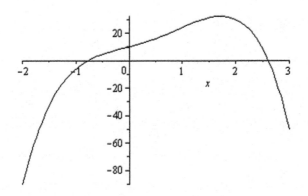

Maple Snippet 4

#*A Maple snippet to show that the marginal cost curve cuts*
#*the average total cost curve at its minimum point*
#*Enter the TC function*

$TC := 0.2\,Q^3 - 7\,Q^2 + 136\,Q + 100$

$$0.2\,Q^3 - 7\,Q^2 + 136\,Q + 100$$

$ATC := \dfrac{TC}{Q}$

$$\dfrac{0.2\,Q^3 - 7\,Q^2 + 136\,Q + 100}{Q}$$

#Differentiate TC function to get the MC
$MC := diff\,(TC, Q)$

$$0.6\,Q^2 - 14\,Q + 136$$

#Find the intersect of ATC and MC curves
$fsolve\,(ATC = MC, Q)$

$$18.2505635$$

#Now find the minimum of the ATC function

$DATC := diff\,(ATC, Q)$

$$\dfrac{0.6\,Q^2 - 14\,Q + 136}{Q} - \dfrac{0.2\,Q^3 - 7\,Q^2 + 136\,Q + 100}{Q^2}$$

$fsolve\,(DATC = 0, Q)$

$$18.2505635$$

#We have exactly the same Q
#Graph the ATC and MC curves

$plot([ATC, MC], Q = 1\,..35, color = black\,)$

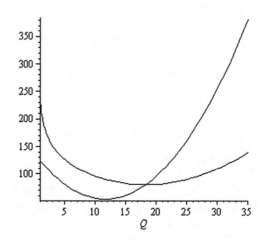

Maple Snippet 5

Maple snippet for determining the long run price in a competitive
market assuming all firms has identical total cost function

$$TC := \frac{1}{12}Q^3 - 2.5\,Q^2 + 30\,Q + 100$$

$$\frac{1}{12}\,Q^3 - 2.5\,Q^2 + 30\,Q + 100$$

Find the average total cost

$$ATC := \frac{TC}{Q}$$

$$\frac{\frac{1}{12}\,Q^3 - 2.5\,Q^2 + 30\,Q + 100}{Q}$$

Determine the minimum of the ATC, take the derivative of ATC
$datc := diff\,(ATC, Q)$

$$\frac{\frac{1}{4}\,Q^2 - 5.0\,Q + 30}{Q} - \frac{\frac{1}{12}\,Q^3 - 2.5\,Q^2 + 30\,Q + 100}{Q^2}$$

Set datc equal to zero and solve for Q
$fsolve\,(datc = 0, Q)$

$$17.0612457$$

#Evaluate ATC at 17.0612457 for the long-run price

$$ATC(Q)\Big|_{Q\,=\,17.0612}$$

$$17.4652978$$

Maple Snippet 6

#Maple snippet to find a competitive firm's supply function
$$TC := Q^3 - 10\,Q^2 + 45\,Q + 100$$

$$Q^3 - 10\,Q^2 + 45\,Q + 100$$

Find the shutdown price by determining the minimum of
average variable cost

$AVC := \dfrac{(TC - 100)}{Q}$

$$\dfrac{Q^3 - 10\,Q^2 + 45\,Q}{Q}$$

$dAVC := diff\,(AVC, Q)$

$$\dfrac{3\,Q^2 - 20\,Q + 45}{Q} - \dfrac{Q^3 - 10\,Q^2 + 45\,Q}{Q^2}$$

$Q1 := fsolve\,(dAVC = 0, Q)$

5.

$AVC(Q)\big|_{Q = Q1}$

20.0000000

#Then the shutdown price is 20
#Determin the marginal cost

$MC := diff\,(TC, Q)$

$$3\,Q^2 - 20\,Q + 45$$

$Qs := solve\,(MC - P = 0, Q)$

$$\dfrac{10}{3} + \dfrac{1}{3}\sqrt{-35 + 3\,P},\ \dfrac{10}{3} - \dfrac{1}{3}\sqrt{-35 + 3\,P}$$

We must reject the second solution, the supply function is

$Qs = \dfrac{10}{3} + \dfrac{\sqrt{3\,P - 35}}{3}$ for $P \geqslant 20$

Maple Snippet 7

From Firm to Industry: A comprehensive
equilibrium model for a competitive market

Assume there are 100 firms with the same cost structure in a
competitive market. If a typical firm's total cost function is

$TC = 0.2\,Q^3 - 6\,Q^2 + 80\,Q + 100$

and the market demand function is
Qmd := 2000 − 5.86 P

We want to determine the market equilibrium price and quantity
and a firm's level of output and profit   in the short run
We first find a firm's supply function

$TC := 0.2\,Q^3 - 6\,Q^2 + 80\,Q + 100$

$$0.2\,Q^3 - 6\,Q^2 + 80\,Q + 100$$

Next we determine the shutdown price by finding the minimum of
the average variable cost   AVC

$AVC := \dfrac{(TC - 100)}{Q}$

$$\dfrac{0.2\,Q^3 - 6\,Q^2 + 80\,Q}{Q}$$

Take the derivative of AVC, set it equal to zero and solve for Q
$dAVC := diff\,(AVC, Q)$

$$\dfrac{0.6\,Q^2 - 12\,Q + 80}{Q} - \dfrac{0.2\,Q^3 - 6\,Q^2 + 80\,Q}{Q^2}$$

$Qsd := fsolve\,(dAVC = 0, Q)$

$$15.00000000$$

Evaluate AVC at Qsd to determine the shutdown price Psd
$AVC(Q)\big|_{Q = Qsd}$

$$35.00000000$$

The shutdown price is 35
 # We next find the marginal cost (MC) function

$MC := diff\,(TC, Q)$

$$0.6\,Q^2 - 12\,Q + 80$$

and derive the supply function for a firm   Qfs
$Qfs := solve\,(MC - P = 0, Q)$

$$10.0 + 0.33333\sqrt{-300 + 15P} \;,\; 10 - 0.33333\sqrt{-300 + 15P}$$

We must reject the second solution, therefore a typical firm's
supply function Qsf is

$Qsf := 10 + \dfrac{1}{3}\text{sqrt}(15\,P - 300)$

$$10 + \frac{1}{3}\sqrt{-300 + 15\,P}$$

Next we aggregate over 100 firms to arrive at the market
supply function Qms

$Qms := 100 \cdot Qsf$

$$1000 + \frac{100}{3}\sqrt{-300 + 15\,P}$$

We can now find market equilibrium price
$Qmd := 2000 - 5.86\,P$

$$2000 - 5.86\,P$$

$Pe := fsolve\,(Qmd = Qms,\, P)$

$$49.9939486$$

Market equilibrium price ≈ 50. The market equilibrium
quantity is

$Qmd(P)\Big|_{P = Pe}$

$$1707.03546$$

$Q^{*} \approx 1707.04$

Each firm's profit maximizing level of output is 17.0704,
which is the total market quantity supplied divided by 100
(number of firms in the market), and the profit is

$Totalcost := TC(Q)\Big|_{Q = 17.0704}$

$$712.098645$$

$Totalrevenue := Pe \cdot 17.0704$

$$853.416700$$

$Profit := Totalrevenue - Totalcost$

$$141.318054$$

Profit > 0 means that firms in this market are making above
"normal" profit. This induces more firms to enter the market
and in the long-run price will be established at the minimum
of the total cost function

$TC := 0.2 \, Q^3 - 6 \, Q^2 + 80 \, Q + 100$
$$0.2 \, Q^3 - 6 \, Q^2 + 80 \, Q + 100$$

$ATC := \dfrac{TC}{Q}$
$$\frac{0.2 \, Q^3 - 6 \, Q^2 + 80 \, Q + 100}{Q}$$

$dATC := diff \, (ATC, Q)$
$$\frac{0.6 \, Q^2 - 12 \, Q + 80}{Q} - \frac{0.2 \, Q^3 - 6 \, Q^2 + 80 \, Q + 100}{Q^2}$$

Firm's normal profit level of output Qn is
$Qn := fsolve \, (dATC = 0, Q)$
$$15.9791167.$$

And the long-run normal profit price Pn is
$Pn := ATC(Q) \big|_{Q = Qn}$
$$41.4499020$$

#We now determine the firms that entered the market.
#Given the long-run price, the longrun market equilibrium quantity
is determined as

$Qmd := 2000 - 5.86 \, Pn$
$$1757.10357$$

#The number of firms in the market Nfirms is then

$Nfirms := \dfrac{Qmd}{Qn}$
$$109.962496$$

round(Nfirms)
$$110$$

#Number of firms in the market ≈ 110
#This means that 10 new firms with the same cost structure entered
#the market

Chapter 9
Logarithmic and Exponential Functions

9.1 Logarithm

The entire algebra of logarithm is based on the following definition:

The *logarithm* of a number a to the base b is a number c such that a can be expressed as b to the power c. Using "log" for logarithm the definition is

$$\log_b a = c \quad \text{such that} \quad a = b^c \tag{9.1}$$

As we will see, there are certain restrictions on values that a and b can take, otherwise a, b, and c are real numbers. The inverse operation of log is *antilogarithm or anti-log*. If c is the log of a, then a is the anti-log of c.

Lets apply the definition of logarithm and find the log of some numbers. What is the log of 10 to the base 10?

$\log_{10} 10 = c$ then $10 = 10^c$ and c clearly must be 1. Hence $\log_{10} 10 = 1$

Similarly $\log_{10} 100 = c$, $100 = 10^c$ and c must be 2. Thus $\log_{10} 100 = 2$.

Notice that $100 = 10^2$ and $\log_{10} 100 = \log_{10} 10^2 = 2$. In the same manner $\log_{10} 1000 = c$ when $1000 = 10^3 = 10^c$ and $c = 3$.

What is $\log_{10} 1$? Let $\log_{10} 1 = c$, then based on the definition (9.1) $1 = 10^c$ and c must be zero. It is easy to verify that Logarithm of 1 at any base is zero.

Lets summarize our findings so far

$$\log_{10} 1 = \log_{10} 10^0 = 0$$
$$\log_{10} 10 = \log_{10} 10^1 = 1$$
$$\log_{10} 100 = \log_{10} 10^2 = 2$$
$$\log_{10} 1000 = \log_{10} 10^3 = 3$$

S. Vali, *Principles of Mathematical Economics*,
Mathematics Textbooks for Science and Engineering 3,
DOI: 10.2991/978-94-6239-036-2_9, © Atlantis Press and the authors 2014

Then it must be the case that $\log_{10} 10^n = n$. This result holds for any base. For base 2

$$\log_2 1 = \log_2 2^0 = 0 \quad \log_2 2 = \log_2 2^1 = 1 \quad \log_2 4 = \log_2 2^2 = 2 \quad \log_2 2^n = n$$

In general

$$\log_a a^n = n \tag{R1}$$

(As before, by R1 I mean Rule number 1. I'll again follow this practice for identifying the log rules)

A logarithm to the base 10 is called the *common logarithm*. It is a widely accepted practice that 'log' is used to denote logarithm to the base 10. In many theoretical and computational areas in science and mathematics, logarithm to the base $e \approx 2.71828$ is used. A logarithm to the base e is called the *natural log* or Napierian log (in honor of Scottish mathematician, physicist and astronomer John Napier who invented the logarithm). It has its own symbol 'ln'.[1] In music, computer science, and information theory logarithm to the base 2 has wide application. 'lg' is a symbol commonly used for the base 2, or binary, logarithm.

The main restriction on b, base of the logarithm, is that it cannot be zero or a negative number. Lets recheck the definition given in (9.1). No number can be expressed as a power of zero, then 0 cannot be the base. As for a negative base, it is true that the logarithm of some numbers to a negative base may exist, but this is not true for all real numbers. The logarithm of 100 to the base -10, for example, is equal to 2, but the logarithm of 10 to the base -10 does not exist. Therefore the base of a logarithm must be strictly positive.

The logarithm for a negative number is not defined; with $b > 0$ in (9.1), no c exists that can produce $b^c < 0$. Therefore, a must also be a positive number. Only positive real numbers have real logarithm.

Numbers between zero and 1 have negative logarithm. If $0 \le a \le 1$ then $\log_b a \le 0$. For example

$$\log 0.1 = \log \frac{1}{10} = \log 10^{-1} = -1 \qquad \text{(applying R1)}$$

$$\text{Similarly} \quad \log 0.01 = \log \frac{1}{100} = \log 10^{-2} = -2$$

$$\log 0 = \log 10^{-\infty} = -\infty$$

All the following rules and identities of "logarithm" are derived from simple application of the log definition stated in (9.1):

---

[1] Swiss mathematician Euler (pronounced "Oiler") introduced the natural logarithm. If you wonder why the symbol for 'natural logarithm' is not 'nl', just remember that for the French speaking mathematician it was "le logarithm natural". In which case, 'ln' makes sense.

1. If X and Y are real numbers, then the log of their product is sum of their logs.

$$\log_a XY = \log_a X + \log_a Y$$

Let

$$\log_a X = b \quad \text{and} \quad \log_a Y = c \tag{9.2}$$

Using the log definition we can write $X = a^b$ and $Y = a^c$ and subsequently $XY = a^b a^c = a^{b+c}$.

Taking log to the base a from both sides of $XY = a^{b+c}$ gives us

$$\log_a XY = \log_a a^{b+c} = b + c \quad \text{(using R1)}$$

and by substituting for b and c from (9.2), we have

$$\log_a XY = \log_a X + \log_a Y \tag{R2}$$

2. By using rule 2 (R2), we can express a more general form of rule 1 (R1), that is

$$\log_a X^n = n \log_a X \tag{R3}$$

Note that X^n is a short hand for $\underbrace{X * X * X * \cdots * X}_{n}$, X multiplied by itself n times.

By rewriting X^n as product of nXs and applying (R2) successively we will arrive at (R3).

$$\log_a X^n = \log_a \underbrace{(X * X * X \cdots * X)}_{n} = \underbrace{\log_a X + \log_a X + \log_a X + \cdots + \log_a X}_{n}$$

then $\quad \log_a X^n = n \log_a X$

3. It is now easy to show that

$$\log_a \frac{X}{Y} = \log_a X - \log_a Y \tag{R4}$$

Rewriting $\log_a \dfrac{X}{Y}$ as $\log_a (XY^{-1})$ and then applying (R2) and (R3) we get

$$\log_a \frac{X}{Y} = \log_a (XY^{-1}) = \log_a X + \log_a Y^{-1} = \log_a X - \log_a Y$$

4. It is also easy to see that

$$\log_a \sqrt{X} = \log_a X^{\frac{1}{2}} = \frac{1}{2} \log_a X$$

In general we can write the log of the nth root of X^m ($\sqrt[n]{X^m}$) as

$$\log_a \sqrt[n]{X^m} = \log_a X^{\frac{m}{n}} = \frac{m}{n} \log_a X \qquad \text{(R5)}$$

5. What is $10^{\log X}$? Let $10^{\log X} = Y$. By taking log from both sides, we get $\log 10^{\log X} = \log X \log 10 = \log X = \log Y$ leading to $X = Y$ which in turn means $10^{\log X} = X$. We can generalize this result as

$$a^{\log_a X} = X \qquad \text{(R6)}$$

As an special case if $a = e$ then $e^{\ln X} = X$

The following examples should help with establishing a better understanding of the rules of logarithms and their application.

1. Find the exact value of $\log_8 4 + \log_8 16$.

Using (R2), we can write

$$\log_8 4 + \log_8 16 = \log_8 (4*16) = \log_8 64 = \log_8 8^2 = 2\log_8 8 \quad \text{(using R3)}$$

Since $\log_8 8 = 1$ then $\log_8 4 + \log_8 16 = 2$. Or alternatively, using (R2) and (R4)

$$\log_8 4 + \log_8 16 = \log_8 \left(\frac{8}{2}\right) + \log_8 (2*8) = \log_8 8 - \log_8 2 + \log_8 2 + \log_8 8 = 2$$

2. Find $\log_7 1715 - \log_7 5$.

Using R4, we have

$$\log_7 1715 - \log_7 5 = \log_7 \frac{1715}{5} = \log_7 343 = \log_7 7^3 = 3$$

3. Find $\lg 5 - \lg 90 + 2\lg 3$ (remember $\lg$ is $\log_2$)

Using a combination of (R2), (R3), and (R4), we have

$$\lg 5 - \lg 90 + 2\lg 3 = \lg \frac{5}{90} + \lg 3^2 = \lg \left(\frac{1}{18}\right) + \lg 9 = \lg \left(\frac{1}{18}*9\right) = \lg \left(\frac{1}{2}\right) = -1$$

4. Expand $\log \left(\frac{w^2 x^3}{y}\right)$.

$$\log \left(\frac{w^2 x^3}{y}\right) = \log w^2 x^3 - \log y = \log w^2 + \log x^3 - \log y = 2\log w + 3\log x - \log y$$

5. Find $\log_6\left(\dfrac{1}{36}\right)$.

$$\log_6\left(\frac{1}{36}\right) = \log_6\left(\frac{1}{6^2}\right) = \log_6 6^{-2} = -2$$

6. The Richter Scale or Magnitude Scale provides a number for quantifying the amount of energy released by an earthquake. The scale is a base 10 logarithm of the severity of an earthquake vibration. On December 26, 2003, the Bam earthquake, which devastated the historic city of Bam in Iran, measured 7.0 on the Richter scale. The 1906 San Francisco earthquake that destroyed most of the city measured at 8.25. How many times more severe was the San Francisco quake compared to the Bam quake?

Denoting the severity of the San Francisco and Bam earthquake by S_{SF} and S_{Bam} respectively, we have

$$\log S_{Bam} = 7.0 \quad \text{and} \quad \log S_{SF} = 8.25 \quad \text{then}$$
$$S_{Bam} = 10^7 \quad \text{and} \quad S_{SF} = 10^{8.25} = 10^7 * 10^{1.25} = 17.78 * 10^7 = 17.78 S_{Bam}$$

That is, the San Francisco earthquake was 17.78 times more severe than the Bam earthquake.

7. Find the value of $\log_9 3$.

We first write $\log_9 3 = \log_9 \sqrt{9} = \log_9(9^{1/2})$ and then apply (R3). Therefore,

$$\log_9 3 = 1/2 \log_9 9 = 1/2$$

8. Find the value of $\log_{16} 8$.

$$\log_{16} 8 = \log_{16}(16/2) = \log_{16} 16 - \log_{16} 2 = 1 - \log_{16} 2$$

Since 2 is the fourth root of 16, we have

$$2 = \sqrt[4]{16} = 16^{\frac{1}{4}} = 16^{0.25}$$

therefore,

$$\log_{16} 8 = 1 - \log_{16} 2 = 1 - \log_{16} 16^{0.25} = 1 - 0.25 = 0.75.$$

9.1.1 Change of Base Rule

We know that $\log 45 = 1.65321$ and $\log 2 = 0.30103$. Can we use this information and determine $\lg 45$ (remember $\lg 45$ means $\log_2 45$)? We can, by using change of base rule.

Let $\lg 45 = X$ so $45 = 2^X$. Now we take log of both sides to the base 10

$\log 45 = \log 2^X = X \log 2$ (using R3). Therefore $X = \dfrac{\log 45}{\log 2} = 5.4919$

That is, $\log_2 45 = 5.4919$. In general, the change of base rule can be expressed as

$$\log_b X = \frac{\log_a X}{\log_a b} \tag{R7}$$

Before the advent of computer and calculator, the change of base rule eliminated need for preparing logarithm tables for different bases. Even today most calculators only offer the 'log' and 'ln' functions, which still make the change of base rule a computational necessity.

9.1.2 Characteristic and Mantissa

By knowing that the log of 5.40 is 0.7324, we can easily compute the log of 540, by writing it as

$$\log 540 = \log(100 * 5.40) = \log 100 + \log 5.40 = 2 + 0.7324 = 2.7324$$

The integer part of $\log 540$, which is 2, is called the *characteristic* and the fractional part, 0.7324, is referred to as the *mantissa*. By knowing the log of 5.40 (the mantissa) the log of any number of the form $10^C * 5.40$, where C is an integer number, is simply $C + 0.7324$. For example

$$\log 5400 = \log(10^3 * 5.40) = \log 10^3 + \log 5.40 = 3 + 0.7324 = 3.7324 \quad \text{or}$$
$$\log 0.540 = \log(5.40/10) = \log(10^{-1} * 5.40) = 0.7324 - 1 = -0.2676$$

Similarly
$$\log 0.054 = \log(5.40 * 10^{-2}) = 0.7324 - 2 = -1.2676$$

9.2 Logarithmic Functions

A function of the form $y = \log f(x)$, where $f(x)$ is any algebraic function of x, is called a *logarithmic function*. $y = \log x$, $y = \ln(x+2)$, and $y = \log(x^2 + 3x) - 5$ are all examples of logarithmic function. Since logarithm is not defined for non-positive

Fig. 9.1 Graph of log(x)

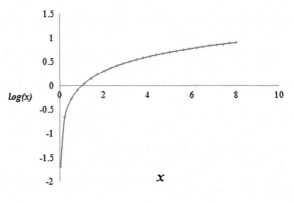

Fig. 9.2 Graph of log(x^2) − 1

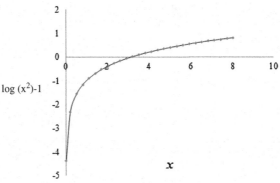

numbers, then the value of algebraic expression of the function that is logged must be positive real number. Figure 9.1 is the graph of a simple log function $y = \log x$. Notice that the curve intersect the x-axes at $x = 1$, where the value of the function is zero. For $x < 1$, the function is negative, and for $x > 1$ it is positive. $x = 1$ is the zero of the function.

Figure 9.2 shows the graph of log function $y = \log x^2 - 1$

At what value of x is the function $y = \log x^2 - 1$ zero? To determine the zero of the function we must solve the logarithmic equation $\log x^2 - 1 = 0$

$$\log x^2 - 1 = 0 \longrightarrow \log x^2 = 1 \quad \text{and} \quad x^2 = 10^1 = 10$$

subsequently $x = \sqrt{10} \approx 3.162$. As the graph indicates $y < 0$ for $0 < x < 3.162$, $y = 0$ for $x \approx 3.162$ and $y > 0$ for $x > 3.162$

9.3 Exponential Functions

A function of the form $y = ca^{f(x)}$ where $a, c > 0$ are constants and $f(x)$ is any function of x is called an *exponential function*. $y = 3^x$ is an example of an exponential function. In exponential functions the variable x, or a function of x, appears as the exponent of positive constant a. We should carefully distinguish between a *power function*, where the base is a variable and the exponent is a constant (e.g. $y = x^3 + 2x$), and an exponential function, where the base is a constant and the exponent is a variable. Generally when a variable and not a function of a variable appears as the exponent, the exponential function is called a *simple exponential function* (e.g. $y = 3^x$). When the number e is used as the base, the function is called a *natural exponential function*.

We should immediately recognize the relationship between exponential and logarithmic functions; they are the inverse of each other. Consider the simple logarithmic function $y = \log x$. From the definition of a logarithm it must be true that $x = 10^y$, which is an exponential function. Figure 9.3 shows the graph of the exponential function $y = 2^x$ for different values of x.

While generally there is no restriction on the domain of an exponential function (unless imposed as a requirement of the model), the range of the function is invariably nonnegative;

$$\text{Domain} = \{x \in R \mid -\alpha < x < \alpha\}$$
$$\text{Range} = \{y \in R \mid 0 \le y < \alpha\}$$

It is clear from Fig. 9.3, the graph of our example $y = 2^x$, that as x takes on larger and larger negative values, 2^x becomes smaller and smaller positive numbers, approaching zero. When, for example, $x = -2$ the value of the exponential function

$$y = 2^{-2} = \frac{1}{2^2} = \frac{1}{4} = 0.25$$

But when $x = -4$

$$y = 2^{-4} = \frac{1}{2^4} = \frac{1}{16} = 0.0625$$

Fig. 9.3 Graph of $y = 2^x$

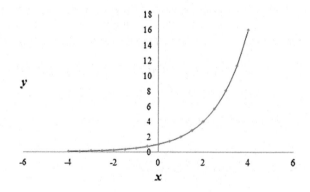

Fig. 9.4 Graph of $y = 2^x$ for $x \geq 0$

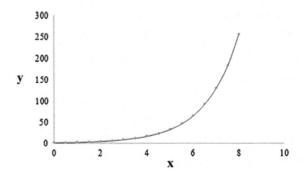

Fig. 9.5 Graph of $y = 0.5^x$

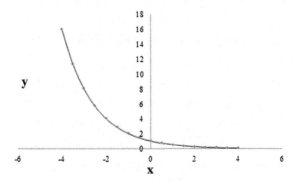

Given the fact that with very few exceptions economic variables are non-negative, it is realistic to restrict the domain of exponential functions to positive real numbers. Figure 9.4 is the graph of the exponential function $y = 2^x$ with values of $x \geq 0$.

When the base of an exponential function is greater than 1, the values of the function rapidly increase as x takes larger values. The graph of $y = 2^x$ in Figs. 9.3 and 9.4 displays rapid grow of the function as x increases. Conversely, when $0 < a < 1$, the values of the function rapidly decline as x increases. The graph of $y = 0.5^x$ in Figs. 9.5 and 9.6 exhibits the rapid decline of the function as x grow larger. The case when $a = 1$ is a trivial and uninteresting one in that $y = 1$ for any value of x.

Note that a, the base of exponential function $y = ca^{f(x)}$ can not be negative. There is a simple reason for this restriction. By taking the log of both sides of $y = ca^{f(x)}$, we get $\log y = \log c + f(x) \log a$. If $a < 0$ then $\log a$ is not defined.

Examples

1. **Growth factor**: It is estimated that the world population grew at an average annual rate of 1.37 % in the twentieth century. The world population in 1900 was estimate to be around 1650 millions (1.65 billions). If we denote the word population in 1900 by N_0, the population in 1901, N_1, is given as

$$N_1 = N_0 + 0.0137 * N_0 = N_0(1.0137)$$

Fig. 9.6 Graph of
$y = 0.5^x$ for $x \geq 0$

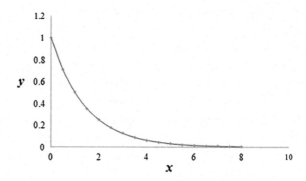

Similarly the word population in 1902, N_2

$$N_2 = N_1 + 0.0137 * N_1 = N_1(1.0137) = N_0(1.0137)^2$$

and in general at time t

$$N_t = N_0(1.0137)^t = 1650(1.0137)^t \qquad t = 0, 1, 2, \ldots$$

which is an example of exponential growth model.

It should be noted that the crucial step in developing this exponential model is the assumption of a constant annual rate of growth of the worlds population. This assumption leads to the estimate of 1.0137 as the annual growth factor. What was the world population at the end of the twentieth century?

$$N_{1999} = N_{100} = 1650(1.0137)^{100} = 6433 \text{ millions or } 6.433 \text{ billions}$$

In what year did the world population grow to twice the population in the year 1900? In this case
$$N_t = 1650(1.0137)^t = 2N_0 = 2 * 1650 = 3300$$

and we have
$$3300 = 1650(1.0137)^t \longrightarrow (1.0137)^t = 2$$

To solve for t we take the log of the both sides

$$t \ln(1.0137) = \ln 2 \longrightarrow 0.01361t = 0.6931 \quad \text{and thus} \quad t = 50.9 \approx 51$$

That is in year 1951 the population grow to twice the size of the population in 1900.

2. **Decay factor**: Assume that the annual inflation rate is 3%. What is the value of $1 after 10 years?

Here due to inflation money loses its purchasing power or value by 3% every year. Let's calculate the value of a dollar $ at the end of the first few years,

$$\$_1 = 1 - 0.03 = 0.97$$
$$\$_2 = \$_1 - 0.03\$_1 = (0.97)\$_1 = (0.97)^2$$
$$\$_3 = \$_2 - 0.03\$_2 = \$_2(0.97) = (0.97)^3 \quad \text{and in general}$$
$$\$_t = (0.97)^t \quad \text{for} \quad t = 1, 2, 3, \ldots$$

Therefore, a dollar $ after 10 years is worth only $(0.97)^{10} \approx 0.74$ or 74 cents.

After how many years does the value of a dollar drops to 50 cents? Here we must determine the value of t that makes

$$\$_t = 0.50 = (0.97)^t$$

By taking the log of the both sides of $0.50 = (0.97)^t$, we have

$$\ln 0.50 = t \ln 0.97 \longrightarrow -0.693 = -0.0305t \quad \text{and} \quad t = 22.7 \text{ years}$$

3. Table 9.1 records the nominal Gross Domestic Product of the United States from 1978 to 2007 (data from the *Economic Report of the President, 2008* and in $billion). Figure 9.7 is the graph or time plot of GDP—GDP plotted against time—and shows the movement of GDP over the 30 year span from 1978 to 2007. The graph's upward curve indicates that an exponential function would provide a reasonable model for the growth of GDP over time.

Table 9.1 US GDP from 1978 to 2007

Year	GDP	Year	GDP
1978	2,294.70	1993	6,657.40
1979	2,563.30	1994	7,072.20
1980	2,789.50	1995	7,397.70
1981	3,128.40	1996	7,816.90
1982	3,255.00	1997	8,304.30
1983	3,536.70	1998	8,747.00
1984	3,933.20	1999	9,268.40
1985	4,220.30	2000	9,817.00
1986	4,462.80	2001	10,128.00
1987	4,739.50	2002	10,469.60
1988	5,103.80	2003	10,960.80
1989	5,484.40	2004	11,685.90
1990	5,803.10	2005	12,433.90
1991	5,995.90	2006	13,194.70
1992	6,337.70	2007	13,843.80

Fig. 9.7 Time plot of GDP

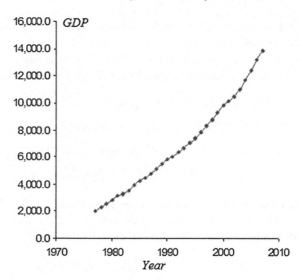

To formulate the model we assume that the US economy grow at a stable annual rate from 1978. To find an exponential model to fit the data in Table 9.1, we must first determine the values of c and a in the formula $GDP_t = ca^t$. Here t denotes time and takes value of 1 for 1978, 2 for 1979, ..., and 30 for 2007. We offer two different methods for calculating the parameters of this model.

Method One: From Table 9.1, we have

$$GDP_1 = ca^1 = 2294.7 \tag{9.3}$$

$$GDP_{30} = ca^{30} = 13843.8 \tag{9.4}$$

By dividing (9.4) by (9.3), we have

$$\frac{ca^{30}}{ca} = \frac{13843.8}{2294.7} = 6.033$$

Parameter c cancels in this ratio and after simplification, we have $a^{29} = 6.033$. To determine a we take the log of both sides,

$$\log(a^{29}) = \log 6.033 \longrightarrow 29 \log a = 0.7805 \longrightarrow \log a = \frac{0.7805}{29} = 0.0269$$

From $\log a = 0.0269$ we can write $a = 10^{0.0269}$ (this is called the *anti-log*) which leads to the value $a = 1.0639$. Now that we have the value of a, we can determine the value of c by substituting for a in (9.3)

$$c(1.0639) = 2294.7 \longrightarrow c = \frac{2294.7}{1.0639} = 2156.88$$

Thus, the estimated exponential function for the nominal GDP from 1978 to 2007 is

$$GDP_t = 2156.88(1.0639)^t \tag{9.5}$$

The growth factor of 1.0639 in (9.5) implies an annual rate of growth of 6.39 % in the nominal GDP.

Method Two: This method is based on estimating the parameters of the exponential function using regression. We must first linearize the exponential function and then apply the technique for estimation of a linear regression model. By taking the log from both sides of the exponential function $GDP_t = ca^t$ we convert it to a linear function.

$$\ln GDP_t = \ln(ca^t) = \ln c + \ln a^t = \ln c + t \ln a$$

By renaming $y_t = \ln GDP_t$; $\beta_0 = \ln c$ and $\beta_1 = \ln a$ we have

$$y_t = \beta_0 + \beta_1 t \qquad t = 1, 2, 3, \ldots, 30 \tag{9.6}$$

Equation in (9.6) is a presentation of a simple linear trend model and its estimate is

$$y_t = 7.7703 + 0.0587t$$

In the estimated equation $\beta_0 = \ln c = 7.7703$ and $\beta_1 = \ln a = 0.0587$. To determine c and a we take the anti-logs

$$\ln c = 7.7703 \longrightarrow c = e^{7.7703} = 2369.18 \quad \text{and}$$
$$\ln a = 0.0587 \longrightarrow a = e^{0.0587} = 1.0605$$

with the estimate of the exponential function finally expressed as

$$GDP_t = 2369.18(1.0605)^t \tag{9.7}$$

Alternatively we can express GDP as a *natural* exponential function of time by writing (9.7) as

$$y_t = \ln GDP_t = 7.7703 + 0.0587t \longrightarrow GDP_t = e^{7.7703+0.0587t}$$
$$GDP_t = (e^{7.7703})(e^{0.0587t}) = 2369.18e^{0.0587t} \tag{9.8}$$

Note that (9.7) and (9.8) are exactly the same since we can write

$$e^{0.0587t} = (e^{0.0587})^t = (1.0606)^t$$

(9.7) and (9.8) are two different expression of the same exponential model, one with the base $a = 1.0606$ and the other with the base $e \approx 2.71828$.

Now we have two competing models, (9.5) and (9.7) (or 9.8). In order to decide which model is more accurate, we must compute a measure of forecasting or estimation accuracy, like *Mean Absolute Percentage Error* (MAPE),[2] for both models. It turned out that MAPE of (9.5) is 9.5 % and that of (9.8) is 6.6 %, leading to the selection of (9.8) as a relatively more accurate exponential model of the US economy. Based on (9.8) the nominal GDP of the United States in 2008 would be

$$GDP_{2008} = GDP_{31} = 2369.18(1.0605)^{31} = 2369.18(6.196) = 14,679.1$$

or about 14.5 trillion dollars.

4. Table 9.2 shows the decline in sales of Long Playing (LP) records, due to rise of Compact Discs (CD) players, from 1982 to 1993.[3] Figure 9.8, a time plot of data represented in Table II, exhibits this dramatic decline in sales of LP records over time. Figure 9.9 is a scatter plot of LP sales against CD sales, indicating an exponential decay of LP sales as CD sales grow.

We can express LP sales as an exponential function of CD sales by

$$LP_t = ca^{CD_t} \tag{9.9}$$

By taking log of both sides of (9.9), we transform the exponential function to a linear function in logarithmic form

$$\ln LP_t = \ln ca^{CD_t} = \ln c + CD_t \ln a$$

Table 9.2 Sales of LP records and CD (in $millions)

Year	LP sales	CD sales
1982	244	0
1983	210	0.8
1984	205	5.8
1985	167	23
1986	125	53
1987	107	102
1988	72	150
1989	35	207
1990	12	287
1991	4.8	333
1992	2.3	408
1993	1.2	495

[2] For a detail discussion of measures of estimation accuracy, see the Appendix to Chap. 11.

[3] This problem is taken from *Functions Modeling Change* by Connaly et al. 2004 John Wiley and Sons. The data is originally from the Recording Industry Association of America, Inc., 1998.

Fig. 9.8 Sales of LP records ($million)

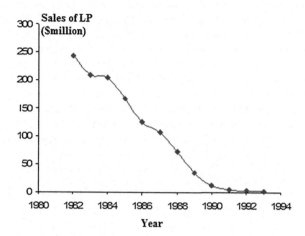

Fig. 9.9 Scatter plot of LP and CD sales

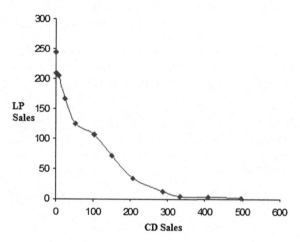

After renaming $y_t = \ln LP_t$, $\beta_0 = \ln c$, and $\beta_1 = \ln a$, we can have a simple linear regression model

$$y_t = \beta_0 + \beta_1 CD_t \qquad \text{for} \quad t = 1982, 1983, \dots, 1993$$

The estimate of the above model gave the following results

$$y_t = 5.5189 - 0.01092 CD_t$$

Writing the model in its original exponential format, we have

$$LP_t = e^{5.5189} - 0.01092CD_t = (e^{5.5189})(e^{-0.01092CD_t})$$

$$LP_t = 249.36e^{-0.01092CD_t} \quad \text{or alternatively}$$

$$LP_t = 249.36(0.989)^{CD_t}$$

It should be noted that exp(x) is widely used as a notational equivalent to e^x. The use of "exp" makes it very convenient to write—and type set—complex exponential/logarithmic expressions. For example, the expression

$$LP_t = e^{5.5189 - 0.01092CD_t}$$

can be more easily written as

$$LP_t = \exp(5.5189 - 0.01092CD_t)$$

$$= \exp(5.5189) \exp(-0.01092CD_t)$$

$$= 249.36 \exp(-0.01092CD_t)$$

The logarithmic and exponential functions have numerous applications in many different fields. A more detailed discussion of average and continuous rates of growth, along with simple and compounding interest rates and their applications will be found in Chap. 12 *Mathematics of Interest Rate and Finance.*

Exercises

1. Find the value of each expression without using a calculator

   (a) $\log 10000$ (b) $\lg 32$ (c) $\log_7 49$

   (d) $\log_8 2$ (e) $\ln e^2$ (f) $\ln e^{-20}$

   (g) $\log_5 \frac{1}{5}$ (h) $\log_6 \frac{1}{36}$ (i) $\log_1 23 + \log_1 248$

   (j) $\log 0.001$ (k) $\log \sqrt{10}$ (l) $\log \frac{1}{\sqrt{10}}$

   (m) $\lg 2^3 \, 2^5$ (n) $e^{\ln 2}$ (o) $\ln(\frac{1}{\sqrt{e}})$

2. If $10^{1.8} = 63.095$, then find the value of $\log 63.095$. What is $\log 6309.5$?

3. If $y = \log x$ what is $\log \sqrt{x}$ in terms of y?

4. Solve the following equations

(a) $20(2)^{2x} = 80$

(b) $e^{3x+1} = 65$

(c) $8e^{0.15x} = 20(1.5)^x$

(d) $\log(x+1) - \log(3x) = 0.2$

(e) $\lg(x-2) + \lg(2x+3) = 2$

(f) $\exp(t+2) \exp(2t) = 125$

(g) $5(3.4)^{x+2} = (2.5)^{2x}$

(h) $\ln(5-3x) = 4$

(i) $\log(3x-1) = 5 - \log x$

(j) $e^{2-6x} = 150$

(k) $\ln(2\sqrt{z}+3) = 5$

(l) $\frac{3e^{2x-1}}{2e^{1-2x}} = 10$

(m) $\log_6(2x+4)^2 = 8$

(n) $e^{4x+6} = \log 125$

5. **Doubling Time**: The population of a city at a certain time is P. If the annual rate of growth of the population is 2 %, after how many years does this city's population doubles? What if the rate is 2.5 %? Write a general expression for the doubling time assuming the rate of growth is g percent. Show that the doubling time is only a function of g.

6. **Half Life**: A company has 25 million dollars worth of equipment. This company depreciates the value of the equipment at the rate of 10 % per year. How long does it takes for the value of the equipment to drop to 50 % of its original value? What would be the value of equipment after 25 years? What would be the value of equipment after 25 years as proportion of the original value?

7. Use either methods described in example 3 and write LP and CD sales, discussed in example 4, as exponential function of time.

8. The sales of a department store grow at the rate of 4.5 % annually. In the year 2000 the store reported $125 million in sales. Write an exponential function for sales. What is the volume of sales in year 2012?

9. Assume that the population of a country grows exponentially. Suppose $N_5 = 20$ million and $N_{10} = 22$ million. Write the expression for N_t. What is the average annual rate of growth of the population in this country? Estimate the population for year 20.

10. The real rate of growth of the US economy is, on average, about 2.5 % annually. The Chinese economy grows, on average, at a rate of 8.5 %. According to World Bank data in 2011 the GDP of the United States and China was 14.99 and 7.31 trillion dollar, respectively. How long will it take for China to reach the current size of the US economy? Is it possible for China to overtake the United States in twenty-first century?

11. The number of cell phones (CP) in the United States has been growing according to the following model

$$CP_t = CP_0(1+g)^t$$

where g is the growth rate and CP_0 is the number of cell phones (in thousands) in 1995. After linearizing this model by a log transformation, we have

$$\ln CP_t = \ln CP_0 + t \ln(1 + g)$$

and we obtain the following result as regression estimate

$$\ln CP_t = 8.5 + 0.23t \qquad\qquad t = 0 \quad \text{for} \quad 1995$$

(a) What is the rate of growth of cell phones?

(b) What was the number of cell phones in use in the year 1995?

(c) Write the exponential equation and estimate the number of cell phones in use for the year 2012.

12. Go to the site of BEA (Bureau of Economic Analysis, US Department of Commerce) www.bea.gov and collect the US nominal GDP figures for years 2008–2012. Now you can extend Table 9.1 in example 3 to include more recent data, specially the years that the economy was in crisis. Use Method One and the extended data and estimate a new exponential function for US GDP similar to (9.5). Is there an appreciable change in the rate of growth of the nominal GDP?

13. [*For those familiar with regression*] Use the extend GDP series from Exercise 12. Implement Method Two of example 3 and estimate the parameters of the GDP growth function.

14. Use your estimated models from Exercises 12 and 13 and forecast the US nominal GDP for 2012. Which model is more accurate?

15. [*For those familiar with regression*] In Method Two of example 3, we offered a procedure for determining c and a in the exponential model of GDP, $GDP_t = ca^t$, fitted to the data in Table 9.1. In the text we denoted t to take the value of 1 for 1978, 2 for 1979, ..., and 30 for 2007. Re-estimate the model by assuming t takes the value of 0 for 1978, 1 for 1979, ..., and 29 for 2007. Do you find any difference between these two models?

9.4 Derivative of Logarithmic and Exponential functions

Consider a simple logarithmic function $y = \ln x$. We are interested in determining the derivative of this function. Recall that the derivative of a function $f(x)$ is defined as

$$f'(x) = \frac{dy}{dx} = \lim_{\Delta x \to 0} \frac{f(x + \Delta x) - f(x)}{\Delta x}$$

Let's apply the definition to our logarithmic function

$$\frac{dy}{dx} = \lim_{\Delta x \to 0} \frac{\ln(x + \Delta x) - \ln x}{\Delta x} \qquad (9.10)$$

Using rules of logarithm, we can write

$$\frac{\ln(x + \Delta x) - \ln x}{\Delta x} = \frac{1}{\Delta x} \ln\left(\frac{x + \Delta x}{x}\right)$$

$$= \frac{1}{\Delta x} \ln\left(1 + \frac{\Delta x}{x}\right)$$

$$= \frac{1}{x} \ln\left(1 + \frac{\Delta x}{x}\right)^{\frac{x}{\Delta x}}$$

By denoting $n = \dfrac{x}{\Delta x}$, we can write

$$\frac{1}{x} \ln\left(1 + \frac{\Delta x}{x}\right)^{\frac{x}{\Delta x}} = \frac{1}{x} \ln\left(1 + \frac{1}{n}\right)^{n}$$

then (9.10) becomes,

$$\frac{dy}{dx} = \lim_{\Delta x \to 0} \left[\frac{1}{x} \ln\left(1 + \frac{1}{n}\right)^{n}\right]$$

But as $\Delta x \to 0$ $n = \dfrac{x}{\Delta x} \to \infty$ and it can be shown that

$$\lim_{n \to \infty} \ln\left(1 + \frac{1}{n}\right)^{n} = e$$

the base of natural logarithm.[4] Therefore,

$$\frac{dy}{dx} = \lim_{\Delta x \to 0} \left[\frac{1}{x} \ln\left(1 + \frac{1}{n}\right)^{n}\right] = \frac{1}{x} \ln e = \frac{1}{x}$$

Thus, if $y = \ln x$ then $y' = \dfrac{dy}{dx} = \dfrac{1}{x}$
The general form of this rule is

---

[4] Try this on your calculator. Start by $n = 10$, change it to 100, then 1000, 10000, and 100000. You'll see that the value of $\left(1 + \frac{1}{n}\right)^{n}$ goes from 2.59374 to 2.70481, 2.71692, 2.71814, and 2.71826. Number e to 5 decimal places is 2.71828. It is clear that larger n becomes closer the result gets to e.

$$y = \ln f(x) \longrightarrow y' = \frac{f'(x)}{f(x)}$$

For example if $y = \ln(x^2 + 3x - 5) \rightarrow y' = \dfrac{2x + 3}{2^x + 3x - 5}$

What is the derivative of the *natural*[5] exponential function $y = e^x$? By taking the natural log of both sides, we have

$$\ln y = x \quad \text{since} \quad \frac{dx}{dy} = \frac{1}{y} \quad \text{then} \quad \frac{dy}{dx} = y = e^x$$

The generalized form of the above rule is

$$y = e^{f(x)} \quad \text{then} \quad \frac{dy}{dx} = f'(x)e^{f(x)}$$

For example

$$y = e^{3x - 2x^2} \longrightarrow y' = (3 - 4x)e^{3x - 2x^2}$$

To find the derivative of $y = \log_a x$ (that is logarithm with base other than e), we first rewrite the function in ln, by applying the change of base rule, and then take the derivative. Recall the change of base rule is

$$\log_a x = \frac{\log_b x}{\log_b a}, \quad \text{therefore}$$

$$y = \log_a x = \frac{\ln x}{\ln a} = \frac{1}{\ln a}\ln x \quad \text{and consequently} \quad y' = \frac{1}{\ln a}\frac{1}{x}$$

In general if $y = \log_a f(x)$ then $y' = \dfrac{f'(x)}{\ln a * f(x)}$

As an example if $y = \log_4(x^2 - 2x)$ then

$$y' = \frac{2x - 2}{\ln 4 \, (x^2 - 2x)} = \frac{2x - 2}{1.386 \, (x^2 - 2x)}$$

Next we find the derivative of general exponential function, an exponential function with base other than e.

What is the derivative of $y = b^x$? By taking the natural log of both sides, we have

$$\ln y = x \ln b \longrightarrow x = \frac{1}{\ln b}\ln y \quad \text{and} \quad \frac{dx}{dy} = \frac{1}{\ln b}\frac{1}{y} \quad \text{then}$$

$$\frac{dy}{dx} = y \ln b = b^x \ln b$$

---

[5] As before, I am using the designation *natural* for exponential function to the base e.

In general if $y = b^{f(x)}$ then $y' = f'(x) b^{f(x)} \ln b$

For example if $y = (15)^{1-2x^2}$ then

$$y' = (-4x)(15)^{1-2x^2} \ln 15 = -10.832x(15)^{1-2x^2}$$

Exercises

1. Differentiate the following functions:

(a) $f(x) = 3\ln(5x)$ (b) $g(x) = \ln(1 - x^2)$

(c) $f(x) = \ln\left(\frac{2x^3 - 2}{x+1}\right)$ (d) $f(y) = \log(y^2 + 3y)$

(e) $f(x) = \log(5x - x^3 + 2)$ (f) $g(z) = \sqrt{z}\ln z$

(g) $f(x) = \sqrt[3]{\ln x^3}$ (h) $w(y) = [\lg(2y + 5)]^3$

(i) $f(z) = \log_8(ax^2 + bx + c)$ (j) $f(x) = \log_{12}\frac{x^2-1}{2x+3}$

(k) $f(x) = 3e^{2x+3}$ (l) $w(z) = (12z + 3)e^{z+5}$

(m) $f(x) = 3(5)^{2x+3}$ (n) $f(x) = 10x^2(6)^{2x+1}$

2. Find the rate of change of the following function at specified value

(a) $f(x) = 5x^2(5)^{2x+1}$ at $x = 5$

(b) $w(y) = [\log(2y + 5)]^3$ at $y = 1$

(c) $f(x) = \ln\frac{2x^2-2}{x+3}$ at $x = 8$

(d) $g(x) = \ln(1 - x^3)$ at $x = 0.1$

(e) $f(x) = 7e^{3x+3}$ at $x = 3$

9.5 Economic Application of Derivative of Logarithm

9.5.1 Exponential Supply and Demand Functions and Market Equilibrium

Example 1

Assume the following market demand and supply functions

$$Q_d = \frac{150}{2P + 5} \qquad\qquad Q_s = Pe^{0.02P}$$

Table 9.3 P and $ED(P)$

P	ED(P)
5.0	4.474
5.5	3.235
6.0	2.059
6.5	**0.931**
7.0	**−0.157**
7.5	−1.214
8.5	−3.257
9.0	−4.253

At equilibrium $Q_d = Q_s$ then

$$\frac{150}{2P+5} = Pe^{0.02P} \longrightarrow \frac{150}{2P+5} - Pe^{0.02P} = 0 \tag{9.11}$$

Root(s) of equation in (9.11) is (are) zero(s) of the excess demand function

$$ED(P) = \frac{150}{2P+5} - Pe^{0.02P} \tag{9.12}$$

(9.12) is a nonlinear function with no obvious zero(s). One approach is the linear or quadratic approximation using Taylor expansion, discussed in Chap. 7. To achieve the highest possible accuracy we must expand the function at a point closest to the zero of the function. Table 9.3 shows various values of P and the corresponding values of $ED(P)$.

When $P = 6.5$ value of the function is positive. At $P = 7$ it is negative. This indicates that the function becomes zero at a value between 6.5 and 7 (at the point that $ED(P)$ curve crosses the horizontal axes or where the supply and demand curves intersect) Fig. 9.10 shows the graph of $ED(P)$.

Let's expand (9.12) around 6.5

$$ED(6.5) = 0.931$$

$$ED'(P) = \frac{-300}{(2P+5)^2} - e^{0.02P} - 0.02Pe^{0.02P}$$

$$ED'(6.5) = -2.2128$$

$$ED''(P) = \frac{1200(2P+5)}{(2P+5)^4} - 0.02e^{0.02P} - 0.02e^{0.02P} - 0.0004Pe^{0.02P}$$

$$ED''(6.5) = 0.15725$$

Fig. 9.10 Graph of excess demand function

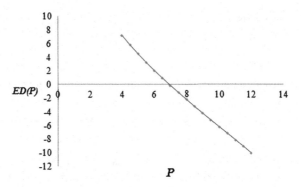

The linear approximation is

$$ED(P) \approx ED(6.5) + ED'(6.5)(P - 6.5)$$
$$ED(P) \approx 0.931 - 2.2128(P - 6.5) \quad \text{and zero of the linear function is}$$
$$ED(P) \approx 0.931 - 2.2128(P - 6.5) = 0 \longrightarrow 15.3142 - 2.2128P = 0$$

and the equilibrium price is $P = 6.92$. By plugging this value for P in the demand function, we get the equilibrium quantity $Q = 7.96$ units. If we use the supply function instead we get a slightly different number for Q. This is partly due to a lot of round off errors and partly to the approximation error. The actual value of zero of the function is 6.926791. The equilibrium quantity for this price calculated from both supply and demand function is exactly the same, 7.956 units.

The quadratic approximation for the excess demand function is

$$ED(P) \approx ED(6.5) + ED'(6.5)(P - 6.5) + \frac{ED''(6.5)}{2}(P - 6.5)^2$$
$$ED(P) = 15.3142 - 2.2128P + \frac{0.15725}{2}(P - 6.5)^2$$
$$ED(P) = 0.07862P^2 - 3.2349P + 18.6361$$

Zero(s) of this function provide the second order (quadratic) approximation,

$$0.07862P^2 - 3.2349P + 18.6361 = 0 \quad \text{or} \quad P^2 - 41.146P + 237.04 = 0$$

This equation has two solutions $P = \dfrac{41.146 \pm 27.292}{2}$. Since the correct solution must be in the vicinity of 6.5, then the acceptable solution is $P = \dfrac{41.146 - 27.292}{2} =$ 6.927, with corresponding $Q = 7.956$. This is clearly an improvement over the linear approximation.

Alternatively, we can use Wolfram-alpha, SAGE, Maple, or Microsoft Mathematics to find zero(s) of (9.12). This function has three zeros, two of them negative numbers that must be discarded. The positive zero is 6.92679, indicating that our quadratic approximation is very good.

Example 2

Consider a market with the following demand and supply functions:

$$Q_d = \frac{5700}{P^{\frac{1}{2}}} \qquad\qquad Q_s = 10P(3)^{0.05P}$$

Here the demand equation is a rational function. The graph of this function is a *rectangular hyperbola*. The supply function is an exponential function.

To find the equilibrium price algebraically we follow, by now, the familiar routine

$$Q_d = Q_s \longrightarrow \frac{5700}{P^{\frac{1}{2}}} = 10P(3)^{0.05P} \longrightarrow 10P^{1.5}(3)^{0.05P} = 5700$$

and finally

$$10P^{1.5}(3)^{0.05P} - 5700 = 0$$

which is equivalent to finding zero of the excess supply function

$$ES(P) = 10P(3)^{0.05P} - \frac{5700}{P^{1/2}}$$

This equation does not yields an analytical solution. Thus we have to turn to approximating methods to find root(s) of the equation. A visual inspection of the Table 9.4 and graph of excess supply function (Fig. 9.11) indicates that the equilibrium price falls in the range of 26–27.

We will drive the linear and quadratic approximation by expanding the function about $P = 26$.

	P	Q_s	Q_d	$ES(P)$
Table 9.4 Values of Q_s, Q_d, and $ES(P)$ for selected values of P	20	1274.559	600.000	−674.559
	21	1243.842	665.574	−578.268
	22	1215.244	736.641	−478.603
	23	1188.532	813.612	−374.920
	24	1163.508	896.926	−266.581
	25	1140.000	987.056	−152.944
	26	1117.862	1084.504	**−33.358**
	27	1096.966	1189.810	**92.844**
	28	1077.199	1303.550	226.352
	29	1058.463	1426.342	367.879
	30	1040.673	1558.846	518.173

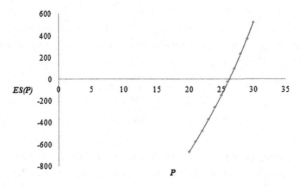

Fig. 9.11 Graph of excess supply function

$$ES(26) = -33.358$$

$$ES'(P) = 10(3)^{0.05P} + 0.5P(3)^{0.05P} \ln 3 + \frac{2850}{P^{1.5}}$$

$$ES'(26) = 122.782$$

$$ES''(P) = 0.5(3)^{0.05P} \ln 3 + 0.5(3)^{0.05P} \ln 3 + 0.025P(3)^{0.05P}(\ln 3)^2 - \frac{4275}{P^{2.5}}$$

$$ES''(26) = 6.615$$

Using the above information, the linear function approximating the nonlinear ES function is

$$ES(P) = -33.358 + 122.782(P - 26) = -3225.69 + 122.782P$$

Zero of this function is

$$-3225.69 + 122.782P = 0 \longrightarrow P = \frac{3225.69}{122.782} = 26.27 \text{ and } Q = 1112.1$$

The quadratic approximation to ES is

$$ES(P) = -33.358 + 122.782(P - 26) + \frac{6.615}{2}(P - 26)^2$$

$$= 3.31P^2 - 49.21P - 989.82$$

$$ES(P) = 0 \longrightarrow 3.31P^2 - 49.21P - 989.82 = 0$$

leading to an acceptable solution $P = 26.26$ and $Q = 1112.3$. Again, the Wolfram-Alpha solution for zero(s) of the excess supply function is $P = 26.2697$.

Example 3

Now let's consider a little more complicated system consisting of the following demand and supply function:

$$Q_d = \frac{100P + 1500}{P^2 + 10}$$
$$Q_s = P(3)^{0.02P} - 10$$

Here demand is a rational function and supply is an *exponential function* with an added negative number. -10 in the supply function ensures that there is a strictly positive price (here $8.33 per unit) at or below which producers will not supply any unit to the market (*See Example 4 below*).

To find the equilibrium price algebraically we follow the familiar routine

$$Q_d = Q_s = \frac{100P + 1500}{P^2 + 10} = P(3)^{0.02P} - 10$$
$$P^3(3)^{0.02P} + 10P(3)^{0.02P} - 10P^2 - 100P - 1600 = 0 \qquad (9.13)$$

Solving (9.13) is equivalent to finding zero(s) of the function

$$f(P) = P^3(3)^{0.02P} + 10P(3)^{0.02P} - 10P^2 - 100P - 1600 \qquad (9.14)$$

As it was demonstrated in Chap. 7, a fairly straight forward strategy for approximating solution to nonlinear supply and demand models is to *linearize* the supply and demand functions over a narrow interval containing the equilibrium price and then solve the resultant linear system for the equilibrium price and quantity. A visual inspection of the Table 9.5 indicates that in this example the equilibrium price falls in the range of 15–16.

	P	Q_s	Q_d	$ED(P)$
Table 9.5 Values of Q_s, Q_d, and $ED(P)$ for selected values of P	11	4.007	19.847	−15.840
	12	5.620	17.532	−11.912
	13	7.298	15.642	−8.344
	14	9.042	14.078	−5.035
	15	10.856	12.766	**−1.910**
	16	12.740	11.654	**1.086**
	17	14.698	10.702	3.996
	18	16.732	9.880	6.852
	19	18.844	9.164	9.680
	20	21.037	8.537	12.500
	21	23.313	7.982	15.330
	22	25.674	7.490	18.184
	23	28.124	7.050	21.074

Note that value of the excess demand *ED* function changes from negative to positive when P moves from 15 to 16, indicating that the excess demand curve cuts the P axis in the interval. At $P = 15$ values of demand, supply, and excess demand functions are 12.766, 10.856, and 1.910 respectively. At $P = 16$ the corresponding values are 11.654, 12.740, and -1.086.

By using this information, we write the linear demand, supply, and excess demand functions over the interval [15, 16] as

$$Q_d = 29.446 - 1.112P$$
$$Q_s = -17.419 + 1.885P$$
$$ED(P) = 46.865 - 2.997P$$

By solving the supply and demand equations simultaneously (or by setting $ED(P) = 0$), we find the equilibrium price $P = 15.6375$, which in turn yields the equilibrium quantity $Q = 12.04$ units.

Next we apply Taylor approximation method to the nonlinear function $f(P)$ in (9.14) above. We first expand $f(P)$ around $P = 15$, where the function is closest to zero, and then find the second degree approximation of zero of the function.

$$f(P) = P^3(3)^{0.02P} + 10P(3)^{0.02P} - 10P^2 - 100P - 1600$$
$$f(15) = -448.878$$
$$f'(P) = 3P^2(3)^{0.02P} + 0.02P^3(3)^{0.02P}\ln3 + 10(3)^{0.02P} + 0.2P(3)^{0.02P}\ln3$$
$$\qquad - 20P - 100$$
$$f'(15) = 660.1052$$

$$f''(P) = 6P(3)^{0.02P} + 0.06P^2(3)^{0.02P}\ln3 + 0.06P^2(3)^{0.02P}\ln3$$
$$\qquad + 0.0004P^3(3)^{0.02P}(\ln3)^2 + 0.2(3)^{0.02P}\ln3 + 0.2(3)^{0.02P}\ln3$$
$$\qquad + 0.004P(3)^{0.02P}(\ln3)^2 - 20$$

which after simplification

$$f''(P) = 6P(3)^{0.02P} + 0.12P^2(3)^{0.02P}\ln3 + 0.0004P^3(3)^{0.02P}(\ln3)^2$$
$$\qquad + 0.4(3)^{0.02P}\ln3 + 0.004P(3)^{0.02P}(\ln3)^2 - 20$$
$$f''(15) = 149.355$$

Denoting the quadratic approximation of $f(P)$ by $f_q(P)$, we have

$$f_q(P) = f(15) + f'(15)(P - 15) + \frac{f''(15)}{2}(P - 15)^2$$
$$= -448.878 + 660.1052(P - 15) + 74.677(P - 15)^2$$
$$= 74.677P^2 - 1580.205P + 6451.869$$

Zero(s) of this quadratic function is

$$74.677P^2 - 1580.205P + 6451.869 = 0$$

From two possible solutions we choose one closer to 15, that is $P = 15.6345$ yielding $Q = 12.04$ units. Given that the actual solution to five decimal places is $P = 15.63345$, the result of linearization method is as good as the quadratic approximation.

Example 4

In the Example 3 the supply function was specified as

$$Q_s = P(3)^{0.02P} - 10$$

For the quantity supplied to be non-negative, we must have

$$P(3)^{0.02P} - 10 \geq 0$$

After moving -10 to the other side of the inequality and taking log of both sides, we have

$$\log P + \log(3)^{0.02P} \geq 1 \longrightarrow \log P + 0.02P \log(3) \geq 1 \text{ or}$$
$$0.009542P + \log P - 1 \geq 0$$

To Solve this inequality we first determine the zero of the function

$$f(P) = 0.009542P + \log P - 1$$

From Table 9.6, zero of the function is between 8 and 9.

We expand $f(P)$ around 8

$$f(8) = -0.0206$$
$$f'(P) = 0.009542 + \frac{1}{\ln 10} \frac{1}{P}$$
$$f'(8) = 0.06383$$

Table 9.6 Values of $f(P) = 0.009542P + \log P - 1$ for selected values of P

P	2	3	4	5	6	7	8	9	10	11
$f(P)$	−0.680	−0.494	−0.360	−0.253	−0.165	−0.088	−0.021	0.040	0.095	0.146

The linear approximation is

$$f(P) = -0.0206 + 0.06383(P - 8)$$
$$f(P) = -0.531 + 0.06383P$$

Zero(s) of this function is

$$-0.531 + 0.06383P = 0 \longrightarrow P = 8.323$$

9.5.2 Profit Maximization for Noncompetitive Firms Facing Semi-Log or Log Linear Demand Functions

Assume a noncompetitive firm with the total cost function $TC = f(Q) + FC$. Further assume that this firm faces a semi-log demand function of the form $P = a - b \ln Q$ in the market. To find the profit maximizing combination of output and price, we write

$$\pi = TR - TC = PQ - f(Q) = aQ - bQ \ln Q - f(Q)$$
$$\frac{d\pi}{dQ} = MR - MC = a - b \ln Q - b - f'(Q) = 0$$

Solution to this equation Q^* is optimal level of output and subsequently $P^* = a - b \ln Q^*$ is the profit maximizing price.

Example 5

Consider a firm with the following cubic cost and semi-log demand function.

$$TC = Q^3 - 10Q^2 + 45Q + 100$$
$$P = 200 - 2 \ln Q$$

The profit function is

$$\pi = TR - TC = 200Q - 2Q \ln Q - Q^3 + 10Q^2 - 45Q - 100$$

and the FOC for a maximum is

$$\frac{d\pi}{dQ} = 200 - 2 \ln Q - \frac{2Q}{Q} - 3Q^2 + 20Q - 45 = 0$$

After simplifying the above equation, we have

$$-3Q^2 + 20Q - 2 \ln Q + 153 = 0$$

Solution to the above mix of quadratic and log function using Wolfram Alpha is $Q = 11.11188$. At $Q = 11.11188$ the SOC for a maximum is satisfied. The optimal price is

$$P = 200 - 2\ln(11.11188) = 195.184$$

This combination generates $11.11188 * 195.185 = 2168.861$ of revenue. The total cost of producing 11.11188 units is 737.323, leaving the firm with 1431.54 of profit.

Now assume that the firm's total cost function is the same as $TC = f(Q) + FC$, but this time firm faces a double log demand function

$$\ln P = \ln a - b\ln Q$$

in the market. This demand function is generally referred to as log-linear demand function and has wide applications in applied macro and micro economics studies. The main problem with linear, quadratic, exponential, and semi-log demand functions is that the elasticity of price or non-price variables, like income, included in the demand equation are not constant. If constancy of own price elasticity and elasticity of non-price variables like income is a desired feature of the model—this is specifically true when a demand model is used for planning and policy formulation—then a log-linear demand function of the form

$$\ln Q = \ln a - b\ln P + c\ln I$$

where the own price and income elasticity of demand are constants '$-b$' and 'c', respectively, must be used. In this case, changes in income 'I' lead to parallel shifts in the demand curve with no change in the income elasticity. We must first show that the price and income elasticity of demand are indeed both constant. We can rewrite our log linear demand function as

$$Q = e^{\ln a - \ln P^b + \ln I^c} = e^{\ln a}\, e^{-\ln P^b}\, e^{\ln I^c} \quad \text{or alternatively} \quad Q = \frac{a\, I^c}{P^b}$$

Recall from Chap. 6 that the own price elasticity of demand $\epsilon(P)$ is

$$\epsilon(P) = \frac{\partial Q}{\partial P}\frac{P}{Q}$$

$$\frac{\partial Q}{\partial P} = \frac{a\, I^c\,(-bP^{b-1})}{P^{2b}} = -\frac{b}{P}\, Q$$

and

$$\epsilon(P) = \frac{\partial Q}{\partial P}\frac{P}{Q} = \left(-\frac{b}{P}\, Q\right)\left(\frac{P}{Q}\right) = -b$$

Similarly, we can show that the income elasticity of demand

$$\epsilon(I) = \frac{\partial Q}{\partial I} \frac{I}{Q} = cI^{c-1} \left(\frac{a}{P^b}\right) \left(\frac{I}{Q}\right) = \frac{c}{I} Q \left(\frac{I}{Q}\right) = c$$

is the constant 'c'. An interesting byproduct of these derivations is that for the log linear demand equation

$$\epsilon(P) = \frac{\partial Q}{\partial P} \frac{P}{Q} = \frac{\partial \ln Q}{\partial \ln P} = -b$$

$$\epsilon(I) = \frac{\partial Q}{\partial I} \frac{I}{Q} = \frac{\partial \ln Q}{\partial \ln I} = c$$

We now turn our attention to profit optimization when a noncompetitive firm faces a log linear demand function. It is easy to show that in order to have a legitimate optimization mode with real solution, the slope of the log linear demand function must be less than 1.

We know that the firm's profit is maximized at the level of output where the marginal cost is equal to the marginal revenue. By rewriting the log linear demand function $\ln P = \ln a - b \ln Q$ as

$$P = e^{\ln a - \ln Q^b} = e^{\ln a} e^{-\ln Q^b} \quad \text{or alternatively} \quad P = \frac{a}{Q^b} = a\, Q^{-b}$$

the total revenue function is

$$TR = PQ = a\, Q^{1-b}$$

and the marginal revenue is

$$MR = \frac{dTR}{dQ} = a(1-b)\, Q^{-b}$$

Now note that when

$$MC = MR = a(1-b)\, Q^{-b}$$

if b is greater than 1 then $1 - b < 0$ and we end up with the absurd result of $MC < 0$. Therefore, for the model to have a real non-trivial solution b must be less than 1. The following is an example of optimization with a log linear demand function.

Example 6

Assume a noncompetitive firm with the cost function

$$TC = 0.04Q^3 - 0.95Q^2 + 10Q + 20$$

faces the following log- linear demand function for its product in the market

$$\ln P = 3.25 - 0.48 \ln Q$$

By rewriting the demand function as

$$P = e^{3.25-0.48\ln Q} = \frac{e^{3.25}}{e^{0.48\ln Q}} = \frac{25.7903}{Q^{0.48}} = 25.7903\ Q^{-0.48}$$

and the profit function as

$$\pi = TR - TC = PQ - TC = (25.7903\ Q^{-0.48})Q - 0.04Q^3 + 0.95Q^2 - 10Q - 20$$
$$\pi = 25.7903\ Q^{0.52} - 0.04Q^3 + 0.95Q^2 - 10Q - 20$$

then the FOC is

$$\frac{d\pi}{dQ} = 0.52\ \left(25.7903\ Q^{-0.48}\right) - 0.12Q^2 + 1.9Q - 10 = 0$$

or

$$13.411\ Q^{-0.48} - 0.12Q^2 + 1.9Q - 10 = 0$$

This equation can be simplified to

$$-0.12Q^{2.48} + 1.9Q^{1.48} - 10Q^{0.48} + 13.411 = 0$$

Using the $f\,solve$ command in Maple

$$f\,solve(-0.12Q^{2.48} + 1.9Q^{1.48} - 10Q^{0.48} + 13.411 = 0)$$

we found a solution to the equation as $Q = 11.62669$. Readers can verify that the second derivative of the profit function evaluated at $Q = 11.62669$ is negative, satisfying the SOC for a maximum. At this level of output the optimal price is

$$\ln P = 3.25 - 0.48\ln(11.62669) = 3.25 - 1.177586 = 2.072414$$
$$P = e^{2.072414} = 7.944$$

This price/quantity combination generate 21.648 units of profit for the firm.

9.5.3 Deriving the Trace Function of a Noncompetitive Firm Facing Semi-Log or Log Linear Demand Functions (Optional)

The method for deriving a noncompetitive firm's trace function discussed for linear and quadratic demand functions in Chap. 8 works equally well when the firm faces other types of nonlinear demand functions. Two examples of nonlinear demand models are semi-log and log-log or log linear functions that we introduced in the last section. In this section we derive the trace function when a firm faces these types of demand equations.

Assume the firm total cost function is $TC = f(Q) + FC$. Further assume that this firm faces a semi-log demand function of the form $P = a - b \ln g(Q)$ in the market. Recall from Chap. 8 that the first step in deriving the trace is to find the shutdown price, by following the routine of denoting the dislocated demand equation by $P = \beta - b \ln g(Q)$ and solving $AVC = P$ and $MC = MR$ simultaneously, that is

$$\begin{cases} \dfrac{f(Q)}{Q} = \beta - b \ln g(Q) \\ f'(Q) = \beta - b \ln g(Q) - \dfrac{bQg'(Q)}{g(Q)} \end{cases} \tag{9.15}$$

By eliminating β in (9.15), we arrive at

$$Qf'(Q)g(Q) - f(Q)g(Q) + bQ^2 g'(Q) = 0 \tag{9.16}$$

with $\tilde{Q}$ as the solution (if exists) and $AVC(\tilde{Q})$ as the shutdown price P_{sd}. We then derive the trace function by using the firm's profit maximizing (loss minimizing) condition $MC - MR = 0$.

$$TR = PQ = [\beta - b \ln g(Q)]Q = \beta Q - bQ \ln g(Q)$$

$$MR = \frac{dTC}{d(Q)} = \beta - b \ln g(Q) - \frac{bQg'(Q)}{(g(Q)}$$

Replacing β by $P + b \ln g(Q)$ from the dislocated demand equation, we have

$$MR = P + b \ln g(Q) - b \ln g(Q) - \frac{bQg'(Q)}{g(Q)} = P - \frac{bQg'(Q)}{g(Q)}$$

Then,

$$MC - MR = f'(Q) + \frac{bQg'(Q)}{(g(Q)} - P = 0$$

$$f'(Q)g(Q) + bQg'(Q) - Pg(Q) = 0 \tag{9.17}$$

As before, if $Q = h(P)$ is the solution to (9.17) then the trace function can be expressed as

$$Q = \begin{cases} 0 & \text{for } P < P_{sd} \\ h(P) & \text{for } P \geq P_{sd} \end{cases}$$

To illustrate, consider the following numerical model

Example 7

Assume

$$TC = Q^3 - 10Q^2 + 45Q + 100$$
$$P = 200 - 2\ln Q$$

Here $f(Q) = Q^3 - 10Q^2 + 45Q$, $b = 2$, and $g(Q) = Q$. We use (9.16) to determine the shutdown price,

$$(3Q^2 - 20Q + 45)Q^2 - (Q^3 - 10Q^2 + 45Q)Q + 2Q^2 = 0$$
$$Q^2 - 5Q + 1 = 0$$

This equation has two nonzero solutions $Q = 4.7913$ and $Q = 0.2087$. For reason that become apparent later, we must discard the second solution. The shutdown price is then $AVC(4.47913) = 20.044$. To derive the trace function we utilize (9.17),

$$(3Q^2 - 20Q + 45)Q + 2Q - PQ = 0$$
$$3Q^2 - 20Q + 45 + 2 - P = 3Q^2 - 20Q + 47 - P = 0$$

And solving for Q in terms of P leads to solution

$$Q = 1/3(10 \pm \sqrt{3P - 41})$$

Since for $P = 20.044$, Q must be 4.7913 then the right answer is

$$Q = 1/3(10 + \sqrt{3P - 41})$$

Note that for the second shutdown solution ($Q = 0.2087$ and $P = 42.957$) to work, we must choose
$$Q = 1/3(10 - \sqrt{3P - 41})$$

which obviously cannot be a supply function. For this reason we discarded this solution. The firm's trace function is then

$$Q = \begin{cases} 0 & \text{for } P < 20.044 \\ 1/3(10 + \sqrt{3P - 41}) & \text{for } P \geq 20.044 \end{cases} \tag{9.18}$$

Let's take our function for a test drive. Using the total cost and demand function, the profit maximizing combination of output and price is $Q = 11.11188$ and $P = 195.184$. The optimal level of output is determined by setting $MC = MR$

$$3Q^2 - 20Q + 45 = 200 - 2\ln Q - 2$$

and solving for Q in

$$3Q^2 - 20Q + 2\ln Q - 153 = 0$$

By plugging 195.184 for price in the trace function (9.18) we get exactly the same answer $Q = 11.11188$. This price/quantity combination generates 1431.54 of profit for the firm. If for any reason demand declines and profit is eliminated, then the firm's normal profit combination (which occurs at the point of tangency between the demand curve and the *ATC* curve) will be $Q = 6.15667$ and $P = 37.58043$. Again, by plugging 37.58043 for price in the trace function (9.18) we get exactly the same answer $Q = 6.15667$. Readers can easily verify that the demand functions associated with shutdown and normal profit situations are $P = 23.178 - 2\ln Q$ and $P = 41.216 - 2\ln Q$.

Now assume the firm's total cost function is $TC = f(Q) + FC$ but this time firm faces a log linear demand function

$$\ln P = \ln a - \ln g(Q) \tag{9.19}$$

in the market. Note that (9.19) can be expressed as

$$P = \frac{a}{g(Q)}$$

and as before, the shutdown price, following the dislocation of demand to

$$P = \frac{\beta}{g(Q)}$$

is found when the following two conditions are simultaneously satisfied

$$\begin{cases} AVC = P \\ MC = MR \end{cases}$$

That is

$$\begin{cases} \dfrac{f(Q)}{Q} = \dfrac{\beta}{g(Q)} \\ f'(Q) = \dfrac{\beta g(Q) - \beta Q g'(Q)}{[g(Q)]^2} \end{cases}$$

or

$$\begin{cases} \dfrac{f(Q)g(Q)}{Q} = \beta \\ \dfrac{f'(Q)[g(Q)]^2}{g(Q) - Qg'(Q)} = \beta \end{cases}$$

Eliminating β leads to

$$\frac{f'(Q)[g(Q)]^2}{g(Q) - Qg'(Q)} - \frac{f(Q)g(Q)}{Q} = 0$$

which after some manipulations results in

$$\left[Qf'(Q) - f(Q)\right]g(Q) + Qf(Q)g'(Q) = 0 \qquad (9.20)$$

As before if $\tilde{Q}$ is a solution to (9.20) then $AVC(\tilde{Q})$ is the shut down price P_{sd}.

To derive the trace function we use the profit optimization condition $MC - MR = 0$ and write,

$$f'(Q) - \frac{\beta g(Q) - \beta Q g'(Q)}{[g(Q)]^2} = 0 \qquad (9.21)$$

By replacing $P\, g(Q)$ for β from the dislocated demand function, we rewrite (9.21) as

$$f'(Q) - \frac{P[g(Q)]^2 - PQg(Q)g'(Q)}{[g(Q)]^2} = 0$$

which after simplification will reduce to

$$f'(Q)g(Q) - Pg(Q) + PQg'(Q) = 0 \qquad (9.22)$$

Solving Eq. (9.22) for Q in terms of P generates the trace function. The following numerical example illustrates the working of the model.

Example 8

Assume a noncompetitive firm with total cost function

$$TC = 0.04Q^3 - 0.95Q^2 + 10Q + 20$$

faces the following log- linear demand function for its product in the market

$$\ln P = 3.25 - 0.48 \ln Q$$

This equation can be expressed as

$$P = e^{3.25 - 0.48 \ln Q} = \frac{e^{3.25}}{e^{0.48 \ln Q}} = \frac{25.79}{Q^{0.48}}$$

By noting that $f(Q) = 0.04Q^3 - 0.95Q^2 + 10Q$ and $g(Q) = Q^{0.48}$ we can use Eq. (9.20) and write

$$\left[Q(0.12Q^2 - 1.9Q + 10) - (0.04Q^3 - 0.95Q^2 + 10Q) \right] Q^{0.48}$$
$$+ Q(0.04Q^3 - 0.95Q^2 + 10Q)(0.48Q^{-0.52}) = 0$$

After simplifying this equation, we have

$$0.0992Q^{3.48} - 1.406Q^{2.48} + 4.8Q^{1.48} = 0$$

or

$$\left(0.0992Q^2 - 1.406Q + 4.8 \right) Q^{1.48} = 0 \tag{9.23}$$

$Q = 8.44099$ and $Q = 5.73239$ are two nonzero solutions to (9.23). For the same reason that we rejected the second solution in Example 7, we must discard the second solution here. The shutdown price is $AVC(8.44099) = 4.831$.

To derive the trace function, we can then utilize (9.22)

$$\left(0.12Q^2 - 1.9Q + 10 \right) Q^{0.48} - PQ^{0.48} + PQ \left(0.48Q^{-0.52} \right) = 0$$

simplified to

$$0.12Q^2 - 1.9Q + 10 - 0.52P = 0$$

By solving this equation for Q in terms of P we arrive at the firm's trace function

$$Q = \begin{cases} 0 & \text{for } P < 4.831 \\ \frac{1}{12}(95 + \sqrt{624P - 2975}) & \text{for } P \geq 4.831 \end{cases} \tag{9.24}$$

As it was shown in Example 6 by using the total cost and original demand function, the profit maximizing combination of output and price is $Q = 11.6267$ and $P = 7.944$. From the trace function (9.24) we get exactly the same answer for Q when we plug in 7.944 for P. This price/quantity combination generates 21.65 of profit for the firm. If demand declines and profit is eliminated, then the firm's normal profit combination will be $Q = 10.5215$ and $P = 6.3335$. Again, by plugging 6.3335 for price in the trace function (9.24) we get exactly the same answer for Q.

9.6 Exercise

1. Given the following market demand and supply functions, determine the market equilibrium price and quantity

$$Q_d = \frac{1200}{0.2P + 0.5}$$

$$Q_s = \frac{P^3}{e^{0.02P+0.1}} = P^3 e^{-(0.02P+0.1)}$$

2. Assume the following market supply and demand functions. Find the market equilibrium price and quantity.

$$D_d = \frac{1200}{0.2P}$$
$$Q_s = P^2 e^{-0.02(P+0.5)}$$

3. Consider a noncompetitive firm with the cubic cost function

$$TC = \frac{1}{3}Q^3 - 20Q^2 + 350Q + 100$$

facing the log linear demand function

$$\ln P = 7.5 - 0.6 \ln Q$$

in the market. What is the firm's profit maximizing combination of price and quantity?

4. In problem (3) determine the firm's break-even price/quantity combination.

5. In problem (3) determine the firm's normal profit price/quantity combination.

6. In problem (3) find the dislocated demand function for shutdown and normal profit situation.

7. Derive the trace function for the firm in problem (3).

8. Assume a noncompetitive firm with cubic cost function

$$Q^3 - 20Q^2 + 150Q + 200$$

facing a semi-log demand function

$$P = 300 - 3 \ln Q$$

in the market. What is the firm's profit maximizing combination of price and quantity?

9. In problem (8) determine the firm's break-even price/quantity combination.

10. In problem (8) determine the firm's normal profit price/quantity combination.

11. In problem (8) find the dislocated demand function for shutdown and normal profit situation.

12. Derive the trace function for the firm in problem (8).

Chapter 10
Production Function, Least-Cost Combinat.~ of Resources, and Profit Maximizing Level of Output

10.1 Production Function

The fundamental concept related to the supply side of an economy is the *production function*. A production function relates the maximum quantity of output that can be produced from given quantities of inputs. It can be defined at various micro and macro economic levels of aggregation: firm, industry, sector, and the whole economy. A generic production function can be expressed as

$$Q = f(x_1, x_2, \ldots, x_n)$$

where Q is the level of output and $x_1, x_2, \ldots, x_n$ are quantities of various inputs— factors or resources—used in the process of production. While some resources like labor and capital are almost universally utilized in any type of productive activity, other inputs (like types of raw materials) vary from sector to sector, industry to industry, and even from firm to firm within the same industry. For example, primary and intermediate inputs used in the furniture industry are widely different from those used in the textile industry. A good example for different inputs used by firms operating in even the same industry is electric-generating firms. Some of the electric-generating power stations burn coal, others oil or natural gas, and still some others (nuclear power plants) use enriched uranium or plutonium to generate the same output, namely electricity.

There are a number of concepts associated with the production function. First we must distinguish between *short run* and *long run* production function. In the short run at least one of the factors of production is fixed. In manufacturing this fixed input is considered to be capital. In agriculture land is more likely to be the fixed input in the short run. In the short run firm's level of output is limited to the firm's plant size or capacity, dictated by the existing capital or land. For example, a steel mill designed and constructed to produce 5 million tons of steel each year cannot produce more than 5 million tons of steel. It can, however, operate at 50 % capacity and produce

S. Vali, *Principles of Mathematical Economics*,
Mathematics Textbooks for Science and Engineering 3,
DOI: 10.2991/978-94-6239-036-2_10, © Atlantis Press and the authors 2014

only 2.5 million tons of steel. This mill has the flexibility of producing output in the range $0 \leq Q \leq 5$ million tons; it just cannot produce more than 5 million tons.

In the long run all inputs are variable. The steel mill in our example can, over time, build additional capacity by adding new electric arc furnaces and expand its plant size to, say, 7 million tons. A short run is then a time period during which a firm maintains the same plant size or productive capacity. As soon as a new and expanded capacity is in place, however, a new short run commences. Now our steel mill is bound by constraint $Q \leq 7$ million tones.

An important concept associated with a production function is the concept of *marginal product*, or more accurately, *marginal physical product MPP*. Assuming a firm operates at a certain capacity producing Q units of output, the marginal physical product of the ith input MPP_i is additional output due to one unit increase in the ith input, *ceteris paribus* (keeping all the other inputs constant at the same level). Mathematically MPP_i is expressed as

$$MPP_i = \frac{\partial Q}{\partial x_i} \qquad\qquad i = 1, 2, \ldots, n$$

It is hard to satisfy the *ceteris paribus* condition if there are *complementary inputs*—inputs that are used together. Coking coal and iron ore are complementary in steel making. So are typists and text processing equipments (computer) in publishing. In our steel-making example the raw materials are complementary. In the publishing example labor (typist) and capital (computer) are.

If the price of a unit of output in the market is P, then $P * MPP_i$ is the *marginal revenue product MRP_i* of the ith input. MRP_i measures the contribution of an additional unit of ith input to a firm's revenue.

Suppose a winery has 10 acres of vineyards. 50 farm workers are hired to harvest the grapes. Assume that these workers pick 1000 pounds of grapes an hour and the market price of grapes is \$3.50 a pound. The winery hires an additional farm worker. If 51 workers collectively pick 1018 lb of grapes, then the *MPP* of the 51st farm worker is 18 lb of grape. The *MRP* of this worker is $3.50 * 18 = \$63$.

Another concept related to the production function is the *average product AP*. The average product of the ith input is

$$AP_i = \frac{Q}{x_i} \qquad\qquad i = 1, 2, \ldots, n$$

In our winery example the AP of labor is $\dfrac{1018}{51} = 19.96$ pounds of grape per worker.

Without loss of generality and to make the presentation more concrete let us assume that in our generic production function the level of output Q is expressed as a function of two resources, capital and labor, that is $Q = f(K, L)$. Associated with this production function we have the marginal products of labor and capital MPP_L and MPP_k, and the average product of labor and capital AP_L and AP_k.

$$MPP_L = f_L = \frac{\partial Q}{\partial L}; \quad MPP_K = f_K = \frac{\partial Q}{\partial K}; \quad AP_L = \frac{Q}{L}; \quad AP_k = \frac{Q}{K}$$

It makes economic sense that a firm would employ additional units of labor or capital as long as f_L and $f_K > 0$, but, alas, the production function must also satisfy

$$f_{LL} = \frac{\partial^2 Q}{\partial L^2} < 0 \text{ and } f_{KK} = \frac{\partial^2 Q}{\partial K^2} < 0$$

This implies that the contribution of labor (capital) to output declines as a firm employs more labor (capital) while amount of capital (labor) remains the same. This is *the law of diminishing marginal product*, which is the source of *the law of diminishing return*, which in turn has been the source of all conflicts in human societies throughout history. If indeed we could have f_{LL} and $f_{KK} > 0$ then it would be possible to produce an infinite number of super computers, five-bedroom houses, Maseraties and Lamborghinies, luxury yachts, etc, by continuously adding labor to a small amount of capital! Unfortunately f_{LL} and $f_{KK} > 0$ must be the mathematical expression of "heaven" or operational procedure in Santa Clause workshop!

The graph of $Q = f(K, L)$ is a solid surface in three-dimensional space, with K and L measured on the X and Y axes and Q on the Z axis. The production surface is a collection of (K, L, Q) triplets. Naturally both the domain and range of the production function are non-negative real numbers.

There are three classes of *optimization* (maximization or minimization) problems in economics associated with production functions. The first class of problems can be stated as finding a firm's least costly combination of resources for producing a given level of output. This is an example of a *constraint optimization* problem. Here the constraint is the predetermined level of output Q_0. In this model the level of output Q, price of the good P, and prices of resources like P_K and P_L for capital and labor are all exogenously determined. The objective is to find values of endogenous variables L and K that produces output of Q_0 at minimum production cost. We label this cost minimization problem as *type I* optimization problem.

In the second class of optimization problems, a firm has a predetermined level of total costs TC_0 and must maximize its profit π. In this model the total cost, prices of resources, and price of the good are exogenous variables. Variables Q, K, L, and π must be endogenously determined. This is another example of constraint optimization. Here the constraint is that the firm cannot spend more than its predetermined level of total cost TC_0. We call this problem the *type II* optimization problem.

The third class, *type III*, of optimization problems seeks to maximize a firm's profit without any predetermined requirement or constraint. This is an example of *unconstrained optimization*. In this case the prices of resources and price of the good are exogenous variables. Variables Q, K, L, and π must be endogenously determined.

Another class of optimizations involving production functions arises from considering the firm's stock of existing capital as fixed (short run production function) or exogenously determined. This assumption reduces the complexity of the problem by converting the production function from a bivariate to a single or univariate function.

Since this special case could arise in all three types of optimizations mentioned above, it is not considered a separate optimization type.

In Sect. 10.1.1 I introduce additional concepts related to production functions and one method of solving type I optimization problems. This section is followed by an extensive discussion of the Cobb–Douglas production function and its properties. Different methods for solving all three types of optimizations, complemented with detailed numerical examples, constitute the rest of this chapter.

10.1.1 Isoquants, MRTS, and the Least-Cost Combination of Resources (Type I Optimization)

We know that the graph of $Q = f(K, L)$ is a solid surface in a three-dimensional space. If we cut this production surface with planes parallel to the KL plane and project the resulting intersections onto the KL plane, we generate what is known as *contour map* or *isoquant map* of the function. A contour map is a collection of contour lines. A contour line for a function of two variables is a curve connecting points on the surface where the function has exactly the same value (height). This contour curve for production function is called an *isoquant* or an "equal quantity" curve (Greek prefix 'iso', meaning 'equal', added to 'quant' for 'quantity'.) For our $Q = f(K, L)$ production function each isoquant represents all combinations of labor and capital that can be used to produce a given amount of output .

Assume a firm's production function is $Q = 10K^{0.5}L^{0.3}$. Assume this firm decides to produce 124 units of output. What is the equation of isoquant for $Q = 124$ units? If $Q = 124$ then $124 = 10K^{0.5}L^{0.3}$ and after solving for K, we have

$$K^{0.5} = \frac{124}{10L^{0.3}} = 12.4L^{-0.3} \longrightarrow (K^{0.5})^2 = (12.4L^{-0.3})^2$$

and the equation of isoquant for 124 units of output is

$$K = 153.76L^{-0.6} \tag{10.1}$$

Figure 10.1 is the graph of isoquant for $Q = 124$ units. Since different levels of output have separate isoquant, we have the isoquant map. Figure 10.2 shows isoquants related to three different levels of output $Q = 100$, 124, and 150 units.

Isoquants exhibit the notion of *substitutability of resources*. In our case a whole set of different combinations of K and L produce the same amount of output. For instance, the following three combinations of labor and capital produce 124 units of output: $(13.31K, 59L)$, $(25.03K, 20.6L)$, and $(46.27K, 7.4L)$. There are, of course, certain limitations in substituting one resource for another. As noted British economist R. J. Hicks once said, "it is not possible to substitute a typewriter for a typist".

Fig. 10.1 Graph of isoquant for $Q = 124$

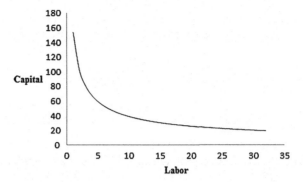

Fig. 10.2 Isoquant map graph of isoquant for $Q = 100,\ 124,\ \text{and}\ 150$ units

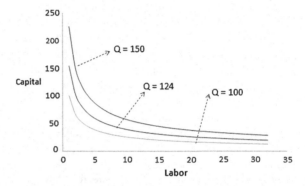

At any point along the isoquant, the slope of the curve measures the rate of substitution between the two resources, here labor and capital. The absolute value of this rate is called the *marginal rate of technical substitution*, or *MRTS*. At any point on the curve, the *MRTS* measures the number of units of capital that could be replaced by one unit of labor, maintaining the same level of output of 124 units.

Alternatively, we could solve $124 = 10K^{0.5}L^{0.3}$ for L and derive the equation of isoquant as

$$L = 4413K^{-1.667}$$

In this case at any point on the isoquant curve, *MRTS* measures the number of units of labor that could be replaced by one unit of capital, maintaining the same level of output at 124 units.

We find slope of the isoquant or *MRTS* associated with the 124 units of output by differentiating $K = 153.76L^{-0.6}$

$$\frac{dK}{dL} = -0.6 * 153.76L^{-1.6} = -92.26L^{-1.6} \quad \text{and} \quad MRST_{LK} = \frac{92.26}{L^{1.6}} \quad (10.2)$$

where $MRST_{LK}$ denotes the marginal rate of technical substitution of "labor" for "capital". Notice that the slope of isoquant is negative for all values of L, indicating the shape of the isoquant. Evaluating the slop at $L = 5$ and 10, and ignoring the sign, we get two different rates

$$\frac{dK}{dL}\bigg|_{L=5} = -92.26(5)^{-1.6} = -7.03 \text{ and } MRTS = 7.03$$

$$\frac{dK}{dL}\bigg|_{L=10} = -92.26(10)^{-1.6} = -2.3 \text{ and } MRTS = 2.3$$

If this firm with the production function $Q = 10K^{0.5}L^{0.3}$ decides to use 5 units of labor to produce 124 units of output it must combine it with about 58.5 units of capital. MRTS of labor for capital at 5 units of labor is 7. This means that the firm can produce the same 124 units of output by substituting one unit of labor for almost 7 units of capital. The new combination is then about $6L$ and $52K$.

But if this firm is utilizing 10 units of L combined with 38.6 units of K to produce 124 units of output, MRTS is 2.3. That is, at this combination one unit of labor can replace only 2.3 units of capital. The new combination would be $11L$ and $38.6 - 2.3 = 36.3K$. This clearly indicates the diminishing MRTS as the firm uses more and more of one resource.

A more fundamental question, of course, is "under what condition does a firm seek to substitute one input for another, and what is the guiding star in searching for the optimal combination of resources for producing a certain level of output"? That guiding star is, simply, the total cost (TC). Given the prices of inputs, a firm must find a least costly combination of inputs that satisfies its production goal.

A numerical example would help us to understand the logic. Let assume that the price of labor P_L and capital P_K in the resource market[1] are $155 and 82.5 per unit, respectively. Based on this assumption I have prepared Table 10.1 below.

Using the equation of isoquant for production of 124 units of output in (10.1), the number of units of capital needed to be combined with units of labor given in the first column are calculated and listed in the second column. The third column gives MRTS for various labor units based on Eq. (10.2). Column 4 gives the approximate amount of saving the firm can realize by substituting the appropriate units of capital, given by MRTS in column 3, by one unit of labor. The firm can substitute, for instance, about 6 units of capital with one unit of labor and move from a $(5L, 58.54K)$ combination to a $(6L, 52.48K)$ combination. This move reduces its resource costs by approximately $82.5 * 6 - 155 = \$345$. The firm can reduce further its production cost by additional substitution of labor for capital.

At $L = 11$ unit the MRTS is 1.99, that is firm can substitute 1 unit of labor for 1.99 units of capital. Our firm moves from 11 L to 11.2 L, increasing labor by only 0.2 units. 0.2, which is 1/5 units of labor, can then be substituted for $1.99/5 = 0.398$ of capital. So the firm moves from combination $(11L, 36.48K)$ to $(11.2L, 36.08K)$, realizing additional $2 of saving in input costs. At combination $11.4L$ and $35.7K$ additional saving almost vanishes. If the firm pushes beyond this combination its input costs start to go up. And our firm has no incentive to do that, so it stays at its

---

[1] Later in the chapter we adopt w as the wage rate and r as the rental cost of a unit of capital.

	1	2	3	4
Table 10.1 $MRTS$ and Saving from Substitution	Units of labor	Units of capital	$MRTS_{LK}$	Approximate saving from substituting L for K
	5	58.54	7.03	–
	6	52.48	5.25	345
	7	47.84	4.10	228
	8	44.16	3.31	149
	9	41.14	2.74	94
	10	38.62	2.32	53
	11	36.48	1.99	22
	11.2	36.08	1.93	2
	11.4	35.70	1.88	0.35
	11.6	35.33	1.83	−0.48
	11.8	34.97	1.78	−1.3
	12.0	34.62	1.73	−2.1
	12.2	34.28	1.69	−2.95
	12.4	33.95	1.64	−3.78
	12.6	33.62	1.60	−3.78
	12.8	33.31	1.56	−5.42
	13.0	33.00	1.52	−5.42

optimal least cost combination of $11.4L$ and $35.7K$. No other combination of labor and capital can produce 124 units of output and cost less than this combination.

$MRTS$ of labor for capital at $(11L, 35.7K)$ combination is 1.88. The ratio of the price of capital to the price of labor $155/82.5$ is also 1.88. Notice that $MRTS * P_K$ is the cost of capital replaced by one unit of labor, which costs P_L. As long as $MRTS * P_K > P_L$ the firm gains by going through with the substitution. The gainful substitution comes to an end when $MRTS * P_K = P_L$, that is $MRTS = \dfrac{P_L}{P_K}$. This is the 'guiding star' for determining the least-cost combination of factors for producing a given level of output.

Let us apply this rule to the above problem: by setting $MRTS$ from (10.2) equal to the ratio of factor prices $\dfrac{P_L}{P_K}$, we have

$$MRTS_{LK} = 92.26L^{-1.6} = \frac{P_L}{P_K} = \frac{155}{82.5} = 1.8$$

$$92.26L^{-1.6} = 1.88 \text{ or } \frac{92.26}{L^{1.6}} = 1.88 \longrightarrow L^{1.6} = \frac{92.26}{1.88} = 49.07$$

$$L^{1.6} = 49.07 \longrightarrow (L^{1.6})^{0.625} = (49.07)^{0.625} \longrightarrow L = 11.4$$

Substituting for L in the equation of isoquant $K = 153.76L^{-0.6}$ we have $K = 35.7$.

Note from (10.2) that if $L = 0$, *MRTS* approaches infinity. Implication is that production without application of labor is impossible.

10.1.2 The Cobb–Douglas Production Function

The production function $Q = 10K^{0.5}L^{0.3}$ used in our discussion of isoquants and *MRTS* is an example of a class of most widely used production functions called the *Cobb–Douglas production function (CDpf)*. Charles Cobb, a mathematician, and Paul Douglas, an economist, first introduced this production function in a paper delivered in March 1928 at the annual meeting of the *American Economic Association*. They formulated the function as $Q = AL^{\alpha}K^{\beta}$ with the assumption that $\alpha + \beta = 1$ or $\beta = 1 - \alpha$ (the original notation used by Cobb and Douglas was $P = AL^{\alpha}C^{1-\alpha}$).

The assumption that the exponents of labor and capital in the function add up to one guarantees that the function is a *linear homogeneous function* (homogeneous of degree one) representing a production process subject to *Constant Returns to Scale (CRS)*. A function is *homogeneous of degree r* if an m-fold change in the values of independent variables leads to m^r-fold change in the value of the function. If r is equal to 1, then the function is homogeneous of degree one, or linear homogeneous. It is not difficult to check that the original Cobb–Douglas production function is linear homogeneous:

Let labor and capital in the CDpf $Q = AL^{\alpha}K^{1-\alpha}$ change by a factor of m (if m is, for example, 1.1 it means an increase in labor and capital by 10%) and denote the new level of output by Q'. We have

$$Q' = A(mL)^{\alpha}(mK)^{1-\alpha} = Am^{\alpha}L^{\alpha}m^{1-\alpha}K^{1-\alpha} = m^{\alpha+1-\alpha}AL^{\alpha}K^{1-\alpha}$$

$$Q' = mAL^{\alpha}K^{1-\alpha} = mQ$$

showing that the level of output changes by the same factor m.

As will be shown shortly, α and β are shares of labor and capital from the total output. If the CDpf is used at a highly aggregated level, it makes sense to assume that proportions of output going to those who supply labor and those who supply capital add up to one. In their original presentation Cobb and Douglas used data from 1899 to 1922 for American manufacturing and estimated the function as $Q = 1.01L^{0.75}K^{0.25}$ which gives share of labor as 75% and capital as 25% of total output.

If $Q = f(K, L)$ is a linear homogeneous function then according to *Euler's theorem* we must have

$$\frac{\partial Q}{\partial L}L + \frac{\partial Q}{\partial K}K = Q \tag{10.3}$$

To show the validity of Euler's theorem for the Cobb–Douglass production function $Q = AL^{\alpha}K^{1-\alpha}$ we have

$$\frac{\partial Q}{\partial L} = \alpha AL^{\alpha-1}K^{1-\alpha} \tag{10.4}$$

$$\frac{\partial Q}{\partial K} = (1 - \alpha)AL^{\alpha}K^{-\alpha} \tag{10.5}$$

Then (10.3) can be expressed as

$$\frac{\partial Q}{\partial L}L + \frac{\partial Q}{\partial K}K = L * \alpha AL^{\alpha-1}K^{1-\alpha} + K * (1 - \alpha)AL^{\alpha}K^{-\alpha}$$
$$= \alpha AL^{\alpha}K^{1-\alpha} + (1 - \alpha)AL^{\alpha}K^{1-\alpha}$$
$$= AL^{\alpha}K^{1-\alpha}[\alpha - (1 - \alpha)]$$
$$= Q$$

Notice that (10.4) and (10.5) are the marginal physical products of labor and capital, so we can write (10.3) as

$$L * MPP_L + K * MPP_k = Q \tag{10.6}$$

Multiplying both sides of (10.6) by P, the price of the product in the market, we have

$$P * L * MPP_L + P * K * MPP_k = PQ \tag{10.7}$$

where $P * MPP_L$, $P * MPP_k$, and PQ are the values of marginal product (*marginal revenue product*) of labor and capital, and the value of the firm's output, respectively. If the factor market is competitive then firms must pay labor and capital the value of their marginal product. Proof of this proposition derives from unconstrained profit maximization, as follows:

Assume a firm with the CDpf operates in competitive product and resource markets. Assume P is the market price of the product, and the firm pays w per unit of labor and r per unit of capital (w is the wage rate and r is the rental price of capital). The firm's total revenue is PQ and its total cost is $wL + rK + FC$, where FC is the Fixed Cost. wL is the total payment to labor (wage bill) and rK is total payment to capital. The profit function of this firm is

$$\pi = TR - TC = PQ - wL - rK - FC$$
$$\pi = PAL^{\alpha}K^{\beta} - wL - rK - FC$$

Here π is a function of L and K. The first-order necessary condition (FOC) for an unconstrained maximization of profit is

$$\frac{\partial \pi}{\partial L} = PA\alpha L^{\alpha-1}K^{\beta} - w = 0$$
$$\frac{\partial \pi}{\partial K} = PA\beta L^{\alpha}K^{\beta-1} - r = 0$$

Solving these two equations leads to

$$PA\alpha L^{\alpha-1}K^{\beta} - w = 0 \longrightarrow w = P * MPP_L = MRP_L$$
$$= \text{the value of marginal product of labor, and}$$
$$PA\beta L^{\alpha}K^{\beta-1} - r = 0 \longrightarrow r = P * MPP_K = MRP_K$$
$$= \text{the value of marginal product of capital}$$

So now we can write (10.7) as

$$wL + rK = PQ$$

Dividing both sides by PQ we have

$$\frac{wL}{PQ} + \frac{rK}{PQ} = 1$$

where $\dfrac{wL}{PQ}$ is the share of labor and $\dfrac{rK}{PQ}$ is the share of capital from output.

Now we need to show that these shares are equal to α and $(1 - \alpha)$ respectively. By substituting $P * MPP_L$ for w and $P * MPP_K$ for r, we have

$$\frac{wL}{PQ} = \frac{LP * MPP_L}{PQ} = \frac{LP\alpha AL^{\alpha-1}K^{1-\alpha}}{PQ} = \frac{\alpha Q}{Q} = \alpha$$

and similarly

$$\frac{rK}{PQ} = \frac{KP * MPP_K}{PQ} = \frac{KP(1 - \alpha)AL^{\alpha}K^{-\alpha}}{PQ} = \frac{(1 - \alpha)Q}{Q} = 1 - \alpha$$

which is what Cobb and Douglas considered in their original formulation.

Also notice that

$$\frac{\partial Q}{\partial L}\frac{L}{Q} = \alpha \text{ and } \frac{\partial Q}{\partial K}\frac{K}{Q} = 1 - \alpha$$

But $\dfrac{\partial Q}{\partial L}\dfrac{L}{Q}$ and $\dfrac{\partial Q}{\partial K}\dfrac{K}{Q}$ are the output elasticity of labor ε_{QL} and output elasticity of capital ε_{QK}, respectively. So α and $1 - \alpha$ are output elasticity of labor and capital.

In a paper published in the Journal of Political Economy, David Durand[2] introduced what is now referred to as the *generalized* Cobb–Douglas production function $Q = AL^{\alpha}K^{\beta}$. In his formulation α and β could add up to more or less than one, indicating production processes that could be subject to Increasing or Decreasing

[2] Durand, D. (1937). Some thoughts on marginal productivity, with special reference to professor Douglas' analysis. *The Journal of Political Economy*, 45(6), 740–758.

Returns to Scale. With this extension it became possible to use the CDpf at a disaggregated level, like for an individual firm or industry. A firm can indeed operate under increasing or decreasing return to scale.

Durand admits that reasons for CD constant return assumption are very interesting:

> If decreasing return prevailed an enterprise consisting of more than one man would be inefficient; society would consists of one-man firms. If increasing return prevailed any firm could expand indefinitely, competition would be impossible and monopoly would be the result. Since industrial society is characterized neither by one-man firms nor by universal monopoly, Professor Douglas concludes that constant returns are essential and that production is expressible as a homogeneous linear function.[3]

But as we will see in Sect. 10.6 neither increasing nor constant returns to scale version of CDpf lend itself to unconstrained profit maximization of a firm or industry.

In $Q = AK^{\alpha}L^{\beta}$, A is a scale factor that could be interpreted as expressing *innovation*.[4] A firm can achieve higher output not just by increasing the level of factors used in the production process but also by improving or rearranging the production process itself (for example by better part distribution or more efficient assembly line setup). A higher value of A reflects this more efficient production arrangement. As an example assume a firm has the following CDpf

$$Q = 1.5K^{0.3}L^{0.4} \qquad (10.8)$$

and produces $Q = 80.77$ units of output by using 200 units of capital and 400 units of labor. If by reorganizing its operation and implementing an innovative production arrangement this firm can increases its output by 5 %, then its new production function should be

$$Q = 1.575K^{0.3}L^{0.4}$$

and the firm using the same units of labor and capital can produce 84.8 units of output.

The exponents α and β represent *degrees of technical and technological sophistication* of the firm. If due to a new *invention* firm can use a more advanced and productive technology, the firm's new level of productivity would be reflected in higher values of α and/or β. If the firm with the original production function given in (10.8) upgrades its equipment, retrains its labor force, and manage to achieve the new production function

$$Q = 1.5K^{0.32}L^{0.41}$$

then with the same level of inputs it can raise its output to $Q = 95.34$ units.

Similar to the case for the classic CDpf, the common interpretation of α and β is that these parameters represent the partial elasticity of output with respect to capital and labor. A different way to show this result is to take the log of both sides of the

---

[3] *Ibid.*

[4] Readers should be aware of the difference between "innovation" and "invention" pertaining to production processes.

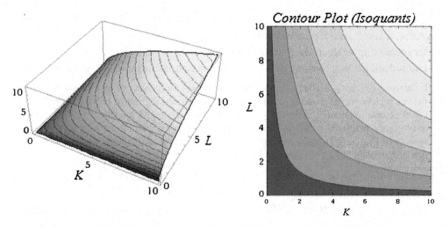

Fig. 10.3 Graph of CD production function $Q = 1.2K^{0.5}L^{0.5}$ and its isoquants

CDpf,

$$\ln Q = \ln A + \alpha \ln K + \beta \ln L$$

By denoting capital and labor elasticity of output by ϵ_{QK} and ϵ_{QL} as before, we have

$$\epsilon_{QK} = \frac{\partial \ln Q}{\partial \ln K} = \alpha$$

$$\epsilon_{QL} = \frac{\partial \ln Q}{\partial \ln L} = \beta$$

Figure 10.3 is the 3D graph of a Cobb–Douglass production function with the contour lines drawn. The projection of these contour lines onto the $K - L$ plane creates the Isoquant map.

10.2 The Cobb–Douglas Production Function and Cost Minimization (Type I Optimization)

Let us assume a firm that operates a plant has the general Cobb–Douglas production function. The price of the firm's product in the market is P dollars per unit. The wage rate is w dollars per day and the rental rate of a unit of capital is r dollars per day. If this firm's management decides to produce Q_0 units of output per day, what would be the least-cost combination of labor and capital?

The total cost of producing Q_0 is the sum of total variable costs—labor and capital costs—and fixed cost. The objective is to determine a combination of labor and capital that minimizes TC given the predetermined level of output:

$$\text{Minimize} \quad TC = wL + rK + FC \quad (10.9)$$
$$\text{Subject to} \quad Q_0 = AK^\alpha L^\beta$$

In this model w, r, Q_0, FC are the exogenous variables and A, α and β are parameters. The endogenous variables are L and K. We discuss two different approaches for solving the constraint optimization problem in (10.9).

We already established that a firm operating in competitive product and factor markets must pay its factors of production the value of their marginal product, marginal revenue product. That is

$$w = MRP_L = P * MPP_L \text{ and } r = MRP_k = P * MPP_k \quad (10.10)$$

where (as before) MPP_L and MPP_K are the marginal physical product of labor and capital. Partially differentiating $Q = AK^\alpha L^\beta$ we have

$$\frac{\partial Q}{\partial L} = MPP_L = A\beta K^\alpha L^{\beta-1} \quad (10.11)$$

$$\frac{\partial Q}{\partial K} = MPP_K = A\alpha K^{\alpha-1} L^\beta \quad (10.12)$$

Using (10.11) and (10.12), we can write the expressions for w and r in (10.10) as

$$w = P * MPP_L = PA\beta K^\alpha L^{\beta-1} \quad (10.13)$$

$$r = P * MPP_K = PA\alpha K^{\alpha-1} L^\beta \quad (10.14)$$

Dividing (10.13) by (10.14) and simplifying the right hand side expression, we have

$$\frac{w}{r} = \frac{PA\beta K^\alpha L^{\beta-1}}{PA\alpha K^{\alpha-1} L^\beta} = \frac{\beta K}{\alpha L} \text{ or}$$
$$\frac{K}{L} = \frac{\alpha w}{\beta r} \quad (10.15)$$

where $\dfrac{K}{L}$ is the capital–labor ratio. Expressing K in terms of L, we have

$$K = \frac{\alpha w}{\beta r}L \quad (10.16)$$

Substituting for K from (10.16) into the production function, we get

$$Q_0 = AK^\alpha L^\beta = A(\frac{\alpha w}{\beta r}L)^\alpha L^\beta = A(\frac{\alpha w}{\beta r})^\alpha L^{\alpha+\beta}$$

Solving for L,

$$L^{\alpha+\beta} = \frac{Q_0}{A(\frac{\alpha w}{\beta r})^\alpha}, \text{ leading to}$$

$$L = \left\{ \frac{Q_0}{\left[A(\frac{\alpha w}{\beta r})^\alpha\right]} \right\}^{\frac{1}{\alpha+\beta}} \tag{10.17}$$

Substituting for L in (10.16), we have

$$K = \frac{\alpha w}{\beta r} \left\{ \frac{Q_0}{\left[A(\frac{\alpha w}{\beta r})^\alpha\right]} \right\}^{\frac{1}{\alpha+\beta}} \tag{10.18}$$

(10.17) and (10.18) give the Least-Cost combination of labor and capital for producing Q_0 units of output.

It should be noted that any combination of capital and labor satisfying the capital-labor ratio in (10.15) would be a least cost combination for the level of output that this combination produces (*see problem 5 in the* **Exercise** *section*).

An alternative method to finding the least-cost combination of labor and capital for producing specified units of output is through a minimization routine. From the production function, we find the equation of the isoquant by solving for K (or L) in terms of L (or K) and substitute the result into the total cost function TC. We then minimize TC as a function of one variable. The process is shown below.

$$Q_0 = AK^\alpha L^\beta \longrightarrow K^\alpha = \frac{Q_0}{AL^\beta} \longrightarrow K = \left(\frac{Q_0}{AL^\beta}\right)^{\frac{1}{\alpha}}$$

or

$$K = \left(\frac{Q_0}{A}\right)^{\frac{1}{\alpha}} L^{\frac{-\beta}{\alpha}}$$

Substituting for K in $TC = wL + rK + FC$ we have

$$TC = wL + r\left(\frac{Q_0}{A}\right)^{\frac{1}{\alpha}} L^{\frac{-\beta}{\alpha}} + FC$$

The first-order condition for minimizing TC is

$$\frac{dTC}{dL} = w - (\frac{\beta}{\alpha})r \left(\frac{Q_0}{A}\right)^{\frac{1}{\alpha}} L^{\frac{-\beta}{\alpha}-1} = 0$$

$$(\frac{\beta}{\alpha})r \left(\frac{Q_0}{A}\right)^{\frac{1}{\alpha}} L^{\frac{-\beta}{\alpha}-1} = w$$

After solving for L and some algebraic manipulation, we get the same answer as in (10.17). Similarly the value of K would be the same as in (10.18).

We must check the SOC for a minimum, which is done by

$$\frac{d^2TC}{dL^2} = -\left(\frac{\alpha+\beta}{\alpha}\right)\left[-\frac{\beta}{\alpha}r\left(\frac{Q_0}{A}\right)^{\frac{1}{\alpha}}L^{\left(-\frac{\alpha+\beta}{\alpha}-1\right)}\right] = \frac{\beta r(\alpha+\beta)}{\alpha^2}\left(\frac{Q_0}{A}\right)^{\frac{1}{\alpha}}L^{-\left(\frac{2\alpha+\beta}{\alpha}\right)} > 0$$

and SOC for a minimum is satisfied.

Example

A competitive firm operates with the following Cobb–Douglas production function

$$Q = 25K^{0.3}L^{0.5}$$

Assume the price of the firm's product in the market is \$75; the wage rate is \$60; and the rental rate of a unit of capital is \$150. Also assume that the firm's fixed cost is \$30,841. If this firm wants to produce 5,000 units per day, what would be the least-cost combination of labor and capital?

In this example $Q_0 = 5000$, $A = 25$, $\alpha = 0.3$, $\beta = 0.5$, $P = 75$, $w = 60$, $r = 150$, and $FC = 30,841$.

The MPP of Labor and Capital are

$$MPP_L = \frac{\partial Q}{\partial L} = 25(0.5)K^{0.3}L^{0.5-1} = 12.5K^{0.3}L^{-0.5}$$

$$MPP_K = \frac{\partial Q}{\partial K} = 25(0.3)K^{0.3-1}L^{0.5} = 7.5K^{-0.7}L^{0.5}$$

By setting the values of marginal product of labor and capital equal to the wage and rental rate, we have

$$75(12.5K^{0.3}L^{-0.5}) = w \quad\longrightarrow\quad 937.5K^{0.3}L^{-0.5} = 60$$

$$75(7.5K^{-0.7}L^{0.5}) = r \quad\longrightarrow\quad 562.5K^{-0.7}L^{0.5} = 150$$

Dividing these two equations, we get

$$1.667\frac{K}{L} = \frac{60}{150} = 0.4 \text{ which leads to a capital labor ratio of } \frac{K}{L} = 0.24$$

Solving for K in terms of L, we have $K = 0.24L$.

Substituting for K in the production function, for the firm to produce 5,000 units we have

$$5000 = 25(0.24L)^{0.3}L^{0.5} = 25(0.24)^{0.3}L^{0.8}$$

$$L^{0.8} = \frac{5000}{16.293} = 306.879 \longrightarrow L = (306.879)^{1.25} = 1284.4 \text{ units}$$

The corresponding capital input is

$$K = 0.24L = 0.24(1284.4) = 308.3 \text{ units}$$

To implement the alternative cost minimization procedure outlined above, we must first use the production function to find the equation of the isoquant for $Q = 5000$, that is

$$5000 = 25K^{0.3}L^{0.5} \longrightarrow K^{0.3} = \frac{5000}{25L^{0.5}} = 200L^{-0.5} \text{ and}$$

$$\left(K^{0.3}\right)^{\frac{1}{0.3}} = \left(200L^{-0.5}\right)^{\frac{1}{0.3}}$$

$$K = 200^{\frac{10}{3}}L^{-\frac{5}{3}}$$

Substituting for K in the cost function, we have

$$TC = wL + rK + FC = 60L + 150\left(200^{\frac{10}{3}}L^{-\frac{5}{3}}\right) + 30841$$

The *FOC* is

$$\frac{dTC}{dL} = 60 - \left(\frac{5}{3} * 150 * 200^{\frac{10}{3}}L^{-\frac{8}{3}}\right) = 60 - 250 * 200^{\frac{10}{3}}L^{-\frac{8}{3}} = 0$$

$$L^{-\frac{8}{3}} = \frac{60}{250 * 200^{\frac{10}{3}}} = \frac{1}{4.16667 * 200^{\frac{10}{3}}}$$

By taking the log of both sides of the above equation, we have

$$-2.6666 \ln L = -\ln 4.166675 - 3.3333 \ln 200 = -19.088$$

$$\ln L = \frac{19.088}{2.6666} = 7.158 \text{ which leads to } L = \exp(7.158) = 1284.4$$

This is the same value we found when we used the method applying the *MPP* of labor and capital on the previous pages. Notice that the *SOC* for minimum is satisfied:

$$\frac{d^2TC}{dL^2} = -2.6666(-1.6667 * 150 * 200^{3.333}L^{-3.6666}) > 0$$

With this level of labor and capital utilization the total variable costs *TVC* and total costs *TC* are

$$TVC = wL + rK = 60 * 1284.4 + 150 * 308.3 = 77064 + 46245 = \$123,309$$

$$TC = TVC + FC = 123309 + 30841 = \$154,150 \qquad (10.19)$$

The firm's Total Revenue TR is equal to $75 * 5000 = \$375,000$. Given the firm's cost structure and the current price of a unit of output, 75, in the market, this firm is profiting $\$220,850$. Also notice that with the given prices of resources established in the factor market, labor and capital are not receiving the value of their marginal products. This situation is generally associated with a temporary deviation from the equilibrium. If the factor and product markets are sufficiently competitive, the dynamics of the market would eliminate these disparities and the markets would reach equilibrium in the long run. At equilibrium the firm's above normal profit (rent) will be eliminated and labor and capital will receive the full value of their marginal product. But how do the markets arrive at their long run equilibrium?

As it was mentioned in previous chapters, a major requirement for establishing a competitive market is the absence of barriers to entry into, and exit from, the market. If this condition is satisfied, the lure of making above normal profit attracts new firms to this market. New firms entering this market bid up the prices of factors suitable for production of this commodity in the resource market, so w and r will start to rise. Additional output supplied to the market by the newly arrived firms will put a downward pressure on the price of the commodity, so P will start to decline. This process—an increase in prices of resources and decline in price of the good—would continue as long as there is opportunity to make above normal profit. The product and factor markets reach their simultaneous equilibrium when the triplet (w, r, P) reaches a point such that the above normal profit is eliminated and labor and capital receive the value of their marginal products.

To exhibit the path of the economic system to equilibrium and the simultaneous determination of equilibrium values of w, r, P, K, and L requires a more expanded and sophisticated dynamic model. At this stage this is beyond our means.

It should be noted that a change in P has no impact on the optimal (the least-cost) combination of K and L [see Eqs. (10.17) and (10.18)]. It is also true, again from (10.17) and (10.18), that as long as w and r change in the same proportions, their changes would not affect the optimal combination. To simplify the situation in order to demonstrate the point, let us make the assumption that w and r at their current values, 60 and 150, are already at equilibrium and only P adjusts in this process. From (10.13) we have

$$w = P * MPP_L = PA\beta K^\alpha L^{\beta-1} \longrightarrow 60 = P * 25(0.5)K^{0.3}L^{-0.5}$$

By solving for P, we have

$$P = \frac{60L^{0.5}}{12.5K^{0.3}} = \frac{60(1284.4)^{0.5}}{12.5(308.3)^{0.3}}$$

$$P = \frac{2150.3}{69.76} = \$30.82$$

We get the same answer for P if we use (10.14) instead of (10.13).

$P = \$30.82$ is the long run equilibrium price of this commodity. When the price is lowered to this level, all the firms operating in this market will earn normal profit, and labor and capital will earn their values of marginal products. To check, $TR = P * Q = 30.82 * 5000 = 154{,}100$ which is equal to TC in (10.19) (the small difference is due to rounding error).

One can use (10.13) and (10.14) to verify that the RHS values are equal to w and r, respectively.

As it was mentioned before, α and β also represent degrees of technical and technological sophistication of the firm. Technical/technological advances can be achieved by new inventions, manifested in new and more advanced and efficient tools and machinery. If the firm with the Cobb–Douglas production function in our example upgrades its capital and operates with a new Cobb–Douglas production function given by

$$Q = 25K^{0.35}L^{0.5}$$

then its optimal least-cost solution would change to

$$L = 860.5 \text{ units of labor and } K = 240.9 \text{ units of capital}$$

Notice that both capital and labor requirement for producing 5,000 units of output are reduced, leading to the reduction of the TVC to \$87,765 and a saving of \$35,544.

One can justifiably argue that more advanced and efficient capital (tools and machinery) may cost more than $\$r$ per unit, as was assumed originally. But as long as the increase in capital costs is overcompensated by reduction in TVC then the capital upgrade is justified.

Let us examine this issue for our current example. Assume that the cost of new capital is r'. Then as long as the new TVC is less than the original TVC of \$123,309 this firm should undertake the capital upgrade. For the current example we can express this as

$$60 * 860.5 + r' * 240.9 < 123309$$

As long as the price of capital is less than \$297.5 per unit, by upgrading its capital this firm can reduce its TVC.

10.3 The Production Function, Isocosts and Cost Minimization (Type I Optimization)

A firm's total cost of production covers several different categories of costs. The two most important financial statements of a firm are the income statement and the balance sheet. The income statement of a company lists two types of costs and expenses: The cost of goods sold and operating expenses. The cost of goods sold includes outlays for raw materials, supplies, utilities, and wages of labor associated with making the goods. Operating expenses cover the administrative, marketing, warehousing, distribution, and research and development costs. These are in fact the costs of factors of production used in the process of producing a good or service.

If a firm with FC dollars of fixed cost uses n different factors $f_1, f_2, \ldots, f_n$, with factor prices $P_1, P_2, \ldots, P_n$ established in competitive factor markets, its total cost is

$$TC = \sum_{i=1}^{n} P_i f_i + FC$$

Amongst factors of production labor L and capital K play the most prominent role. Limiting our analysis of factor costs to these two factors, the firm's total cost can be expressed by the familiar equation

$$TC = wL + rK + FC$$

where, as usual, w and r are the wage and rental rate of a unit of labor and capital, respectively. If the firm plans to spend TVC_0 (Total Variable Cost) dollars on labor and capital, then TVC must be written as

$$TVC_0 = wL + rK \longrightarrow rK = TVC_0 - wL$$

By dividing through by r, we have

$$K = \frac{TVC_0}{r} - \frac{w}{r}L \qquad (10.20)$$

Equation (10.20) is the equation of an *isocost* associated with TVC_0. The graph of (10.20) is an *isocost line*. The isocost line shows all the combinations of capital and labor that are available to the firm for a given total variable cost TVC_0. If the price of labor and capital are \$10 and \$20, and our firm plans to spend \$3,000 then the isocost equation is

$$K = \frac{3000}{20} - \frac{10}{20}L = 150 - 0.5L$$

Notice that the ratio of input prices is the absolute value of slop of isocost line. The derivative of K with respect to L,

$$\frac{dK}{dL} = -\frac{w}{r} = -\frac{10}{20} = -0.5$$

measures the rate of change in capital per unit change in labor and indicates the number of units of capital that can be substituted by one unit of labor. In our example

Fig. 10.4 Graph of Isocosts

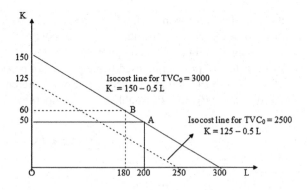

one unit of labor can be substituted for 0.5 unit of capital (or one unit of capital by 2 units of labor). We call this rate the *Marginal Rate of Factor Substitution (MRFS)*.

Different values of *TVC* generate different isocosts. Figure 10.4 shows two isocosts associated with total costs of $3,000 and $2,500. Points on the isocost lines give all the combinations of units of capital and labor that the firm can buy for $3,000 (solid line) or $2,500 (dashed line). At point A the firm can buy 200 units of labor and 50 units of capital for $3,000. Alternatively, this firm can chose combination B, buying 180 units of labor and 60 units of capital (substituting 10 units of capital for 20 units of labor) for the same value of $3,000.

It is clear that any change in the prices of labor or capital in the resource market leads to changes in the *L*- or *K*-intercept and slope of the isocost.

With these principles in mind, we now offer another approach to solving Type I optimization problems. We must find the lowest possible cost for a level of output Q_0 that is exogenously determined. For a firm with the production function $Q = f(K, L)$ and *total factor cost* function[5] $TFC = wL + rK$, the model can be formulated as

$$\text{Minimize } TFC = wL + rK$$
$$\text{Subject to } Q_0 = f(K, L)$$

From the production function, we derive the equation of the isoquant by solving for *K* in terms of *L* and Q_0. Recall that the slope of the isoquant $\dfrac{dK}{dL}$ is the marginal rate of technical substitution *MRTS* of capital for labor. From the total factor cost function we solve for *K* in terms of *L* and *TFC* to derive the isocost $K = \dfrac{TFC}{r} - \dfrac{w}{r}L$. The slope of this isocost, $\dfrac{dK}{dL} = -\dfrac{w}{r}$, is the marginal rate of factor substitution *MRFS* of capital for labor.

The optimal solution for the firm's minimization problem is reached at a point where the isoquant is tangent to an isocost, that is where *MRTS* = *MRFS*. Graphically the firm reaches equilibrium at point E in Fig. 10.5.

---

[5] A slightly fancier term than total variable cost *TVC*.

Fig. 10.5 Isocost meets isoquant

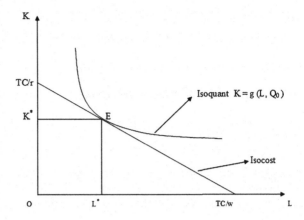

The following numerical example should help in understanding the isoquant–isocost Type I optimization method.

Consider that a firm with the Cobb–Douglas production function $Q = 2K^{0.5}L^{0.4}$ plans to produce 130 units of output. Assume the factor prices for labor and capital are $w = \$10$ and $r = \$20$ per unit. What is the firm's optimal combination of factors for minimizing its cost?

From the production function we derive the isoquant associated with 130 units as

$$130 = 2K^{0.5}L^{0.4} \longrightarrow K^{0.5} = 65L^{-0.4}$$

which leads to $K = 4225L^{-0.8}$

From the *TFC* function we derive the isocost equation

$$TFC = wL + rK \longrightarrow TFC = 10L + 20K$$
$$K = \frac{TFC}{20} - \frac{10}{20}L = 0.05TFC - 0.5L$$

Differentiating $K = 4225L^{-0.8}$ and $K = 0.05TFC - 0.5L$ with respect to L, we have

$$MRTS = -0.8(4225L^{-1.8}) = -3380L^{-1.8} \text{ and } MRFS = -0.5$$

At equilibrium *MRTS* must be equal to *MRFS*, therefore

$$-3380L^{-1.8} = -0.5 \longrightarrow L^{1.8} = 6760 \text{ so } L^* = 134.2 \text{ units}$$

Substituting for L in isoquant equation, we have

$$K = 4225(134.2)^{-0.8} \quad K = \frac{4225}{50.373} \text{ and } K^* = 83.9 \text{ units}$$

At the combination $(L^*, K^*) = (134.2, 83.9)$ the firm's cost of producing 130 units, or its total factor cost, is $TFC = 1342 + 1678 = 3020$

10.4 Type II Optimization Models

10.4.1 Substitution Method

In a Type II optimization model a firm's objective is to maximize profit subject to a total cost constraint. In this case the firm has a limited expenditure outlay, with its total cost TC exogenously determined. The model is

$$\text{Maximize} \quad \pi = TR - TC$$
$$\text{Subject to} \quad TC_0 = wL + rK + FC$$

Given that the total cost is fixed at TC_0, profit maximization is indeed achieved by revenue maximization. Since the price for a competitive firm is also exogenously determined, the revenue maximization becomes a matter of output maximization. For a firm with the production function $Q = f(K, L)$, the model is

$$\text{Maximize} \quad Q = f(K, L)$$
$$\text{Subject to} \quad TC_0 = wL + rK + FC$$

We will look at two different methods for solving this problem. The first method (generally called the *substitution method*) is rather straightforward. We solve for K (or L) in the cost function in terms of the other variable. We then substitute for K (or L) in the production function. This converts the bivariate production function into a univariate function. Finally, we simply apply the maximization routine.

Lets consider a firm with the same production function as in the previous example

$$Q = 2K^{0.5}L^{0.4}$$

Also assume that the price of labor and capital are the same as the above example, namely \$10 and \$20 per unit. Further assume that the firm's fixed cost is \$20 and its predetermined level of total cost is \$3,040. What would be the profit maximizing level of output? Using the total cost function,

$$TC_0 = wL + rK + FC \quad \text{or} \quad 3040 = 10L + 20K + 20$$

Solving for K, we have

$$K = \frac{3020}{20} - \frac{10}{20}L \quad \longrightarrow \quad K = 151 - 0.5L \qquad (10.21)$$

Substituting for K in the production function,

$$Q = 2(151 - 0.5L)^{0.5}L^{0.4} = 2\left[(151 - 0.5L)L^{0.8}\right]^{0.5} = \left(604L^{0.8} - 2L^{1.8}\right)^{0.5}$$

The *FOC* for a maximum is

$$\frac{dQ}{dL} = 0.5\left(0.8 * 604L^{-0.2} - 2 * 1.8L^{0.8}\right)\left(604L^{0.8} - 2L^{1.8}\right)^{-0.5} = 0$$

which is satisfied if

$$0.8 * 604L^{-0.2} - 2 * 1.8L^{0.8} = 0 \quad \longrightarrow \quad 483.2L^{-0.2} - 3.6L^{0.8} = 0$$

and the optimal value of L is $L^* = 134.2$ units. Substituting for L in (10.21), we have

$$K^* = 151 - 0.5(134.2) = 83.9 \text{ units}$$

The profit maximizing output is then

$$Q^* = 2(83.9)^{0.5}(134.2)^{0.4} = 130 \text{ units}$$

The second-order necessary condition for a maximum is also satisfied:

$$\frac{d^2Q}{dL^2} = -0.2 * 483.2L^{-1.2} - 0.8 * 3.6L^{-0.2} < 0$$

10.4.2 The Method of the Lagrange Multiplier

Another method for solving a constraint optimization problem is the Lagrange Multiplier method. The first step is to write the *Lagrangian Function* Y^6 as

$$Y = f(K, L) + \lambda(TC - wL - rK) \tag{10.22}$$

where the Greek letter λ (lambda) is called the *Lagrange multiplier*.

The Lagrangian function for our example is

$$Y = 2K^{0.5}L^{0.4} + \lambda(3020 - 10L - 20K)$$

Expression $TC - wL - rK$ in (10.22) is simply the model's cost constraint. If we can identify some values of L and K that satisfy this constraint then the last term

---

[6] Commonly L or script L ($\mathcal{L}$) is used for the Lagrangian function. Since we use L for labor it is denoted here by Y.

in (10.22) will become zero regardless of the value of λ. We must consider Y as a function of three variables K, L and λ and maximize it. The *FOC* for an optimum will consist of the following set of three equations with three unknowns K, L, and λ:

$$\begin{cases} \dfrac{\partial Y}{\partial K} = \dfrac{\partial f}{\partial K} - r\lambda = 0 \longrightarrow f_K - r\lambda = 0 \\[2ex] \dfrac{\partial Y}{\partial L} = \dfrac{\partial f}{\partial L} - w\lambda = 0 \longrightarrow f_L - w\lambda = 0 \\[2ex] \dfrac{\partial Y}{\partial \lambda} = TC - wL - rK = 0 \end{cases} \qquad (10.23)$$

In our example,

$$\begin{cases} \dfrac{\partial Y}{\partial K} = K^{-0.5}L^{0.4} - 20\lambda = 0 \\[2ex] \dfrac{\partial Y}{\partial L} = 0.8K^{0.5}L^{-0.6} - 10\lambda = 0 \\[2ex] \dfrac{\partial Y}{\partial \lambda} = 3020 - 10L - 20K = 0 \end{cases} \qquad (10.24)$$

The third equation in (10.23) and (10.24) is a restatement of the cost constraint. Rewriting the first two equations in (10.24) as

$$K^{-0.5}L^{0.4} = 20\lambda$$
$$0.8K^{0.5}L^{-0.6} = 10\lambda$$

and dividing the first one by the second, we have

$$\frac{K^{-0.5}L^{0.4}}{0.8K^{0.5}L^{-0.6}} = \frac{20\lambda}{10\lambda}$$

After simplifying, the expression becomes

$$\frac{L}{0.8K} = 2 \longrightarrow L = 1.6K$$

Substituting $1.6K$ for L in the third equation (the cost constraint) in (10.24), we have

$$3020 - 10(1.6K) - 20K = 0 \longrightarrow 36K = 3020$$

which leads to the same answer we obtained by the substitution method, $K^* = 83.9$ units. Similarly the optimal values of L and Q are $L^* = 134.2$ units and $Q^* = 130.02$

units. By substituting K^* and L^* for K and L in the first equation of (10.24) we obtain the optimal value of λ as $\lambda^* = 0.0387$.

It is clear that both the substitution and Lagrange multiplier methods lead to the same optimal values of the endogenous variables. But the method of Lagrange multiplier has two advantages over the substitution method. It is not always possible or easy to solve for one variable in terms of the other if the constraint is a complicated function; in that case the substitution method stalls. The second advantage is calculation of the Lagrange multiplier λ. The optimal value of the multiplier λ^* measures the sensitivity of Y—and by extension the optimal solution—to small changes in the constraint. In the above example the multiplier is 0.03875. This means that if the firm changes the total cost outlay by a small amount the level of output would changes by roughly λ times that amount, that is $\Delta Q = \lambda^* * \Delta TC$. For example, if the firm increases its cost outlay, net of fixed cost, from 3,020 by 30.2 units (1 %) to 3,050.2, its new level of output would be 131.19. The change in the level of output of $131.19 - 130.02 = 1.17$ is about $0.0378 * 30.2 = 1.17$ units. The optimal value of λ sometimes is interpreted as a measure of marginal cost.

There is an important relationship between the Type I and Type II constraint optimizations. To highlight this connection we summarize the results of the numerical examples we solved for each case. In the Type I optimization the firm with the production function $Q = 2K^{0.5}L^{0.4}$ and prices of L and K given as 10 and 20 sets the objective of producing 130 units of output at the minimum total cost. Here the objective is cost minimization and the constraint is the exogenously determined level of output. We found the optimal solution for the endogenous variables L^*, K^*, TC^* as 134.2, 83.9, and \$3,020, respectively.

In the Type II optimization the same firm sets its total cost outlay (net of fixed cost) at \$3,020 with the objective of maximizing its output. We found the optimal solution for the endogenous variables L^*, K^*, Q^* as 134.2, 83.9, and 130, respectively.

It is clear that we arrive at the same solution whether the objective is the cost minimization or output maximization. Commonly one of the optimization problems is called *Primal*, and the other problem its *Dual*.

Graphically, the optimal factor use (L^*, K^*) for both problems occurs at the point of tangency between an isoquant and an isocost. If the level of output is given, we must search for the lowest possible isocost tangent to the isoquant associated with the given output. If the total cost is given, we must search for the highest possible isoquant tangent to this isocost. Whatever problem is considered as primal, the other is its dual counterpart.

In the next example we will consider a more comprehensive case, where a firm with given set of production and cost functions searches for the optimal level of output and factor inputs.

10.5 Production Function, Cost Function, and Profit Maximization

10.5.1 Isoquant Meets Isocost, Again

Assume a firm in a competitive market has the following Cobb–Douglas production function:

$$Q = 1.5K^{0.5}L^{0.4}$$

Also assume that the firm has the following cubic cost function:

$$TC = Q^3 - 24Q^2 + 2000$$

If the price of this good in the market, P, is 1260 and the prices of labor and capital are 100 and 80 per unit, respectively, what is the firm's optimal level of output and its profit maximizing factor utilization?

Determining the optimal level of output is a simple routine: the firm produces at the level where the marginal cost is equal the market price, that is,

$$MC = \frac{dTC}{dQ} = 3Q^2 - 48Q$$

$$3Q^2 - 48Q = 1260 \longrightarrow Q^2 - 16Q - 420 = 0$$

An acceptable solution to this quadratic equation is $Q^* = 30$. We now must determine the least-cost combination of factors that produces this output and generates the maximum profit for the firm.

At the level of output $Q^* = 30$ the total cost is

$$TC = (30)^3 - 24 * (30)^2 + 2000 = 7400$$

The isocost associated with this total cost is

$$TC = wL + rK + FC = 7400 \text{ or } 100L + 80K + 2000 = 7400$$

and the isocost equation can be expressed (in terms of K) as

$$K = 67.5 - 1.25L$$

The isoquant associated with 30 units of output is

$$30 = 1.5\, K^{0.5}\, L^{0.4} \longrightarrow K^{0.5} = \frac{20}{L^{0.4}} = 20L^{-0.4}$$

and after squaring both sides, we have

Table 10.2 Values of $f(L)$ for selected values of L

L	11	13	15	29	33	35	37	39	41
$f(L)$	33.9	1.1	−25.4	−62.1	−30.5	−8.2	18.1	48.6	82.9

$$K = 400\,L^{-0.8}$$

Minimum cost/Maximum profit combination of factors occurs at the point of tangency of the isoquant and the isocost, Therefore

$$K = 67.5 - 1.25\,L = 400\,L^{-0.8} \text{ or}$$
$$1.25L^{1.8} - 67.5L^{0.8} + 400 = 0$$

To solve this 1.8-degree polynomial we use the Taylor expansion method discussed in Chap. 7 and linear approximate the function

$$f(L) = 1.25L^{1.8} - 67.5L^{0.8} + 400$$

by expanding it around an estimate of its zero(s).

Table 10.2 shows that this function has two zeros: one between 10 and 15 and the other between 35 and 37.

For the first zero we linear approximate the function by expanding it around 13, and for the second zero by expanding it around 36. We assemble the required elements

$$f(13) = 1.25(13)^{1.8} - 67.5(13)^{0.8} + 400 = 1.114$$
$$f(36) = 1.25(36)^{1.8} - 67.5(36)^{0.8} + 400 = 4.429$$
$$f'(L) = 2.25L^{0.8} - 54L^{-0.2}$$
$$f'(13) = 2.25(13)^{0.8} - 54(13)^{-0.2} = -14.818$$
$$f'(36) = 2.25(36)^{0.8} - 54(36)^{-0.2} = 13.186$$

The first linear approximation is then

$$f(L) = 1.114 - 14.818(L - 13) = 193.748 - 14.818L$$
$$193.748 - 14.818L = 0 \longrightarrow L = \frac{193.748}{14.818} = 13.075$$

By substituting for L in the equation $K = 67.5 - 1.25L$ or $K = 400\,L^{-0.8}$ we find that $K = 51.157$ units.

The second linear approximation is

$$f(L) = 4.429 + 13.186(L - 36) = -470.267 + 13.186L$$

$$-470.267 + 13.186L = 0 \longrightarrow L = \frac{470.267}{13.186} = 35.66 \text{ units}$$

and $K = 67.5 - 1.25(35.66) = 22.92$ units.

Both combinations of labor and capital (13.075, 51.157) and (35.66, 22.92) produce 30 units of output and cost \$5,400. The firm's profit in both cases is maximized at \$30,400.

10.6 The Cobb–Douglas Production Function and Profit Maximization

10.6.1 Type III Optimization and Restrictions on α and β

In general the main objective of a firm is to find its optimal level of output, i.e. the level of output that maximizes its profit. Consider a firm with a CDpf operating in a competitive market. As before, P is the market price of the product and all firms pay w per unit of labor and r per unit of capital. Our firm's total revenue is PQ and its total cost is $wL + rK + FC$, where FC is the fixed cost. The profit function of this firm is

$$\pi = \pi(Q, L, K) = TR - TC = PQ - wL - rK - FC$$
$$\pi = P * f(L, K) - TC = PAK^\alpha L^\beta - wL - rK - FC$$

Here π is a function of L and K. The first-order necessary condition (FOC) for an unconstrained maximization of π is

$$f_L = \frac{\partial \pi}{\partial L} = \frac{\partial f}{\partial L} = \beta PAK^\alpha L^{\beta-1} - w = 0$$
$$f_K = \frac{\partial \pi}{\partial K} = \frac{\partial f}{\partial K} = \alpha PAK^{\alpha-1} L^\beta - r = 0$$

which are the same as Eqs. (10.13) and (10.14) introduced in Sect. 10.2. The second-order sufficient condition requires that

$$f_{LL} \text{ and } f_{KK} < 0 \text{ and } f_{LL} * f_{KK} > (f_{LK})^2$$

$$f_{LL} = \frac{\partial^2 \pi}{\partial L^2} = \beta(\beta - 1)PAK^\alpha L^{\beta-2} < 0 \tag{10.25}$$

$$f_{KK} = \frac{\partial^2 \pi}{\partial K^2} = \alpha(\alpha - 1)PAK^{\alpha-2} L^\beta < 0 \tag{10.26}$$

$$f_{LK} = f_{KL} = \frac{\partial^2 \pi}{\partial L \partial K} = \frac{\partial^2 \pi}{\partial K \partial L} = \alpha\beta PAK^{\alpha-1}L^{\beta-1} \tag{10.27}$$

It is clear from conditions (10.25) and (10.26) that we must have $\beta - 1 < 0$ and $\alpha - 1 < 0$, leading to the first set of restrictions on α and β:

$$\alpha < 1 \text{ and } \beta < 1$$

Next we must check $f_{LL} * f_{KK} > (f_{LK})^2$. Using (10.25), (10.26), and (10.27), we have

$$[\beta(\beta - 1)PAK^{\alpha}L^{\beta-2}][\alpha(\alpha - 1)PAK^{\alpha-2}L^{\beta}] > [\alpha\beta PAK^{\alpha-1}L^{\beta-1}]^2$$
$$\alpha\beta(\alpha - 1)(\beta - 1)P^2A^2K^{2\alpha-2}L^{2\beta-2} > \alpha^2\beta^2 P^2 A^2 K^{2\alpha-2}L^{2\beta-2}$$

Simplifying the above expression we get $(\alpha-1)(\beta-1) > \alpha\beta$, which upon further simplification lead us to the restriction that $\alpha+\beta < 1$. This is a very interesting result: a firm with a Cobb–Douglas production function can maximize its profit if and only if it operates under decreasing returns to scale! Is there any economic explanation for this result?

Yes, indeed there is. As we learned earlier, if a firm operates under increasing returns to scale it can systematically and constantly expand its size or capacity,[7] producing larger and larger volumes of output and making more profit. In this case the profit function of the firm is a *monotonically increasing* function of output, with its graph extending upward into the space, like a mountain without a peak. In short, the maximization routine is irrelevant in this case. However, if this firm chooses to operate at a certain level of output Q_0 then the optimization routine for determining the least-cost combination of labor and capital is applicable.

For a firm with constant returns to scale, its profit increases as the firm expands its output, but beyond a certain output level the profit plateaus at a constant, flat rate. However, for a firm with decreasing returns to scale, as the firm expands its output the profit first increases, reaches a peak and then begins to decline. Profit maximization is naturally most suitable in this case.

The following three graphs depict the trajectory of *unit profit* (profit/output) for three CDpfs:

$$Q = 1.5K^{0.5}L^{0.7} \qquad \text{for increasing returns to scale (Fig. 10.6)}$$
$$Q = 1.5K^{0.5}L^{0.5} \qquad \text{for constant returns to scale (Fig. 10.7)}$$
$$Q = 1.5K^{0.5}L^{0.4} \qquad \text{for decreasing returns to scale (Fig. 10.8)}$$

To calculate unit profit, the price of a unit of output was assumed to be \$50 and the unit prices of labor and capital assumed to be \$15 and \$20, respectively. The fixed cost was set at \$100.

---

[7] By building more facilities or through merger and acquisition.

Fig. 10.6 Unit profit: increasing returns to scale

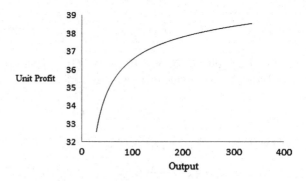

Fig. 10.7 Unit profit: constant returns to scale

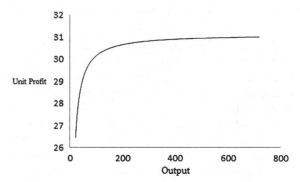

Fig. 10.8 Unit profit: decreasing returns to scale

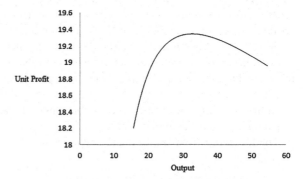

10.6.2 An Example of Type III Unconstrained Maximization

Consider a firm with the CDpf $Q = 10K^{0.5}L^{0.3}$. Assuming P, w, r, and FC are 50, 155, 82.5, and 200 respectively. We want to determine the firm's profit maximizing level of output, total profit, and the optimal combination of labor and capital input. The firm's profit function is

$$\pi = \pi(Q, L, K) = TR - TC = PQ - wL - rK - FC$$

After substituting for Q, we have

$$\pi = P * f(L, K) - TC = PAK^{\alpha}L^{\beta} - wL - rK - FC$$
$$\pi = 50\,(10K^{0.5}L^{0.3}) - 155L - 82.5K - FC$$
$$\pi = 500\,K^{0.5}L^{0.3} - 155L - 82.5K - FC$$

The first-order necessary condition for optimization requires that

$$f_L = 150\,K^{0.5}L^{-0.7} - 155 = 0 \tag{10.28}$$

$$f_K = 250\,K^{-0.5}L^{0.3} - 82.5 = 0 \tag{10.29}$$

We can write (10.29) as

$$250\,K^{-0.5}L^{0.3} = 82.5 \longrightarrow K^{-0.5}L^{0.3} = 0.33 \longrightarrow K^{-0.5} = 0.33L^{-0.3}$$

or $K^{0.5} = 3.0303\,L^{0.3}$.

Substituting for $K^{0.5}$ in (10.28), we get

$$150\,(3.0303\,L^{0.3})\,L^{-0.7} - 155 = 0 \longrightarrow 454.545\,L^{-0.4} = 155 \text{ and}$$
$$L^{-0.4} = \frac{155}{454.545} = 0.341 \longrightarrow L^{0.4} = 2.9325 \longrightarrow (L^{0.4})^{2.5} = (2.9325)^{2.5}$$

which gives us the profit optimizing level of labor input $L^* = 14.727$ units. By substituting for L in $K^{0.5} = 3.0303L^{0.3}$ we get the profit optimizing level of capital input $K^* = 46.115$ units.

The second-order sufficient condition for a maximum requires that

$$f_{LL} \text{ and } f_{KK} < 0 \text{ and } f_{LL} * f_{KK} > (f_{LK})^2$$

Differentiating (10.28) and (10.29) partially, we have

$$f_{LL} = -105\,K^{0.5}L^{-1.7} \text{ and } f_{KK} = -125\,K^{-1.5}L^{0.3},$$
$$\text{and } f_{LK} = f_{KL} = 75K^{-0.5}L^{-0.7}$$

Evaluating f_{LL}, f_{KK}, and f_{LK} at L^* and K^*,

$$f_{LL} = -105\,(46.115)^{0.5}\,(14.727)^{-1.7} < 0$$
$$f_{KK} = -125\,(46.115)^{-1.5}\,(14.727)^{0.3} < 0$$

$$f_{LL} * f_{KK} = \left[-105(46.115)^{0.5}(14.727)^{-1.7}\right]\left[-125(46.115)^{-1.5}(14.727)^{0.3}\right]$$
$$= 13125\,(46.115)^{-1}\,(14.727)^{-1.4}$$

$$(f_{LK})^2 = (75 \, K^{-0.5}L^{-0.7})^2 = 5625 \, (46.115)^{-1}(14.727)^{-1.4}$$

Clearly $f_{LL} * f_{KK} > (f_{LK})^2$. Therefore all three $SOC's$ for a maximum are satisfied. The firm's profit maximizing level of output and the total profit are

$$Q^* = 10 \, (46.115)^{0.5}(14.727)^{0.3} = 152.179 \text{ units}$$

$$\pi^* = PQ - wL - rK - FC$$
$$= 50 * 152.179 - 155 * 14.727 - 82.5 * 46.115 - 200 = 1321.79$$

10.7 Exercises

1. Assume a firm has the production function $Q = 2.5K^{0.4}L^{0.5}$. If this firm employs 15 units of labor and 20 units of capital and the price of the good in the market is $100 per unit, find the following:

   (a) Marginal product of labor and capital.

   (b) Marginal revenue product of labor and capital.

   (c) Average product of labor and capital.

   Should the fact that in part (a) the marginal product of labor is larger than the marginal product of capital encourage this firm to substitute labor for capital?

2. Assume a firm has the production function $Q = 2.5K^{0.4}L^{0.5}$.

   (a) Write the firm's Isoquant equation associated with the production of 20 units of output.

   (b) Calculate $MRTS$ for labor input $L = 6$ and $L = 8$.

   (c) Assume labor costs $80 per unit and capital rental costs $200 a unit. What is the cost of producing 20 units of output when the firm employs 6 units of labor? What about 8 units of labor? Which one costs more? Could this be anticipated by the $MRTS$s you calculated in part(b)?

   (d) What is the firm's least-cost combination of labor and capital for producing 20 units of output?

   (e) What is the firm's total cost at the least-cost combination of labor and capital?

   (f) What is the firm's profit if the price of this good in the market is $200 per unit?

   (g) Write the firm's Isocost equation.

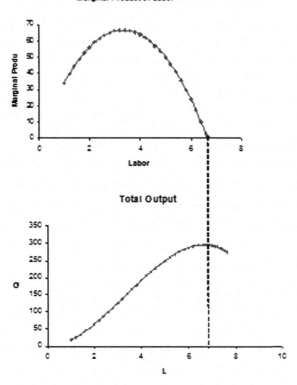

3. A firm has the following short run production function

$$Q = 20L^2 - 2L^3$$

(a) Write the firm's marginal product of labor.

(b) At what level of labor input does the MPP_L reach its maximum? At what level does it become zero?

(c) Show that the output reaches its maximum at the level of labor input where MPP_L is equal zero (use the following graphs as a guide).

4. A firm with a CD production function $Q = 2K^{0.5}L^{0.4}$ plans to produce 500 units of output per day. Assume the price of labor and capital are $w = \$50$ and $r = \$100$ a day.

(a) What is the firm's optimal combination of factors that minimizes its cost?

(b) What is the market price of this good if this firm makes a $2,500 profit each day?

5. Consider a firm with the production function $Q = 1.5K^{0.5}L^{0.5}$.

   (a) Write equations of three isoquants associated with 30, 40, and 60 units of output.

   (b) If the prices of labor and capital are $40 and $160 per unit, respectively, find the least-cost combination of labor and capital for 30, 40, and 60 units of output by setting *MRTS* equal to the factor price ratio.

   (c) Show that the three combinations in part (b) fall on a straight line (use the following graph as a guide). In general, if the production function is homogeneous then the least-cost combinations of all different levels of output fall on a straight line. This line is called the *Output Expansion Path*. Note that the capital–labor ratio in all three cases in this problem is $\frac{1}{4}$, which is equal to the ratio of prices of labor and capital: $\frac{10}{40} = \frac{1}{4}$. Equation (10.16) in Sect. 10.4, which relates K to L by

$$K = \frac{\alpha w}{\beta r} L$$

   is indeed the equation of the linear output expansion path.

   (d) Write the equation of the linear expansion path for this production function.

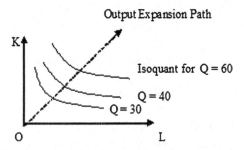

Output Expansion Path

6. For the production function in problem (3), $Q = 20L^2 - 2L^3$, write the equation for the average product of labor. At what level of labor does input the *AP* of labor reach its peak? At what level does it drop to zero?

7. A farmer uses land and labor to grow soybeans. He has the following production function $Q = 126.2A^{0.6}L^{0.4}$ where Q is the amount of soybeans in bushels, A is land in acres and L is number of farm hands.

   (a) If this farmer has 100 acres of land, how many farm workers must he employ in order to produce 10,000 bushels of soybeans? What is the marginal product of labor at this level of employment?

   (b) Assume this farmer pays each farm worker $1,000 over the growing season and spends $2,000 for seed and fertilizer. If the price of soybeans is $14.75 a bushel, what is the farmer's profit?

(c) Is this farmer paying his workers the value of their marginal product?

(d) If this farmer doubles the amount of land and labor, how many bushels of soybeans he can produce?

8. A firm operates with the production function $Q = 5.4K^{0.3}L^{0.5}$.

   (a) If the price of labor and capital are $10 and $20, respectively, and the firm's management has decided to spent only $7,000 daily on production costs, how many units of output can the firm produce? (Use the substitution method).

   (b) How many units of labor and capital does firm employ?

   (c) If the price of the good is $20 per unit, what is the daily profit for this firm?

9. Redo part (a) of problem (8) using the method of Lagrange Multiplier.

10. Assume the firm in problem (8) has a fix 213.71 units of capital.

   (a) Write the firm's short run production function.

   (b) If the price of the good, labor, and capital are $20, $10, and $20 respectively, what is the profit maximizing level of output and employment ?

   (c) What is this firm's daily profit?

11. Consider a firm with the production function $Q = KL$.

   (a) Is this production function homogeneous? If yes, of what degree?

   (b) If labor and capital cost $100 and $200 per unit, what is the equation of the output expansion path?

   (c) If the firm wants to spend $8,000 on labor and capital, what is the maximum output it can produce? How many units of L and K does it employ?

12. For the production function $Q = K^{0.35}L^{0.5}$ write the marginal product of K and L and determine whether they are downward sloping. Let $K = 50$ and write the equations of total, average, and marginal product of L and graph them.

13. Assume a firm's output is a function of capital, labor, and energy (E) given by

$$Q = AK^{\alpha}L^{\beta}E^{\gamma}$$

   (a) Show that this function is homogeneous and determine the degree of homogeneity of the function.

   (b) What should the relations between α, β, and γ be in order to have a constant returns to scale? Increasing or decreasing returns to scale?

14. In an interesting paper published in the journal *American Economic Review* in September 1970, Professor Martin Weitzman from Yale University estimated a production function for Russia (then the Soviet Union) based on a large amount

of macroeconomic data from 1950 to 1969. The production function he used in his estimate

$$Q = \gamma \left[\delta K^{-\rho} + (1 - \delta) L^{-\rho} \right]^{-\frac{1}{\rho}}$$

is known as the *Constant Elasticity of Substitution* (CES) production function. Weitzman estimated the parameters of the function and expressed it as

$$Y = 0.8(0.64K^{-1.5} + 0.36L^{-1.5})^{-0.67} \text{ (where } Y \text{ is output)}$$

(a) Is this function homogeneous?

(b) Write the marginal physical product of labor MPP_L and capital MPP_K.

(c) If $K = 5{,}000$ units and $L = 8{,}000$ units, what is Y, MPP_L, and MPP_K?

15. Assume a firm with the production function $Q = AK^{\alpha} L^{\beta}$. Show that α is the output elasticity of labor, ϵ_{QL}, and β is the output elasticity of capital, ϵ_{QK}.

Chapter 11
Economic Dynamics

11.1 Economic Dynamics and Difference Equations

Most macro- and micro-economic variables are time series variables. We can track changes of time series variables over time in two ways, *continuously* or *discretely*. In continuous case values of a variable are measured at every moment of time. Recall from Chap. 2 that periodicity of a time series variable is the time interval over which values of the variable are measured or computed. In continuous case measurement is made over infinitesimally small time intervals to the extent that periodicity of the variable approaches zero. However, values of most time series variables, particularly in economics, are recorded or measured at predetermined equal-interval periods. Variables with weekly, monthly, quarterly, semi-annual and annual periodicity are common in economics. Also recall that some variables are *accumulated* or "stock" variables and cannot be measured at every moment of time. Rainfall, volume of production, and sales revenue are example of these types of variables. In general, only "flow" variables could be measured continuously.

Study of how economic variables evolve over time is called *Economic dynamics*. Study of behavior of discreet time series variables is known as *period analysis*. While the mathematical techniques for analyzing time series variables with continuous periodicity are differential equation and integral calculus, researchers use *difference equations* for study and analysis of variables with discreet periodicity. First and higher order difference equations often arise in period analysis as part of formulation and solution of economic models relating two or more time series variables.

Static analysis is study of economic entities—households, firms, markets, sectors, and the whole economy—at *rest* or equilibrium. In static analysis our task is limited to determining values of the model's endogenous variables that satisfy certain equilibrium condition(s). If change in parameters or exogenous variables of the model perturbs the system to the extent that it cannot sustain its equilibrium, then it moves to a new position. In *comparative static analysis* economists compare and contrast the general properties of old and new equilibrium and values of the model's endogenous variables. The objective of *dynamic analysis* is to trace the time

S. Vali, *Principles of Mathematical Economics*,
Mathematics Textbooks for Science and Engineering 3,
DOI: 10.2991/978-94-6239-036-2_11, © Atlantis Press and the authors 2014

path of endogenous variables in their journeys from one "equilibrium" to another. In comparative static we make the assumption that the endogenous variables, after a series of adjustment, inevitably *reach* a new equilibrium. In contrast, part of dynamic analysis task is to determine whether, or under what condition, the endogenous variables reach equilibrium.

We first encountered an example of a difference equation in our presentation of general linear demand and supply model. Recall that we introduced an alternative formulation to supply function in which instead of relating quantity supplied and price *contemporaneously*, it was assumed that producers make their supply decision based on price they received in the previous period. That is

$$Q_{st} = -c + dP_{t-1}$$

With the demand function specified as

$$Q_{dt} = a - bP_t$$

and invoking the equilibrium condition $Q_{dt} = Q_{st}$ we obtain

$$a - bP_t = -c + dP_{t-1} \quad or$$

$$P_t + \frac{d}{b}P_{t-1} = \frac{a+c}{b} \tag{11.1}$$

Equation (11.1) is an example of *first order difference equation with constant coefficients*.

11.1.1 Basic Definitions

Consider a time series variable with a discreet periodicity. We denote this variable by y_t, where the time index t takes values $0, 1, 2, \ldots, T$. For $t = 0$, y_0 is the *initial value* of the variable. For $t = 1$, y_1 denotes the value of the variable after one period (at the end of the first or beginning of the second period). If the variables periodicity is weekly, then y_1 denotes value of the variable at the end of the first week (or begging of the second week). If periodicity is hourly, monthly, quarterly, or annually, then y_1 is the value of the variable at the end of the first hour, month, quarter, or year. Similarly, if $t = 2, 3, \ldots, T$ then $y_2, y_3, \ldots, y_T$ are the values of the variable at the end of the second, third,..., and Tth period. T represents the *terminal period* and y_T represents the *terminal value* of the variable.

Suppose y_t denotes the GDP of the United States from 1960 to 2006. Here $y_{1960}, y_{1961}, \ldots, y_{2006}$ stand for the GDP of the United States from 1960 to 2006, a 47-year worth of data. A more compact presentation is y_t, $t = 1960, 1961, \ldots, 2006$. If we consider year 1960 the initial year, then expression y_t, $t = 0, 1, 2, \ldots, 46$

would convey exactly the same information. Notice that if y_t represent the GDP for a generic year, say 1987, then y_{t+1} and y_{t+2} represent the GDP for 1988 and 1989. Similarly y_{t-1} represents the GDP for 1986. y_t and y_{t-1} represent two consecutive values of the variable, so do y_{t+2} and y_{t+1}, or y_{t+6} and y_{t+7}. In all these cases values of the variable are "one period" apart, or one is one period *lagged* of the other. In general y_{t+n-1} is one period lagged value of y_{t+n}. We can similarly express "two period" apart or lagged values of y as y_{t-2} and y_t (or y_t and y_{t+2}).

Consider a time series variable y_t, $t = 0, 1, 2, \ldots, T$. The *first difference* of y_t denoted by Δy_t is defined as difference of y at time $t + 1$ and its one period lagged, i.e.

$$\Delta y_t = y_{t+1} - y_t \qquad\qquad t = 0, 1, 2, \ldots, T - 1 \qquad (11.2)$$

If t in (11.2) is, for instance, 7 then $\Delta y_7 = y_8 - y_7$.

The *second difference* of y_t denoted by $\Delta(\Delta y_t)$ or $\Delta^2 y_t$ is

$$\Delta^2 y_t = \Delta(\Delta y_t) = \Delta(y_{t+1} - y_t) = \Delta y_{t+1} - \Delta y_t = y_{t+2} - y_{t+1} - (y_{t+1} - y_t)$$
$$= y_{t+2} - 2y_{t+1} + y_t$$

Similarly we can define the higher order differences of y.

We are now in a position to define a *difference equation*. An equation relating the values of a time series variable to one or more of its differences is called a difference equation. Since this book is an introductory text, we will limit ourselves to only the first difference and hence first order difference equations. A *first order linear difference equation* can be expressed as

$$f_1(t)\, y_{t+1} + f_2(t)\, y_t = g(t) \qquad\qquad t = 0, 1, 2, \ldots$$

where $f_1(t), f_2(t)$ and $g(t)$ are functions of t and y_{t+1} and y_t appear in linear form. If y_{t+1} or y_t appear with powers different than 1, or as cross product, then the difference equation would be *nonlinear*. If $f_1(t)$ and $f_2(t)$ are nonzero constants, then the first order linear difference equation is a first order linear difference equation with *constant coefficients*. If $g(t)$ is also a constant then we have an *autonomous* first order difference equation, otherwise the difference equation would be a *non-autonomous*. If $g(t)$ is zero, then the difference equation is a *homogeneous* difference equation, otherwise it is *non-homogeneous*.

Examples

1. $(t + 1)y_{t+1} - y_t = t^2$ is a non-autonomous linear first order difference equation.
2. $5y_{t+1} - 3y_t^2 = 5$ is an autonomous but nonlinear first order difference equation.
3. $3y_{t+1} + 5y_t = 10$ is an autonomous linear first order difference equation.

4. $3y_t + 5y_{t-1} = 10$ is an autonomous linear first order difference equation. This equation is the same as example 3.
5. $8y^{t+2} + 5y_{t+1} - 5y_t = 6$ is a second order linear autonomous difference equation.
6. $y_t - 5y_{t-1} = 0$ is a homogeneous linear first order difference equation.

All equations in examples 1–5 are non-homogeneous equations.

11.2 Solution of a Difference Equation

In what follows we will focus on solution to a first order linear difference equation with constant coefficient. Consider the following homogeneous difference equation

$$y_{t+1} - 4y_t = 0 \tag{11.3}$$

A solution to a difference equation like (11.3) is a function which makes this equation a true statement for all values of t. Let's try equation $y_t = 4^t$. Notice that if $y_t = 4^t$ then $y_{t+1} = 4^{t+1}$. Substituting for y_{t+1} and y_t in (11.3) we have

$$y_{t+1} - 4y_t = 4^{t+1} - 4(4^t) = 4^{t+1} - 4^{t+1} = 0$$

Since $y_t = 4^t$ makes the left hand side (LHS) of the equation equal to its right hand side (RHS), it is a solution to the difference equation.

But $y_t = 4^t$ is not the only solution, $y_t = C4^t$, where C is any constant, is also a solution. To check

$$y_{t+1} - 4y_t = C4^{t+1} - 4(C4^t) = C4^{t+1} - C4 * 4^t = C4^{t+1} - C4^{t+1} = 0$$

so the function $y_t = C4^t$ satisfies the difference equation. This function is called the *general solution* of the difference equation (11.3). If we assign an arbitrary value, say 3, to C, we get what is called a *particular solution*. For $C = 3$ we have

$$yt = C4^t = 3 * 4^t \quad \text{as a particular solution}$$

It is clear that different values of C lead to different particular solutions. There are, indeed, infinitely many particular solutions. But if the initial value for y is specified, C can be uniquely determined. Assume, for example, that $y_0 = 12$. In that case

$$y_0 = C4^0 = C \quad \text{therefore} \quad C = 12 \quad \text{and} \quad y_t = 12(4^t) = 3 * 4(4^t) = 3(4^{t+1})$$

$y_t = 3(4^{t+1})$ is a unique solution that satisfies both the difference equation and the initial condition.

11.2.1 Solution to First Order Homogeneous Difference Equation

The general form of the first order homogeneous difference equation is

$$y_{t+1} - ay_t = 0 \quad \text{where } a \text{ is a constant and} \quad t = 0, 1, 2, \ldots.$$

We can write this equation as

$$y_{t+1} = ay_t \quad t = 0, 1, 2, 3, \ldots \tag{11.4}$$

At $t = 0$ the initial value of y is y_0. For $t = 1$ we have

$$y_1 = ay_0$$

and for $t = 2$

$$y_2 = ay_1 = a(ay_0) = a^2 y_0$$

Similarly for $t = 3$ and $t = 4$

$$y_3 = ay_2 = a(a^2 y_0) = a^3 y_0$$
$$y_4 = ay_3 = a(a^3 y_0) = a^4 y_0$$

A pattern emerges from using this iterative method. This pattern leads us to conjecture that the general solution to (11.4) can be expressed as

$$y_t = a^t y_0 \tag{11.5}$$

By replacing y_0 with C, a generic symbol for constant, the general solution for (11.4) is

$$y_t = Ca^t$$

Notice that if a in difference equation in (11.4) is 2, then the general solution is $y_t = C2^t$.

11.2.2 Behavior of Solution of First Order Homogeneous Difference Equation

It is clear from (11.5) that values of y, as it evolves through time, depends on a (y_0 being a positive constant, simply amplifies the effect of a). To visually capture the impact of a, the following four graphs are prepared. The first graph, Fig. 11.1, is created by assuming that $a = 2$ and y_0 is 10 (solution to difference equation $y_{t+1} - 2y_t = 0$ with $y_0 = 10$). Figure 11.2 is the graph of the difference equation

Fig. 11.1 Graph of
$y_t = 10(2)^t$

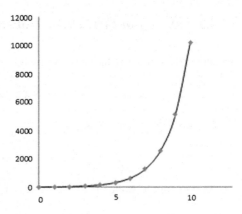

Fig. 11.2 Graph of
$y_t = 10(-2)^t$

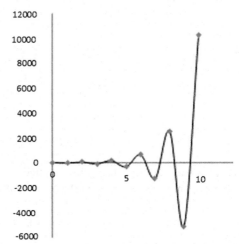

when $a = -2$. Figure 11.3 is the graph of the difference equation when $a = 1/2$ and Fig. 11.4 is the graph of the equation when a is set to be $-1/2$. All the graphs were generated by letting t takes values from 0 to 10.

The behavior of solution to the homogeneous first order difference equation depends on the sign and magnitude of a. For different values of a, $y_t = y_0 a^t = 10 a^t$ exhibits four different behaviors as t increases, they are summarized bellow:

Examples

1. Consider a close economy with no government (we can relax this assumption without any significant complication). Denoting the National Income, Personal Consumption Expenditure, Gross Private Domestic Investment, and Gross Domestic Saving at time t by $Y_t, C_t, I_t,$ and S_t we will develop two macroeconomic models who's

Fig. 11.3 Graph of
$y_t = 10(\frac{1}{2})^t$

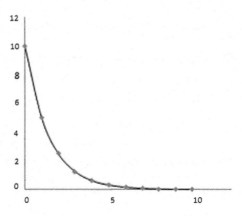

Fig. 11.4 Graph of
$y_t = 10(-\frac{1}{2})^t$

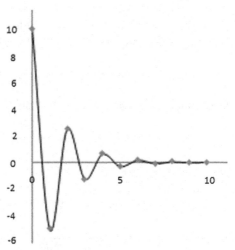

If $a > 1$	y_t increases steadily and without limit, or increases *monotonically* and *diverges* to extremely large number. This case signifies the *exponential growth* of y over time. (Fig. 11.1)
If $a < -1$	y_t alternates between positive and negative values and increases without limit, it shows an *oscillatory divergent* (Fig. 11.2)
If $0 < a < 1$	y_t declines steadily towards 0, i.e. *monotonically decreases* and *converges* to 0. This case exhibits the *exponential decay* of the time series variable y (Fig. 11.3)
If $-1 < a < 0$	y_t alternates between positive and negative values and decreases towards 0. *Damp oscillatory, converging* to 0 (Fig. 11.4)
Case of $a = 1$	is a trivial case, leading to $y_t = y_0$ for all values of t.

solution requires solving a homogeneous first order difference equation. We first write the national income identity as

$$Y_t = C_t + I_t \tag{11.6}$$

We next make two assumptions:

(a) Aggregate consumption is proportional to aggregate income, that is

$$C_t = \alpha Y_t \tag{11.7}$$

α, of course, is the familiar marginal propensity to consume

(b) Private investment in any period is proportional to increase in consumption in that period over the preceding period

$$I_t = \beta(C_t - C_{t-1}) \tag{11.8}$$

(11.8) is the formulation of the *acceleration principle* and β is called the *relation*.[1]
Substituting for C_t and C_{t-1} from (11.7) in (11.8), we have

$$I_t = \beta(C_t - C_{t-1}) = \beta(\alpha Y_t - \alpha Y_{t-1}) = \alpha\beta Y_t - \alpha\beta Y_{t-1} \tag{11.9}$$

Substituting for C_t and I_t from (11.7) and (11.9) in (11.6), we get

$$Y_t = \alpha\ Y_t + \alpha\beta\ Y_t - \alpha\beta\ Y_{t-1}$$

and after rearrange terms

$$(1 - \alpha - \alpha\beta)Y_t = -\alpha\beta\ Y_{t-1}$$

$$Y_t = \frac{-\alpha\beta}{1 - \alpha - \alpha\beta}\ Y_{t-1}\ =\ \frac{\alpha\beta}{\alpha\beta + \alpha - 1}\ Y_{t-1} \tag{11.10}$$

which is a first order homogeneous equation similar to (11.4). Here a is

$$a = \frac{\alpha\beta}{\alpha\beta + \alpha - 1}$$

Solution to (11.10) is

$$Y_t = \left(\frac{\alpha\beta}{\alpha\beta + \alpha - 1}\right)^t Y_0 \tag{11.11}$$

According to the Department of Commerce, Bureau of Economic Analysis α was 0.938 in 1999 [see Table B-30 of Economic Report of the President (ERP), Council

---

[1] The model under discussion is a modified version of the *multiplier-accelerator* model proposed by Paul Samuelson, "Interaction Between the Multiplier and the Principle of Acceleration", *Review of Economic Statistics* 1939. We will discuss the original model, which leads to a second order difference equation, later in the chapter.

Table 11.1 Actual and estimated GDP based on model 1

Year	GDP actual	GDP estimated
1999	9470.3	9470.3
2000	9817.0	9707.1
2001	9890.7	9949.7
2002	10048.8	10198.5
2003	10301.0	10453.4
2004	10703.5	10714.8
2005	11048.6	10982.6

of Economic Advisors, 2009] and β estimated to be around 2.6. Using these values we have

$$\frac{\alpha\beta}{\alpha\beta + \alpha - 1} = \frac{2.44}{(0.938 * 2.6 + 0.938 - 1)} = 1.025$$

and (11.11) will be

$$Y_t = (1.025)^t Y_0$$

By using 1999 real GDP = 9470.3 billion dollar (from Table B-2 of ERP-2009) as the initial value

$$Y_t = 9470.3 \ (1.025)^t \qquad\qquad t = 0, 1, 2, \ldots. \qquad\qquad (11.12)$$

The above Table 11.1, shows the actual real GDP from 1999 to 2005 and the estimated GDP, based on (11.12), for the same period. Based on the model the average rate of growth of the economy over this period was about 2.5 %. This model is surprisingly accurate, generating less than 1 % error on the average—the *Mean Absolute Percentage Error (MAPE)*[2] of the model over 6 forecasting years is less than 1 %.

2. A variation of the model in Example 1 suggested by W.J. Baumol (based on model introduced originally by the British Economist Roy Forbes Harrod) is to write the induced investment and saving function as

$$I_t = \lambda(Y_t - Y_{t-1})$$

$$S_t = sY_{t-1}$$

That is the desired investment in any period t is a constant proportion, λ, of increase in income in that period over income in period $t - 1$, and saving in period t is a constant proportion, s, of income in the previous period.

Utilizing the equilibrium condition $I_t = S_t$, we have

---

[2] See the Appendix for a short discussion of MAEP and other measures of prediction accuracy.

$$\lambda (Y_t - Y_{t-1}) = sY_{t-1}$$
$$\lambda Y_t - \lambda Y_{t-1} = sY_{t-1} \quad \longrightarrow \quad \lambda Y_t = \lambda Y_{t-1} + sY_{t-1} = (\lambda + s)Y_{t-1} \quad \text{and}$$
$$Y_t = \frac{\lambda + s}{\lambda} Y_{t-1} = \left(1 + \frac{s}{\lambda}\right) Y_{t-1}$$

This is a first order homogeneous difference equation with solution

$$y_t = \left(1 + \frac{s}{\lambda}\right)^t Y_0 \qquad (11.13)$$

Since $\lambda > 0$ and $0 < s < 1$ then $1 + \frac{s}{\lambda} > 1$ and (11.13) must be monotonically increasing, which indicates the growth of the economy. In order to operationalize the model we need estimate of λ and s. Real GDP for years 1998 and 1999 are 9066.9 and 9470.3 billion dollars and real gross private domestic investment for year 1999 is 1642.6. From these figures we get an estimate for λ

$$1642.6 = \lambda (9470.3 - 9066.9) \quad \longrightarrow \quad \lambda = \frac{403.4}{1642.6} = 4.07$$

For year 1999 constant $s = 1674.3/9066.9 = 0.18$ (this is gross saving rate, otherwise the personal saving rate for the same year is only 0.024)

Substituting for λ and s in (11.13) and using 1999 GDP = 9470.3 for Y_0 as the initial value, the model would be

$$Y_t = 9470.3 \, (1.044)^t \qquad t = 0, 1, 2, \ldots. \qquad (11.14)$$

The following Table 11.2, shows the actual GDP from 1999 to 2005 and the estimated GDP, based on (11.14), for the same period. Based on this model the average rate of growth of the economy over this period was about 4.4 %. This model is not as accurate as the model in Example 1. It generates about 8 % error on the average—the Mean Absolute Percentage Error (MAPE) of the model over 6 forecasting years is 7.8 %.

Table 11.2 Actual and estimated GDP based on model 2

Year	GDP actual	GDP estimated
1999	9470.3	9470.3
2000	9817.0	9886.9
2001	9890.7	10322.0
2002	10048.8	10776.2
2003	10301.0	11250.3
2004	10703.5	11745.4
2005	11048.6	12801.7

3. Another application of first order homogeneous difference equation is driving the general formula for compound interest (more in Chap. 12). If P dollar is invested at $100i$ percent interest, compounded annually then value of this investment after t years is

$$P_t = P_{t-1} + iP_{t-1} = (1+i)P_{t-1} \qquad t = 1, 2, \ldots.$$

which is a first order homogeneous difference equation with the solution

$$P_t = (1+i)^t P_0 \qquad t = 1, 2, \ldots. \qquad \text{where } t \text{ is in } \underline{\text{year}}$$

Since we assumed the initial investment of P dollar, so $P_0 = P$ and we rewrite the equation as

$$P_t = (1+i)^t P \qquad t = 1, 2, \ldots$$

If interest is compounded 4 times a year (quarterly compounding) then the value of investment after t quarters is

$$P_t = P_{t-1} + \frac{i}{4} P_{t-1} = (1 + \frac{i}{4})P_{t-1} \qquad t = 1, 2, \ldots.$$

So after t quarters, we have,

$$P_t = (1 + \frac{i}{4})^t P \qquad t = 1, 2, \ldots \qquad \text{where } t \text{ is in } \underline{\text{quarter}}$$

If interest is compounded monthly, value of investment after t month is

$$P_t = (1 + \frac{i}{12})^t P \qquad t = 1, 2, \ldots \qquad \text{where } t \text{ is in } \underline{\text{month}}$$

In general if interest is compounded m times a year the solution to the difference equation

$$P_t = P_{t-1} + (\frac{i}{m}) P_{t-1} = (1 + \frac{i}{m}) P_{t-1} \qquad (11.15)$$

will be

$$P_t = (1 + \frac{i}{m})^t P$$

where t is appropriately expressed as a function of m.

If, for example, the initial investment $P = \$1,000$ and rate of interest is 6% per year compounded monthly, the value of investment after 14 month is $P_t = (1 + i/m)^t P$, where number of annual compounding $m = 12$ and t, expressed in month, is 14. Therefore,

$$P_t = (1 + \frac{i}{m})^t P = (1 + \frac{0.06}{12})^{14} * 1000 = \$1,072.32$$

Value of this investment after two years is

$$P_t = (1 + \frac{i}{m})^t P = (1 + \frac{0.06}{12})^{24} * 1000 = \$1,127.16$$

Alternatively, we can write the general solution to (11.15) as

$$P_t = (1 + \frac{i}{m})^{mt} P \qquad \text{where } t = 1, 2, \ldots \text{ is expressed in } \underline{\text{year.}}$$

11.2.3 First Order Non-Homogeneous Difference Equation

First order difference equation with constant coefficients has the form

$$y_{t+1} = ay_t + b \qquad\qquad t = 0, 1, 2, 3, \ldots \qquad\qquad (11.16)$$

where a and b are constant and $a \neq 0$ (if b is zero we have homogeneous equation).

Suppose y_0, the initial value of y, is known. Then from (11.16) with $t = 0$ we have

$$y_1 = ay_0 + b \qquad\qquad (11.17)$$

With $t = 1$ we find

$$y_2 = ay_1 + b \quad \longrightarrow \quad y_2 = a(ay_0 + b) + b = a^2 y_0 + ab + b = a^2 y_0 + b(1 + a) \quad \text{then}$$

$$y_2 = a^2 y_0 + b(1 + a) \qquad\qquad (11.18)$$

We now let $t = 2$ in (16.16) and substitute for y_2 the above expression

$$y_3 = ay_2 + b \longrightarrow y_3 = a[a^2 y_0 + b(1 + a)] + b = a^3 y_0 + ab(1 + a) + b$$
$$y_3 = a^3 y_0 + b(1 + a + a^2) \qquad\qquad (11.19)$$

In order to get a better picture of the pattern which is emerging, we continue the iterative process one more time with $t = 3$

$$y_4 = ay_3 + b \longrightarrow y_4 = a[a^3 y_0 + b(1 + a + a^2)] + b = a^4 y_0 + ab(1 + a + a^2) + b$$
$$y_4 = a^4 y_0 + b(1 + a + a^2 + a^3) \qquad\qquad (11.20)$$

By comparing (16.17), (16.18), (16.19), and (16.20) we can safely guess that for any period t

$$y_t = a^t y_0 + b(1 + a + a^2 + a^3 + a^4 + \cdots + a^{t-1}) \qquad\qquad (11.21)$$

Table 11.3 U.S. real GDP
from 1990 to 2005

Year	Real GDP actual
1990	7112.5
1991	7100.5
1992	7336.6
1993	7532.7
1994	7835.5
1995	8031.7
1996	8328.9
1997	8703.5
1998	9066.9
1999	9470.3
2000	9817.0
2001	9890.7
2002	10048.8
2003	10301.0
2004	10703.5
2005	11048.6

The expression in the bracket in (16.21) is the sum of t terms of a geometric series[3] with first term 1 and common ratio of a. If $a = 1$ this sum is simply equal to t, otherwise it is equal to $\dfrac{(1 - a^t)}{(1 - a)}$. Then depending on value of a the general solution to (16.16) is

$$(a) \quad y_t = y_0 + bt \qquad\qquad \text{if } a = 1 \qquad t = 0, 1, 2, \ldots$$
$$(b) \quad y_t = a^t y_0 + b\left(\tfrac{1-a^t}{1-a}\right) \qquad \text{if } a \neq 1 \qquad t = 0, 1, 2, \ldots$$

Case (a)

In case (a) y is expressed as a linear function of time t. This equation is simply a linear trend equation. Here y_0 is the intercept and b is the slope of the trend line. If $b > 0$ then y has an *upward* trend, otherwise its trend is *downward* . Larger the magnitude of b relative to magnitude of y, steeper would be the trend, indicating rapid growth or decline of y. As an example consider the U.S. real GDP from 1990 to 2005 given in Table 11.3.

A time plot of GDP exhibits an upward trend over time, indicative of steady growth of US economy. Change in GDP over time can be fairly accurately approximated by the following trend equation[4]

$$GDP_t = 6542.6 + 276.8\,t \qquad t = 1, 2, 3, \ldots \quad \text{where } t = 1 \text{ is } 1990$$

---

[3] See Chap. 12 for a detailed presentation of *geometric series*.

[4] Simple linear regression in Microsoft Excel is used to estimate this equation.

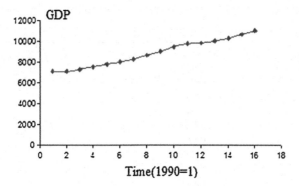

Fig. 11.5 Time plot of GDP from 1990 to 2005

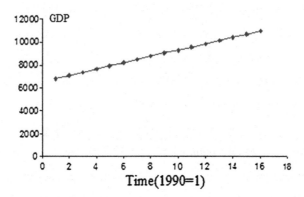

Fig. 11.6 Graph of GDP linear trend equation

or alternatively

$$GDP_t = 6542.6 + 276.8\,(t - 1989) \qquad t = 1990, 1991, \ldots$$

Figures 11.5 and 11.6 show the time plot of GDP and graph of its linear trend equation.

While U.S. GDP shows an upward trend, there are components of GDP exhibiting downward trend. Data (see Table B-12 of Economic Report of the President, Council of Economic Advisors, 2009) indicates a steady decline in the share of manufacturing of total output in the United States. In the graph below (11.7) M is the percentage of value added in the manufacturing sector from the GDP.

The linear trends equation for share of manufacturing from total GDP is

$$M_t = 17.1 - 0.293\,t \qquad t = 1, 2, 3, \ldots \quad \text{where } t = 1 \text{ is } 1990$$

where negative slope (-0.293) clearly indicates the downward trend in M. Figure 11.7 shows the decline in the share of manufacturing form the GDP of the United States.

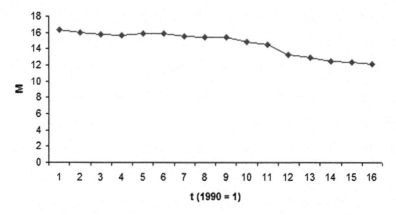

Fig. 11.7 Share of manufacturing form the US GDP

Case(b)

Case (b) of solution to a first order difference equation $y_{t+1} = ay_t + b$

$$y_t = y_0 a^t + b \left(\frac{1 - a^t}{1 - a} \right) \quad a \neq 1 \quad t = 0, 1, 2, \ldots$$

is more complicated. Here different combinations of values of y_0, a, and b lead to different values for y. We consider a few of these combinations and ask you to analyze others as homework.

We first eliminate possibility of $a = 0$ or $b = 0$ in $y_{t+1} = ay_t + b$. If $a = 0$ we no longer have a first order difference equation, we have a constant function $y_{t+1} = b$. $b = 0$ turns the equation into a homogeneous first order difference equation which we have already discussed.

Now lets consider other values for y_0, a and b. Since few economic variables take negative values (the most likely are profit and change in value of dollar against other currencies), it is not so unrealistic to assume y_0 to be a positive number. We make this assumption in the rest of our discussion. b could be either positive or negative. Assuming $b > 0$, we consider four different alternatives, listed below:

And the same four alternatives when b is negative We also assume following numerical values for y_0, a and b. We set $y_0 = 5$ and assume 3 and -6 for positive and negative b. We let $a = 2.5$ for $a > 1$; $a = 0.7$ for $0 < a < 1$; $a = -0.5$ for $-1 < a < 0$; and $a = -2$ for $a < -1$. With these assumptions we will study the behavior of the eight cases

(1) $b > 0$ $a > 1$

(2) $b > 0$ $0 < a < 1$

(3) $b > 0$ $-1 < a < 0$

(4) $b > 0$ $a < -1$

(5) $b < 0$ $a > 1$

(6) $b < 0$ $0 < a < 1$

(7) $b < 0$ $-1 < a < 0$

(8) $b < 0$ $a < -1$

Case 1: $y_0 = 5$, $a = 2.5$, and $b = 3$

$$y_t = 5\,(2.5)^t + 3\left(\frac{1 - 2.5^t}{1 - 2.5}\right) = 7\,(2.5)^t - 2 \qquad t = 0, 1, 2, \ldots.$$

A sketch of Case 1 for different values of t generates a graph similar to Fig. 11.1, which indicates a monotonically increasing y diverging to larger and larger numbers as the process moves through time.

Case 2: $y_0 = 5$, $a = 0.7$, and $b = 3$

$$y_t = 5\,(0.7)^t + 3\left(\frac{1 - (0.7)^t}{1 - 0.7}\right) = -5\,(0.7)^t + 10 \qquad t = 0, 1, 2, \ldots.$$

In Case 2, y monotonically increases and converges to a limit, which in this case is 10. Here as we move through time and farther away from the initial situation as t becomes larger, $(0.7)^t$ approaches zero and y_t converges to $3/(1 - 0.7) = 10$. Movement of y over time and its convergence to 10 is depicted in Fig. 11.8.

Fig. 11.8 Graph of $y_t = -5(10)^t + 10$

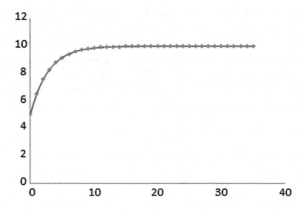

Case 3: $y_0 = 5$, $a = -0.5$, and $b = 3$

$$y_t = 5\,(-0.5)^t + 3\left(\frac{1 - (-0.5)^t}{1 - (-0.5)}\right) = 3\,(-0.5)^t + 2 \qquad t = 0, 1, 2, \ldots.$$

In Case 3, y fluctuates or *oscillates*. But after each oscillation its *amplitude*, the difference between two successive values of y, declines and y ultimately converges to a limit. As we move through time, $(-0.5)^t$ changes from positive (negative) to negative (positive), becomes smaller in magnitude and approaches zero at large values of t. In our case here y converges to $3/(1 + 0.5) = 2$. This pattern of behavior is called *damped oscillatory*. Figure 11.4 is a sample graph of a damped oscillatory equation.

Case 4: $y_0 = 5$, $a = -2$, and $b = 3$

$$y_t = 5\,(-2)^t + 3\left(\frac{1 - (-2)^t}{1 - (-2)}\right) = 4\,(-2)^t + 1 \qquad t = 0, 1, 2, \ldots.$$

In Case 4, y also *oscillates*, but after each oscillation its amplitude increases and it diverges to a very large number in magnitude. As we move through time, $(-2)^t$ changes from positive (negative) to negative (positive) and becomes larger and larger in absolute value. This pattern of behavior of y is called *explosive oscillatory*. Figure 11.2 depicts the graph of a difference equation which is explosive oscillatory.

Case 5: $y_0 = 5$, $a = 2.5$, and $b = -6$

$$y_t = 5\,(2.5)^t - 6\left(\frac{1 - 2.5^t}{1 - 2.5}\right) = (2.5)^t + 4 \qquad t = 0, 1, 2, \ldots.$$

In this case, y *monotonically increases* in magnitude and does not converge to any number.

Case 6: $y_0 = 5$, $a = 0.7$, and $b = -6$

$$y_t = 5\,(0.7)^t - 6\left(\frac{1 - 0.7^t}{1 - 0.7}\right) = 25(0.7)^t - 20 \qquad t = 0, 1, 2, \ldots.$$

In this case, y *monotonically declines* and converge to -20. Decline of y overtime and its convergent to -20 is shown in Fig. 11.9.

Case 7: $y_0 = 5$, $a = -0.5$, and $b = -6$

$$y_t = 5\,(-0.5)^t - 6\left(\frac{1 - (-0.5)^t}{1 - (-0.5)}\right) = 9(-0.5)^t - 4 \qquad t = 0, 1, 2, \ldots.$$

In this case, y exhibit *damped oscillatory* behavior and converge to -4.

Fig. 11.9 Graph of
$y_t = 25(0.7)^t - 20$

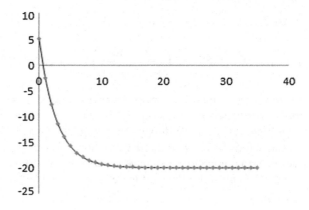

Case 8: $y_0 = 5$, $a = -2$, and $b = -6$

$$y_t = 5\,(-2)^t - 6\left(\frac{1 - (-2)^t}{1 - (-2)}\right) = 7(-2)^t - 2 \qquad t = 0, 1, 2, \ldots .$$

In this case, y displays an *explosive oscillatory* behavior.

If a difference equation comes to rest by converging to a specific value then this value, let's denote it by y^*, is called the *stationary value* or the *steady state* of the difference equation; or more accurately, the stationary value of variable y. In the eight cases that we considered only cases 2, 3, 6, and 7 reach a steady state. The stationary value of y in case 2 is $y^* = 10$. Similarly $y^* = 2, y^* = -20$, and $y^* = -4$, are the steady states of cases 3, 6, and 7.

Note that the behavior of solution to the first order difference equation

$$y_t = a^t y_0 + b\,\frac{1 - a^t}{1 - a} \qquad a \neq 0 \qquad t = 0, 1, 2, \ldots .$$

to a large extend depends on magnitude of a. If $|a| > 1$ (that is either $a > 1$ or $a < -1$) the solution would display monotonic increase (decrease) or explosive oscillations. Only when $|a| < 1$ (that is $-1 < a < 1$) the solution converges to a stationary value, as is the case for alternatives 2, 3, 6, and 7. When $|a| < 1$, a^t goes to zero as t goes to infinity and y_t converges to

$$y_t = y^* = \frac{b}{(1 - a)}$$

Thus the convergent requirement is $-1 < a < 1$ (11.22)

Examples

1. Equilibrium in a single market

We arrived at equation (11.1) by writing the supply function for a commodity as

$$Q_{st} = -c + dP_{t-1}$$

and the demand function as

$$Q_{dt} = a - bP_t$$

where $a, b, c,$ and $d > 0$

We then invoked the equilibrium condition $Q_{st} = Q_{dt}$ and arrived at

$$-c + dP_{t-1} = a - bP_t \qquad \text{or alternatively}$$

$$P_t = \frac{-d}{b} P_{t-1} + \frac{a+c}{b}$$

which is a first order difference equation with the solution

$$P_t = (\frac{-d}{b})^t P_0 + \frac{(a+c)}{b}\left[\frac{1 - (-d/b)^t}{1 - (-d/b)}\right] \qquad t = 0, 1, 2, \dots . \qquad (11.23)$$

This market reaches equilibrium if P in (11.23) converges to a positive stationary value. This can only occurs if the convergent requirement of (11.22) is satisfied, that is $-1 < -\dfrac{d}{b} < 1$. With the assumption that d and b are positive numbers $-\dfrac{d}{b} < 1$ is always satisfied, leaving us with the other part of the requirement that $-1 < -\dfrac{d}{b}$ or $\dfrac{d}{b} < 1$. Stressing again that b and d are positive numbers, the only condition for the market to reach an equilibrium is now reduced to $\dfrac{d}{b} < 1$ or $d < b$.

This condition is satisfied as long as d (slope of the supply function) is less than b (slope of the demand function). Or saying it differently, as long as the demand curve is steeper (relative to price axis) than the supply curve.

As we observed earlier, out of the eight cases we studied cases 2, 3, 6 and 7 reach a steady state or converge, with only cases 2 and 3 converging to positive numbers. Since price cannot be negative, this raises the possibility that we may need additional condition(s) in order to make sure that the market reaches an economically meaningful equilibrium. We will show that $d < b$ is the only condition we need.

After simplifying (11.23) we can rewrite it as

$$P_t = (-d/b)^t P_0 + \left(\frac{a+c}{b+d}\right)\left[1 - (-d/b)^t\right] \qquad t = 0, 1, 2, \dots . \qquad (11.24)$$

As we move through time and t becomes larger P_t in (11.24) shows a damped oscillatory behavior. When condition $d < b$ is satisfied as t gets larger terms $(-d/b)^t$ and $(-d/b)^t P_0$ approach zero and P_t eventually converges to the stationary value $P^* = \dfrac{(a+c)}{(b+d)}$, which is the equilibrium price. Given the assumption that a, b, c, and d are all positive numbers, the equilibrium price is safely positive. A quick numerical example should help to better understand our discussion.

Assume the following supply and demand equations

$$Q_{dt} = 100 - 10P_t$$
$$Q_{st} = -20 + 5P_{t-1}$$

What is the equilibrium price in this market? Notice that the slope of the demand function $|b| = 10$ and that of the supply function $d = 5$, satisfying the convergence condition $d < b$. From $Q_{dt} = Q_{st}$ we get

$$100 - 10P_t = -20 + 5P_{t-1} \quad \longrightarrow \quad -10P_t = 5P_{t-1} - 120$$

$$P_t = -0.5P_{t-1} + 12 \qquad t = 0, 1, 2, \ldots. \quad \text{with the solution}$$

$$P_t = (-0.5)^t P_0 + 12 \, \frac{1 - (-0.5)^t}{[1 - (-0.5)]}$$

$$P_t = (-0.5)^t P_0 + 8 \, [1 - (-0.5)^t] \qquad t = 0, 1, 2, \ldots.$$

When t is very large P_t converges to a limit

$$P_t = P^* = 8$$

By substituting 8 for P in either supply or demand equation we get $Q^* = 20$ units.

It should be noted that the initial price P_0 has no impact on the ultimate equilibrium price. The initial price could be widely deviant from the equilibrium price. But through a series of adjustment in excess demand (supply), market gradually moves toward the equilibrium price. The graph in Fig. 11.10 depicts the movement of price toward the equilibrium.

Here the initial price is set at $12, 50\%$ above the equilibrium price. It is clear from the supply equation that at the offered price of $12 per unit producers are willing to supply 40 units of the good to the market

$$Q_{st} = -20 + 5P_{t-1} = -20 + 5 * 12 = 40$$

But based on the demand function $Q_{dt} = 100 - 10P_t$ consumers are willing to buy 40 units if price is $6 per unit

$$40 = 100 - 10P_t \quad \longrightarrow \quad 10P_t = 60 \quad \longrightarrow \quad P_t = 6$$

Fig. 11.10 Path of price to equilibrium

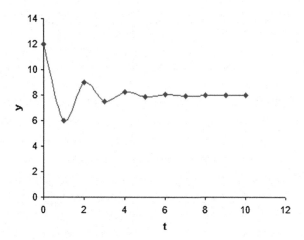

At the price of $6 producers are only willing to produce 10 units of output. Consumers are willing to pay $9 per unit to buy these 10 units. At price of $9 per unit producers are ready to supply 25 units. Consumers would offer to pay $7.5 to clear the market from 25 units. At this price suppliers can produce only 17.5 units. Consumers are willing to pay $8.25 to buy these 17.5 units. At this price suppliers adjust their output to 21.25 units. This process continues till price gets closer and closer to $8 and the quantity supplied and demanded converge to 20 units and market reaches its equilibrium.

The market would go through exactly the same process if the initial price were set below the equilibrium price. This model is famously known as the *Cobweb Model*.

2. A Simple Dynamic National Income Model
A simple dynamic Keynesian national income model includes a national income identity and a consumption function

$$Y_t = C_t + I_o + G_o$$
$$C_t = C_o + b\, Y_{t-1}$$

where Y and C are the national income and consumption expenditure. Here consumption is expressed as a linear function of previous period income. In the consumption function b is the marginal propensity to consume from national income and C_o is the exogenously determined autonomous consumption. I_o and G_o are the exogenously given investment and government expenditure. Y and C are the endogenous variables of the model.

Substituting for C_t from the consumption function into the national income identity, we have

$$Y_t = C_o + b\, Y_{t-1} + I_o + G_o = b\, Y_{t-1} + C_o + I_o + G_o$$

which is a first order difference equation with solution

$$Y_t = (b)^t Y_0 + (C_o + I_o + G_o) \left[\frac{(1 - b^t)}{(1 - b)} \right] \qquad (11.25)$$

Since $0 < b < 1$, Y will increase monotonically but converges to a limit. This limit will be

$$\frac{1}{1 - b} (C_o + I_o + G_o)$$

where $\dfrac{1}{1 - b}$ is, of course, the simple Keynesian multiplier.

Assume we have an estimate of the consumption function as

$$C_t = 400 + 0.81 \, Y_{t-1}$$

Given that for year 1990 private domestic investment and government expenditure are 861 and 1530 billion dollars, we can write (20) as

$$Y_t = (0.81)^t Y_0 + (400 + 861 + 1530) \left[\frac{1 - (0.81)^t}{1 - (0.81)} \right]$$

$$Y_t = (0.81)^t Y_0 + 2791 \left[\frac{1 - (0.81)^t}{0.19} \right]$$

$$Y_t = (0.81)^t Y_0 + 14689.5 \left[1 - (0.81)^t \right]$$

As t increases $(0.81)^t$ declines and approaches zero and Y_t converges to $14,689.5$ billion or 14.689 trillion dollars.

3. Peak Oil[5]

US crude oil well productivity peaked in 1972. That year 501000 active oil wells in the United States produced a total of 9.441 million barrels of oil per day, an average productivity of 18.6 barrels per day per well. In year 2006 with the roughly the same number of wells in production, the US daily output declined to 5.136 million barrels, an average productivity of 10.25 barrels per day per well. Figure 11.11 provides a time series plot of crude oil well productivity in the US from 1972 to 2006. A simple statistical analysis of data indicates that over the last 33 years well productivity has declined by an average of 2.4 percent per year.[6]

In spite of the fact that there are evidences pointing to acceleration of decline in productivity of oil wells, let's assume the rate of decline remains at 2.4 percent per year. We also assume that the number of active wells stays the same at 501000, as

---

[5] Data discussed in this section are from Energy Information Administration, US Department of Energy, *Annual Energy Review* 2006, specially Table 5.2 "Crude Oil Production and Crude Oil Well Productivity 1954-2006". See http://www.eia.doe.gov.

[6] A linear trend estimate for well productivity (WP) produced the trend equation $WP_t = 17.7 - 0.24t$.

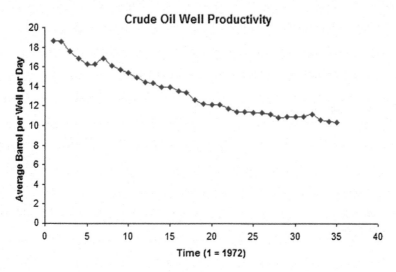

Fig. 11.11 US oil well productivity from 1972 to 2006

in 2006. It is roughly estimated that operating an oil well requires 0.15 barrels of oil energy input per day (a barrel of oil is about 42 gallons. 0.15 barrel is about 6 gallons of oil. We can think of this as using roughly 6 gallons of fuel to run pumps or other equipments used for oil extraction).

Given the above information and assumptions, how long it takes for the US oil production to decline by 50% from 2006 level of 5.136 million barrels a day to 2.568 million barrels?

Let us denote the oil daily output at year t by Q_t. Let κ denote the rate of productivity decline. We can formulate the problem by writing the following first order difference equation

$$Q_t = (1 - \kappa)\, Q_{t-1} - K \qquad t = 0, 1, 2, \ldots. \tag{11.26}$$

where K is the total amount of oil energy used in the process of operating the oil wells. Substituting 0.024 for κ and 0.15 * 501000 = 75150 barrels, or 0.075 million barrels, for K in (21), we have

$$Q_t = (1 - 0.024)\, Q_{t-1} - 0.075 = 0.976\, Q_{t-1} - 0.075 \qquad t = 0, 1, 2, \ldots.$$

Solution to this difference equation is

$$Q_t = (0.976)^t\, Q_0 - 0.075 \left[\frac{1 - (0.976)^t}{1 - 0.976} \right] \tag{11.27}$$

Treating 2006 output of 5.136 million barrels as initial value Q_0, we must determine t, number of years, such that Q_t becomes equal to 2.568 million barrels. Substituting these numbers in (11.27) and simplifying it, we get

$$2.568 = 5.136(0.976)^t + 3.125(0.976)^t - 3.125$$
$$5.693 = 8.261(0.976)^t \quad \longrightarrow \quad (0.976)^t = 0.718$$

By taking the log of both sides, we have

$$\ln[(0.976)^t] = \ln 0.718 \quad \longrightarrow \quad t\ln(0.976) = -0.3313$$

$$-0.024\, t = -0.3313 \quad \longrightarrow \quad t = \frac{0.3313}{0.024} \approx 14$$

That is, in about 14 years (year 2020) the output of existing and operating oil wells will drop to half.

11.3 Second and Higher Order Difference Equation

In the original multiplier-accelerator model, the national income identity is expressed as

$$Y_t = C_t + I_t + G_0 \qquad\qquad (11.28)$$

and aggregate consumption function and private investment are expresses as

$$C_t = \alpha\, Y_{t-1}$$

$$I_t = \beta(C_t - C_{t-1})$$

In this version government is added to the model, with its expenditure exogenously determined, and consumption is expressed as a function of previous year income rather than the current income. Then

$$I_t = \beta(C_t - C_{t-1}) = \beta\,(\alpha\, Y_{t-1} - \alpha\, Y_{t-2}) = \alpha\beta\, Y_{t-1} - \alpha\beta\, Y_{t-2}$$

By substituting for C_t and I_t in (23), we get

$$Y_t = \alpha\, Y_{t-1} + \alpha\beta\, Y_{t-1} - \alpha\beta\, Y_{t-2} + G_0$$

and after rearrange terms, we have

$$\alpha\beta\, Y_{t-2} - (\alpha + \alpha\beta)\, Y_{t-1} + Y_t \;=\; G_0$$

which is an example of a second order difference equation. Solving this or higher order difference equations may involve complex numbers and their presentation in the polar coordinate system, which is beyond the scope of this introductory text. Interested reader should consult some of the sources cited at the end of the chapter.

11.4 Exercises

1. Determine the order of the following difference equations.

   (a) $4y_{t+2} + 2y_{t+1} - 5y_t = 8$

   (b) $y_{k+2} - y_{k+1} - 5y_k = k$

   (c) $y_{t+1} = 5y_t - 6t$

   (d) $12y_{t+2} - 5y_t = \log t$

   (e) $y_{t+2} - 3y_{t+1} + 2y_t = 0$

   (f) $(1 + t) y_{t+2} + 2t y_{t+1} - 5y_t = 6t + 1$

   (g) $3y_{t+2} - 2y_{t+1} + 5y_{t-1} = 0$

2. Classify the difference equations in problem (1) as

   (a) linear or nonlinear.

   (b) with or without constant coefficients.

   (c) autonomous or non-autonomous.

   (d) homogeneous or non-homogeneous.

3. Find the general solution of the following equations. Explain the behavior of y. Find the particular solutions for the given initial values. Determine whether y converges to the steady state.

   (a) $y_{t+1} = 5y_t - 6$ $\qquad\qquad$ $y_0 = 10$

   (b) $4y_{t+2} + 3y_{t+1} = 2$ $\qquad\qquad$ $y_0 = 4$

   (c) $y_{t+1} = 0.34y_t + 4$ $\qquad\qquad$ $y_0 = 2.5$

   (d) $2y_{t+1} = -0.34y_t - 4$ $\qquad\qquad$ $y_0 = 5$

4. Let K_t denote the market value of capital stock of a firm at the beginning of year t. Assume λ is the rate of firm's capital depreciation each year. Further assume that the firm makes I dollar of new capital expenditure or investment each year.

   (a) Write the first order difference equation relating the capital stock of the firm at the beginning of the year K_{t+1} to the capital stock and investment in year t.

   (b) Solve the equation. What would happen to the firm's capital stock if

(I)   Investment is more than depreciation?

(II) Investment is only for replacement of the depreciated stock?

(III) Investment is less than depreciation?

5. In problem (4) assume that the firm's initial capital stock is $5 million. Also assume the rate of capital depreciation is 10% a year. What would be the firm's capital stock after 5 years if annual investment are

(I) $600,000?

(II) $400,000?

(III) $500,000?

6. Assume the firm in problem (4) has the following production function

$$Q_t = 0.45 \, K_t^{0.5}$$

What is the firm's level of output after 7 years if annual investment were the same as I, II, or III in problem (5)?

7. Let P_t denote the aggregate price level at time t. The inflation rate from time $t-1$ to t, denoted by τ, is defined as

$$\tau_t = \frac{P_t - P_{t-1}}{P_{t-1}}$$

(a) Write the first order homogeneous difference equation for the price level and solve it.

(b) If the annual inflation rate remains stable at 3.5%, what is the aggregate price index after 10 years? (Hint: the initial price level is the price index at the *base year*).

(c) After how many years the price level doubles?

8. **Inflationary Spiral and Hyperinflation**

When prices increase rapidly then the inflation rate grows from period to period. A rapid growth of inflation rate creates what is known as the *inflationary spiral*, which if not contained would lead to a *hyperinflation*.

If the inflation rate τ in problem (4) grows at the rate of 100λ percent per period, then

$$\tau_t = \tau_{t-1} + \lambda \, \tau_{t-1}$$

expresses the relations between inflation rates in two consecutive periods.

(a) Assuming τ_0 is the initial inflation rate, write the general solution for the inflation rate.

(b) Redo part (a) in problem (7) after substituting τ_t by its general solution from part (a).

(c) Assume that the initial inflation rate is 35% and grows at the rate of 10% each month. What is the aggregate price index after one year?

[**Note**: Solution to the price level in part (b) is a crude approximation; otherwise the actual aggregate price index in period t has a more complicated structure. If P_0 and τ_0 are the initial values of the price index and inflation rate, and λ is the rate of growth of inflation, then the price index at period t is

$$P_t = P_0(1 + \tau_0)[1 + (1 + \lambda) \tau_0][1 + (1 + \lambda)^2 \tau_0][1 + (1 + \lambda)^3 \tau_0] \ldots$$
$$[1 + (1 + \lambda)^t \tau_0] \quad \text{or}$$

$$P_t = P_0 \prod_{i=1}^{t} [1 + (1 + \lambda)^i \tau_0]$$

where $\prod$ is product notation, similar to $\sum$ which is summation notation. Students are encourage to try and drive this expression.]

9. *Econometricians/Time Series Analysts* refer to linear difference equations with constant coefficient as *Autoregressive models*. The first and second order difference equations are called the first and second order autoregressive models and generally denoted by AR(1) and AR(2). AR(1) and AR(2) are widely used for estimating economic time series, notably macroeconomic series, with upward or downward trend. The following is an example of AR(1) estimate of US money supply.

The narrowest measure of money that the Federal Reserve Bank reports is M1. M1 includes currency, checking account deposits, and the traveler's checks. A broader measure of monetary aggregate is M2. M2 consists of M1 plus saving deposits, small denomination (less than $100,000) time deposits, money market mutual fund deposits, and several other liquid financial assets. Using monthly data from January 2005 to September 2006, the following AR(1) model for M2 was estimated

$$M2_t = 1.0043 \, M2_{t-1} - 5.05$$

(a) Solve the first order difference equation for the money supply and analyze its behavior over time.

(b) If in January 2005 the money supply was $6,415.1 billion, what is your estimate of October, November, and December 2006 money supply?

(c) If the actual money supply for October, November, and December 2006 are $6,936.2, $6,977.0, and $7,021.0 billion respectively, what is the models average estimation error for the three month?

10. Assume the market supply and demand for a perfectly competitive market are given by the following equations

$$Q_t^s = -c + dP_t$$
$$Q_t^d = a - bP_t + \delta\, Q_{t-1}^d$$

where $a, b, c, d > 0$; $-1 < \delta < 1$; and Q_{t-1}^d is the quantity demanded at time $t - 1$.

Derive the first order difference equation for P by assuming that the market reaches equilibrium in each period. Find the general solution for the equation and determine the behavior of P over time.

Appendix A: Measures of Estimation Accuracy

In several places in this and other chapters we referred to forecasting accuracy of a model and used the Mean Absolute Percentage Error (MAPE) as a measure of estimation accuracy. MAPE is one of many measures of forecasting error. Two other widely used are Mean Absolute Forecasting Error (MAFE) and Root Mean Square Error (RMSE). These measures are valuable in evaluating the strength of a model in terms of generating more accurate estimations or forecasts of values of a variable compared to other competing mathematical/statistical models. In the text we introduced two macroeconomic models of national income (example 1 and 2) leading to two different first order difference equations. Solutions to these difference equations were consequently used to predict values of the US GDP from 1999 to 2005. It was then judged, based on MAPE, that the first model was more accurate compared to the second model. What follows is a short description of three most widely used measure of estimation accuracy.

Assume a model estimate or forecast n values of a variable. Let E_i $i = 1, 2, \ldots, n$ denote the estimated values and A_i $i = 1, 2, \ldots, n$ denote the actual or observed values of the variable. The difference between actual and estimated values $A_i - E_i$ $i = 1, 2, \ldots, n$ are n estimation or forecasting errors. A model that generates quality forecast should minimize these errors. One measure of forecasting accuracy could be the average of these errors, *Mean Forecasting Error* (MFE)

$$MAF = \frac{\sum_{i=1}^{n}(A_i - E_i)}{n}$$

The main problem with MFE is that a model could some times over-estimate and some other times under-estimate the actual values [that is $(A_i - E_i < 0)$ or $(A_i - E_i > 0)$], and by summing these values positive and negative numbers may canceled each other out, leading to a small and misleading number for MFE. To avoid this problem we can compute *Mean Absolute Forecasting Error* (MAFE), by calculating the average of absolute values of forecasting errors

$$MAFE = \frac{\sum_{i=1}^{n} |A_i - E_i|}{n}$$

An alternative way of measuring estimation error is to express it in percentage terms. A large estimation error in absolute terms might be very small in percentage terms. For example, a model may overestimate the United States GDP by \$200 billion. In absolute term \$200 billion is a huge number, but not in relative terms. \$200 billion overestimation of a \$14 trillion economy amounts to only about 1.4 % error. To measure the relative size of the estimation error the *Mean Absolute Percentage Error* (MAPE) is computed as

$$MAPE = \frac{\sum_{i=1}^{n} |A_i - E_i|/A_i}{n} * 100$$

The third measure of estimation error is the *Root Mean Square Error* (RMSE). RMSE is defined as

$$RMSE = \sqrt{\frac{\sum_{i=1}^{n} (A_i - E_i)^2}{n}}$$

This measure penalize larger errors more heavily, make it more appropriate for use in a situation where the costs of making errors proportionately larger than the size of error.

Regardless of which measure of estimation error is used, the model generating the smallest MAFE, MAPE, or RMSE is the most accurate.

Example

The following table for the US GDP (in Billion) is constructed from Example 1 discussed in the chapter.

Year	A	E	$|A_i - E_i|$	$\frac{|A_i - E_i|}{A_i}$	$(A_i - E_i)^2$
	Actual	Est.			
2000	9817.0	9707.1	109.9	0.0112	12078.0
2001	9890.7	9949.7	59.0	0.0059	3481.0
2002	10048.8	10198.5	149.7	0.0149	22410.0
2003	10301.0	10453.4	152.4	0.0148	23227.8
2004	10703.5	10714.8	11.3	0.0011	127.7
2005	11048.6	10982.6	66.0	0.0059	4356.0
Total			548.3	0.0538	65680.5

Using the figures from the last three columns of the table we compute MAFE, MAPE, and RMSE for the model in Example 1 as

$$MAFE = \frac{548.3}{6} = 91.38$$

$$MAPE = \frac{0.0538}{6} * 100 = 0.897$$

$$RMSE = \sqrt{\frac{65680.5}{6}} = 104.63$$

The following table is constructed from Example 2 discussed in the chapter. US GDP (in billion)

Year	A	E	$|A_i - E_i|$	$\frac{|A_i - E_i|}{A_i}$	$(A_i - E_i)^2$
	Actual	Est.			
2000	9817.0	9886.9	69.9	0.007	4886.0
2001	9890.7	10322.0	431.3	0.044	186019.7
2002	10048.8	10776.2	727.4	0.072	529110.8
2003	10301.0	11250.3	949.3	0.092	901170.5
2004	10703.5	11745.4	1041.9	0.097	1085556.0
2005	11048.6	12801.7	1753.1	0.159	3073360.0
Total			4972.9	0.471	5780102.0

Again using the figures from the last three columns of the above table we compute MAFE, MAPE, and RMSE for the model in Example 2 as

$$MAFE = \frac{4972.9}{6} = 828.8$$

$$MAPE = \frac{0.471}{6} * 100 = 7.8$$

$$RMSE = \sqrt{\frac{5780102.0}{6}} = 981.5$$

In the text the two models were compared based on MAPE. But comparing MAFE and RMSE for the two models also indicates that the model 1 is superior to the model 2 base on these measures.

Exercise

Redo macroeconomic Examples 1 and 2 by taking year 2003 to be the initial year Y_0 and forecast the US real GDP for period 2004 to 2006. Use the following actual real GDP and calculate MAFE, MAPE, and RMSE for both models.

Year	Actual GDP
2003	10301.0
2004	10703.5
2005	11048.6
2006	11319.4

References

Allen, R.G.D. 1963. *Mathematical Economics*. London: Macmillan and Co., Ltd.

Baumol, William. 1971. *Economic Dynamics, an Introduction*. Third Edition, Macmillan and Company.

Chiang, Alpha C. 1984. *Fundamental Methods of Mathematical Economics*. Third Edition, McGraw-Hill Inc.

Goldberg, Samuel. 1986. *Introduction to Difference Equations*. John Wiley & Sons.

Hoy, Michael, et al. 2001 *Mathematics for Economics*. Second Edition, MIT Press.

Marotto, Frederick R. 2006. *Introduction to Mathematical Modeling Using Discrete Dynamical Systems*. Thomson Books.

Chapter 12
Mathematics of Interest Rates and Finance

12.1 Sequences and Series

A *sequence* is an ordered set of numbers. Consider the sequence of real numbers expressed as

$$a_1, a_2, a_3, a_4, \ldots, a_{n-2}, a_{n-1}, a_n, \ldots \tag{12.1}$$

We can express this sequence in a more compact manner as

$$a_i \qquad\qquad i = 1, 2, 3, \ldots, n, \ldots$$

where a_i are *terms* of the sequence. a_1 is the first term and a_n is the nth term of the sequence. If n is a finite number, then the sequence in (12.1) is a *finite sequence*; otherwise it is an *infinite sequence*. The sum of terms of a sequence is called a *series*. $\sum_{i=1}^{n} a_i$ is a series formed by sum of the first n terms of the above sequence. If a sequence is finite its associated series is also finite, in other words the sum of any number of terms *converges* to a real number. An infinite series may or may not be convergent.

Two sequences are of particular interest; *Arithmetic* and *Geometric* sequences. The series associated with these sequences, Arithmetic and Geometric series, are also of particular interest.

12.1.1 Arithmetic Sequences

A sequence is arithmetic sequence when the difference between any two consecutive terms is constant. Each term in an arithmetic sequence differs from the preceding term by a constant amount, called the *common difference*, denoted by d. For the above sequence to be arithmetic it must be the case that

S. Vali, *Principles of Mathematical Economics*,
Mathematics Textbooks for Science and Engineering 3,
DOI: 10.2991/978-94-6239-036-2_12, © Atlantis Press and the authors 2014

$$a_2 = a_1 + d; \quad a_3 = a_2 + d; \quad a_4 = a_3 + d; \quad \ldots; \quad a_n = a_{n-1} + d$$

Notice that the difference between the $(i + 1)$th and the ith term of this sequence is the constant d, the common difference.

$$(a_{i+1} - a_i) = d \qquad i = 1, 2, 3, \ldots, n - 1, \ldots \quad \text{which simply means that}$$
$$a_{i+1} = a_i + d \qquad i = 1, 2, 3, \ldots, n - 1, \ldots$$

Given this, we can simplify our exposition by dropping the index and denoting the first term of an arithmetic sequence by a and writing the entire sequence as

$$a, \ (a + d), \ (a + 2d), \ (a + 3d), \ \ldots, \ [a + (n - 1)d], \ldots$$

where $a + (n-1)d$ is the nth term or *general term* of the sequence. Since $a + (n-1)d$ for $n = 1, 2, 3, \ldots$ generates all terms of the sequence, it is also known as the *term generating function* of the sequence.

Examples 1

Clearly the following sequence is not an arithmetic sequence

$$-10, \ -5, \ 2, \ 5, \ 7, \ 12, \ 20, \ \ldots$$

but all of the following sequences are:

1. $1, 2, 3, 4, 5, \ldots$ sequence of positive integers $\qquad\qquad$ a $= 1$; d $= 1$
2. $2, 4, 6, 8, 10, \ldots$ sequence of even positive integers $\qquad$ a $= 2$; d $= 2$
3. $6, 10, 14, 18, 22, \ldots$ $\qquad\qquad\qquad\qquad\qquad\qquad\qquad$ a $= 6$; d $= 4$
4. $700, 400, 100, -200, -500, \ldots$ $\qquad\qquad\qquad$ a $= 700$; d $= -300$

The nth or general term (or term generating function) of the above sequences are:

1. $1 + (n - 1) * 1 = n$ $\qquad\qquad\qquad\qquad\qquad$ $n = 1, 2, 3, \ldots$
2. $2 + (n - 1) * 2 = 2n$ $\qquad\qquad\qquad\qquad\quad$ $n = 1, 2, 3, \ldots$
3. $6 + (n - 1) * 4 = 2 + 4n$ $\qquad\qquad\qquad\quad$ $n = 1, 2, 3, \ldots$
4. $700 + (n - 1) * (-300) = 1000 - 300n$ $\qquad$ $n = 1, 2, 3, \ldots$

There are occasions when we are interested in determining the sum of n terms of an arithmetic sequence, that is express an arithmetic series. Denoting this sum by S_n we want

$$S_n = a + (a + d) + (a + 2d) + (a + 3d) + \cdots + [a + (n - 2)d] + [a + (n - 1)d]$$

which can be written as

$$S_n = na + [d + 2d + 3d + \cdots + (n-2)d + (n-1)d]$$

If we find the sum of ds in the square bracket, then S_n is determined. Let's denote this sum by S

$$S = d + 2d + 3d + \cdots + (n-2)d + (n-1)d \qquad (12.2)$$

We can rewrite the same sum backward as

$$S = (n-1)d + (n-2)d + \cdots + 3d + 2d + d \qquad (12.3)$$

If we add both sides of (12.2) and (12.3) we get

$$2S = d + 2d + 3d + \cdots + (n-2)d + (n-1)d + (n-1)d + (n-2)d + \cdots + 3d + 2d + d$$

which can be written as

$$2S = d + (n-1)d + 2d + (n-2)d + 3d + (n-3)d + \cdots + (n-2)d + 2d + (n-1)d + d$$

After simplifying the right hand side of the above expression, we have the sum of $(n-1)$ terms of nd. Thus

$$2S = n(n-1)d \quad \longrightarrow \quad S = \frac{n(n-1)d}{2}, \quad \text{and subsequently}$$

$$S_n = na + S = na + \frac{n(n-1)d}{2} = \frac{n[2a + (n-1)d]}{2}$$

Examples 2

Using sequences from Example 1:

1. The sum of the first 100 terms of Sequence 1 (positive integers) is

$$S_{100} = \frac{100[2 * 1 + (100 - 1) * 1]}{2} = \frac{100 * 101}{2} = 5050$$

Notice that since in this example $a = 1$ and $d = 1$ the formula for the series is reduced to

$$S_n = \frac{n[2 + (n-1)]}{2} = \frac{n(n+1)}{2}$$

You were asked to show this result a result in problem 6 of the Appendix to Chap. 1.

2. The sum of the first 50 terms of Sequence 2 (even positive integers) is

$$S_{50} = \frac{50[2 * 2 + (50 - 1) * 2]}{2} = \frac{50 * 102}{2} = 2550$$

3. The sum of the first 20 terms of Sequence 3 is

$$S_{20} = \frac{20[2 * 6 + (20 - 1) * 4]}{2} = \frac{20 * 88}{2} = 880$$

4. The sum of the first 25 terms of Sequence 4 is

$$S_{25} = \frac{25[2 * 700 + (25 - 1) * (-300)]}{2} = \frac{25(1400 - 7200)}{2} = -72500$$

Exercises 1

1. List the first 5 terms of each sequence with the following general term:

   (a) $a_n = 1 - (0.2)^n$ (b) $a_n = 3(-1)^n$

   (c) $a_n = \dfrac{(n + 1)}{(3n - 1)}$ (d) $a_1 = 2$ and $a_{n+1} = 3a_n - 1$

2. Find a formula for the general term a_n of the sequence if the pattern of the given terms continues.

   (a) $\dfrac{1}{2}, \dfrac{1}{4}, \dfrac{1}{8}, \dfrac{1}{16}, \dfrac{1}{32}, \ldots$ (b) $\dfrac{1}{2}, \dfrac{1}{4}, \dfrac{1}{6}, \dfrac{1}{8}, \ldots$

   (c) $2, 7, 12, 17, 22, \ldots$ (d) $\dfrac{-1}{4}, \dfrac{2}{9}, \dfrac{-3}{16}, \dfrac{4}{25}, \ldots$

12.1.2 Geometric Sequences

A sequence is geometric when the ratio between any two consecutive terms is constant. Each term in a geometric sequence increases or decreases from the preceding term by a constant factor, called the *common ratio*, denoted by r. For (12.1) to be a geometric sequence it must be the case that

$$a_2 = ra_1; \quad a_3 = ra_2 = r^2 a_1; \quad a_4 = ra_3 = r^3 a_1; \quad \ldots; \quad a_n = ra_{n-1} = r^{n-1} a_1, \ldots$$

Notice that the ratio between the $(i + 1)$th and the ith term of this sequence is the constant r, the common ratio. Given that all terms of a geometric sequence can be expressed as a factor (or multiple) of its first term, we can simplify our presentation by denoting the first term a and writing the sequence as

$$a, \ ar, \ ar^2, \ ar^3, \ ar^4, \ \ldots, \ ar^{n-1}, \ \ldots$$

with the general term or term generating function being

$$a_n = ar^{n-1} \qquad\qquad n = 1, 2, 3, \ldots$$

The geometric series associated with the above geometric sequence is

$$a + ar + ar^2 + ar^3 + ar^4, + \cdots + ar^{n-1} + \cdots$$

If r is numerically less than 1, the higher terms of the sequence tend to zero. In this case the geometric series is convergent. If the absolute value of r is greater than 1, the series either tends to infinity or *oscillates*.

There are occasions when we are interested in the partial or finite sum of terms of a geometric sequence. Denoting the sum of n terms of the series by S_n we have

$$S_n = a + ar + ar^2 + ar^3 + ar^4 + \cdots + ar^{n-1} = \sum_{i=0}^{n-1} ar^i \qquad (12.4)$$

Multiplying both sides of (12.4) by r, we have

$$rSn = r(a + ar + ar^2 + ar^3 + ar^4 + \cdots + ar^{n-1}) \qquad (12.5)$$
$$= ar + ar^2 + ar^3 + ar^4 + \cdots + ar^{n-1} + ar^n$$

Subtracting (12.5) from (12.4) we have

$$S_n - rS_n = a - ar^n = a(1 - r^n)$$
$$(1 - r)S_n = a(1 - r^n)$$

Dividing through by $(1 - r)$ we get the general formula

$$S_n = \frac{a(1 - r^n)}{(1 - r)} \qquad (12.6)$$

If $|r| < 1$ as n gets larger and larger, r^n gets smaller and smaller and as $n \to \infty$ then $r^n \to 0$. In this case for relatively large values of n (12.6) becomes

$$S_n = \frac{a}{(1 - r)} \qquad (12.7)$$

Examples 3

1. $1, \dfrac{1}{2}, \dfrac{1}{4}, \dfrac{1}{8}, \dfrac{1}{16}, \cdots$ is a geometric sequence with the first term 1 and the common ratio $\dfrac{1}{2}$. The nth term or the general term of the sequence is $\dfrac{1}{2^{n-1}}$. The sum of the first 10 and 20 terms of this sequence are, respectively

$$S_{10} = \frac{1\left[1 - \left(\frac{1}{2}\right)^{10}\right]}{1 - \frac{1}{2}} = 1.998$$

$$S_{20} = \frac{1\left[1 - \left(\frac{1}{2}\right)^{20}\right]}{1 - \frac{1}{2}} = 1.9999$$

For a very large n the series converges to 2. From (12.7) we have

$$S_n = \frac{1}{1 - \frac{1}{2}} = 2 \quad \text{as} \quad n \longrightarrow \infty$$

2. Find the geometric series of the sequence 5, $\frac{-10}{3}$, $\frac{20}{9}$, $\frac{-40}{27}$, ... The first term is 5 and the common ration r is $\frac{-2}{3}$. Since $|r| = \frac{-2}{3} < 1$ the series is convergent. By (12.7) the sum is

$$S_n = \frac{5}{1 - \left(\frac{-2}{3}\right)} = 3 \quad \text{as} \quad n \text{ becomes very large}$$

3. Express the number 3.4353535 ... as a rational number (that is as ratio of two integers). In number 3.4353535 ... I am using ellipsis ... to mean that the cycle 35 repeat itself forever.[1]

$$3.4353535... = 3.4 + \frac{35}{10^3} + \frac{35}{10^5} + \frac{35}{10^7} + \cdots$$

After the first term we have a geometric series with $a = \dfrac{35}{10^3}$ and $r = \dfrac{1}{10^2}$. Therefore using (12.7), we have

$$3.4353535... = 3.4 + \frac{\frac{35}{10^3}}{1 - \frac{1}{10^2}} = 3.4 + \frac{\frac{35}{1000}}{1 - \frac{1}{100}}$$

$$3.4353535... = 3.4 + \frac{\frac{35}{1000}}{\frac{99}{100}} = \frac{34}{10} + \frac{35}{990} = \frac{3401}{990}$$

---

[1] Actually the notation used to indicate repeating decimal is putting a −, called overline or overbar, over the repeating decimal, like $3.4\overline{35}$.

Exercises 2

1. Find the general term of the following sequences:

(a) $\dfrac{1}{8}, \dfrac{-1}{4}, \dfrac{1}{2}, -1, \ldots$ (b) $-2, \dfrac{5}{2}, \dfrac{-25}{8}, \dfrac{125}{32}, \ldots$

2. Find the sum of the first 10 terms of the sequences in problem 1.
3. Find the infinite series of problem 1.
4. Find $1 + 0.4 + 0.16 + 0.064 + \cdots$

12.2 Simple and Compound Interest

12.2.1 Simple Interest

Suppose you deposit P dollars in a savings account which gives you i percent interest per year. The initial P dollar investment is called the "principal." Assuming no withdrawal, at the end of the first year your account balance will be the principal plus the interest applied to the principal. Denoting your account balance at the end of the first year by P_1, we have

$$P_1 = P + iP = (1 + i)P$$

If in the following periods interest is paid only to the principal, then we have a *simple interest* model. Otherwise, if the principal and interest earned and accumulated from preceding periods both earn interest, then we have a *compound interest* model. Let's consider simple interest first.

Your account balance at the end of the second year, P_2, is

$$P_2 = P_1 + iP = (1 + i)P + iP = (1 + 2i)P$$

For the third period we have

$$P_3 = P_2 + iP = (1 + 2i)P + iP = (1 + 3i)P$$

The pattern is clear. Your balance at the end of t years or periods, denoted by P_t, is

$$P_t = (1 + it)P \qquad (12.8)$$

$P_1, P_2, P_3, P_4, \ldots, P_t$—the balance of the account at the end of the first, second, third, $\ldots$, tth year–form an arithmetic sequence with the first term $P_1 = (1 + i)P$ and common difference iP. Equation (12.8) is the term generating or general term of this sequence. There are four variables in (12.8): P_t, P, i, and t. If the values for any three of these variables are given, we should be able to determine the value of the fourth variable.

Examples 4

1. You invest $10,000 at the rate of 5.0% a year with simple interest. How much money you will have accumulated after 5 years? After 10 years? Using (12.8), we have

$$P_5 = (1 + 5 * 0.05)10000 = \$12,500$$

$$P_{10} = (1 + 10 * 0.05)10000 = \$15,000$$

2. How long does it take for $5,000 invested at 5.0% simple interest to double in value?

$$10000 = (1 + 0.05t)5000 \quad \longrightarrow \quad 1 + 0.05t = 2 \quad \longrightarrow \quad 0.05t = 1$$

and $t = 20$ years.
Analytically, investment growing n−fold can be written as

$$nP = (1 + it)P \quad \longrightarrow \quad 1 + it = n \quad \longrightarrow \quad it = n - 1, \quad \text{and finally } t = \frac{n - 1}{i}$$

Therefore if the interest rate is 5.0%, it takes t $= 1/0.05 = 20$ years for your money to double ($n = 2$). The interest rate must be 10% in order to double your investment over 10 years. If the interest rate is 5%, it takes 60 years for an investment to grow four fold: $t = \dfrac{4 - 1}{0.05} = 60.$

12.2.2 Compound Interest

A more interesting and common version of interest rate model in the financial market is *compound interest*. As it was mentioned above, in this version both the principal and accumulated interest earn interest. Assuming compounding is done annually, at the end of the first year of investing P dollars at the rate of i percent per year, the balance in the account is

$$P_1 = P + iP = (1 + i)P$$

At the end of the second year the balance is

$$P_2 = P_1 + iP_1 = (1 + i)P_1 = (1 + i)(1 + i)P = (1 + i)^2 P$$

And at the end of third year

$$P_3 = P_2 + iP_2 = (1 + i)P_2 = (1 + i)(1 + i)^2 P = (1 + i)^3 P$$

A clear pattern emerges here. We can express the value of the account after t years as

$$P_t = (1 + i)^t P \tag{12.9}$$

As an example, if you invest \$10,000 at the rate of 5.0% compounded annually, after 5 years you will have

$$P_5 = (1 + 0.05)^5 10000 = (1.05)^5 10000 = \$12{,}762.82$$

and after 10 years the amount will be

$$P_{10} = (1 + 0.05)^{10} 10000 = (1.05)^{10} 10000 = \$16{,}288.95$$

Comparing these values with the above values calculated using simple interest, clearly exhibit the power of compounding—interest earning interest. In this case over 10 years an additional \$1,288.95 is earned due to annual compounding.

Notice that $P_1, P_2, P_3, P_4, P_5, \ldots, P_t,$—the balance of the account at various periods—form a geometric sequence with the first term $P_1 = (1 + i)P$ and the common ration $(1 + i)$. Equation (12.9) is in fact the general term of this sequence.

With annual compounding the "periodicity" of compounding—the number of times interest is compounded each year, denoted by m—is one. The periodicity of compounding could be more than one, however. If m is 2 then compounding is done twice a year or semiannually. Compounding could also be done quarterly ($m = 4$), monthly ($m = 12$), weekly ($m = 52$), daily ($m = 365$), hourly ($m = 8760$), each minute, each second, or for that matter continuously ($m \to \infty$, i.e. an infinite number of times each year).

Let's develop the general formula for semiannual compounding. In this case interest on the account balance is computed every six months and added to the account to earn interest the succeeding periods. Given that i is annual, the interest earned for 6 months is $\frac{i}{2}$. At $t = \frac{1}{2}$ year the balance in the account is

$$P_{\frac{1}{2}} = P + \frac{i}{2}P = (1 + \frac{i}{2})P$$

At the end of the first year

$$P_1 = P_{\frac{1}{2}} + \frac{i}{2}P_{\frac{1}{2}} = (1 + \frac{i}{2})P_{\frac{1}{2}} = (1 + \frac{i}{2})(1 + \frac{i}{2})P = P(1 + \frac{i}{2})^2$$

Halfway through the second year

$$P_{1.5} = P_1 + \frac{i}{2}P_1 = P_1(1 + \frac{i}{2}) = (1 + \frac{i}{2})(1 + \frac{i}{2})^2 P = P(1 + \frac{i}{2})^3$$

At the end of the second year

$$P_2 = P_{1.5} + \frac{i}{2}P_{1.5} = (1 + \frac{i}{2})P_{1.5} = (1 + \frac{i}{2})(1 + \frac{i}{2})^3 P = P(1 + \frac{i}{2})^4$$

Recognizing the pattern, we can express the balance in the account after t years with semiannual compounding as

$$P_t = P(1 + \frac{i}{2})^{2t} \qquad\qquad\qquad (12.10)$$

It is easy to deduce that the most general formula for compounding can be expressed as

$$P_t = P(1 + \frac{i}{m})^{mt} \qquad\qquad \text{for} \quad t = 1, 2, 3, \ldots, T \qquad (12.11)$$

where m is the number of compounding periods per year and T is the *investment horizon*, the length of time the investor holds the investment. If compounding is annual then $m = 1$ and we have Eq. (12.9). For semiannual compounding $m = 2$ and we get Eq. (12.10). Similarly if compounding is done monthly we have $P_t = P(1 + \frac{i}{12})^{12t}$.

The following table gives the values of \$10,000 invested at 5.0 % interest rate after 5 years for various compounding periodicity:

Compounding	Periodicity	Account balance
Annually	1	\$12,762.82
Semi-annually	2	\$12,800.85
Quarterly	4	\$12,820.37
Monthly	12	\$12,833.59
Weekly	52	\$12,838.71
Daily	365	\$12,840.03
Hourly	8760	\$12,840.24
Minutely	525600	\$12,840.24

As the number of compounding periods increases from 1 to 2 (from annual to semiannual) there is a relatively noticeable increase in the account balance, about \$38. But as the number of compounding periods continue to increase the additional earning rapidly diminishes. From compounding daily to compounding by minutes, a 1440-fold increase, the additional money earned is only 26 cents!

The number of compounding periods could be taken to the extreme, to the *continuous compounding*. This is when m grows infinitely large. In that case the value of the account at time t must be expressed as

$$P_t = \lim_{m \to \infty} P(1 + \frac{i}{m})^{mt}$$

$$P_t = P \lim_{m \to \infty} (1 + \frac{i}{m})^{mt}$$

$$P_t = P \lim_{m \to \infty} \left[(1 + \frac{i}{m})^{\frac{m}{i}} \right]^{it} \tag{12.12}$$

By denoting $\dfrac{m}{i} = k$ so $\dfrac{i}{m} = \dfrac{1}{k}$, we can express (12.12) as

$$P_t = P \lim_{m \to \infty} \left[(1 + \frac{1}{k})^k \right]^{it}$$

But as we saw in Chap. 9, "Logarithm and Exponential Functions",

$$\lim_{m \to \infty} (1 + \frac{1}{k})^k = e, \quad \text{the base of the natural logarithm}$$

With this result we can write (12.12) as

$$P_t = Pe^{it} \tag{12.13}$$

which is the formula for continuous compounding.[2]

If we use (12.13) to determine the value of a $10,000 investment over 5 years when $i = 5\%$ and compounding is continuous, we get

$$P_t = 10000\ e^{0.05*5} = \$12,840.25$$

This is an improvement of only 21 cents in 5 years over daily compounding!! Even the improvement over monthly compounding is negligible. For this reason the monthly compounding is most common in the banking industry and the financial market.

There are 5 variables in the general formula for discrete compounding (12.11), namely P, i, m, t, and P_t. In general, if the values of any four variables is given, the value of the remaining variable can be determined analytically or with the help of numerical methods.

Examples 5

1. How long would it take for an amount of money invested at a 5.0% interest rate, compounded monthly, to double? Using (12.11)

---

[2] To be mathematically correct, (12.13) must be expressed as $P(t) = Pe^{it}$ to indicate P as a function of the continuous variable t (time). We are ignoring these subtleties.

$$P_t = P(1 + \frac{i}{m})^{mt} \quad \text{and} \quad 2P = P(1 + \frac{0.05}{12})^{12t}$$

$$(1 + \frac{0.05}{12})^{12t} = 2 \quad \longrightarrow \quad (1.00417)^{12t} = 2$$

Taking the logarithm of both sides, we have

$$\ln(1.00417)^{12t} = \ln 2$$

$$12t \ln(1.00417) = 0.69315 \quad \longrightarrow \quad 0.04994t = 0.69315 \quad \text{and} \quad t = 13.89$$

At the interest rate of 5.0 % compounded monthly, it takes 13.89 years or about 13 years and 11 months for any investment to double. This is a substantial improvement over the case with simple interest (20 years).

2. What interest rate compounded monthly doubles an investment over a period of 10 years?

$$P_t = P(1 + \frac{i}{m})^{mt} \quad \longrightarrow \quad 2P = P(1 + \frac{i}{12})^{12*10}$$

$$(1 + \frac{i}{12})^{120} = 2$$

By taking the logarithm of both sides, we have

$$\ln(1 + \frac{i}{12})^{120} = \ln 2 \quad \longrightarrow \quad 120 \ln(1 + \frac{i}{12}) = 0.69315 \quad \text{and}$$

$$\ln(1 + \frac{i}{12}) = 0.0057763$$

Now we take the anti-log

$$1 + \frac{i}{12} = e^{0.0057763} = 1.005793 \quad \longrightarrow \quad \frac{i}{12} = 0.005793 \quad \text{and finally}$$

$$i = 0.0695 \quad \text{or} \quad 6.95 \%$$

3. How much money do you need to invest at 5.6 % compounded quarterly in order to have $20,000 after 5 years?
Here we have $P_t = 20000$, $i = 0.056$, $m = 4$, and $t = 5$. We want to determine P.

$$P_t = P(1 + \frac{i}{m})^{mt} \quad \longrightarrow \quad 20000 = P(1 + \frac{0.056}{4})^{20} = P(1.014)^{20}$$

$$P = \frac{20000}{1.321} = \$15,145.06$$

Table 12.1 Comparing simple and compound interest

End of period	Simple interest	δ for simple interest	Compound interest	δ for compound interest
1	1050	5.000	1050.000	5
2	1100	4.762	1102.500	5
3	1150	4.545	1157.625	5
4	1200	4.348	1215.506	5
5	1250	4.167	1276.282	5
6	1300	4.000	1340.096	5
7	1350	3.846	1407.100	5
8	1400	3.704	1477.455	5
9	1450	3.571	1551.328	5
10	1500	3.448	1628.895	5
11	1550	3.333	1710.339	5
12	1600	3.226	1795.856	5

As we will see shortly, the amount $15,145.06 is called the *Present Value* or the *Discounted Value* of $20,000.

For simple interest the actual amount of interest that the account earns from year to year declines. Another way of expressing this idea is that the rate of growth of balance of the account is a decreasing function of time t. Let's denote the rate of change or growth in the account balance by $δ$. Note that

$$δ = \text{Rate of change in account balance} = \frac{P_t - P_{t-1}}{P_{t-1}}$$

Using (12.8) for simple interest we have

$$δ = \frac{(1 + it)P - [(1 + i(t - 1))]P}{[(1 + i(t - 1))]P} = \frac{i}{1 + i(t - 1)} \quad \text{for } t = 1, 2, 3, \ldots$$

Only for year 1 ($t = 1$) this rate is equal to the stated interest rate i. Beyond the first year the actual interest rate (rate of change or growth rate) that the account realizes from year to year is a decreasing function of time. Table 12.1 is prepared to provide a better understanding of this idea. The third column in this table shows $δ$ for $1,000 invested at 5.0 % simple interest over 20 years. As the figures in this column clearly indicate, the rate of change for the account balance, called the *effective* interest rate, shows systematic decline over time.

The story is different for compound interest. Using (12.11) we have $δ$ for general compound interest as

$$δ = \frac{P(1 + \frac{i}{m})^{mt} - P(1 + \frac{i}{m})^{m(t-1)}}{P(1 + \frac{i}{m})^{m(t-1)}} = (1 + \frac{i}{m})^m - 1$$

Table 12.2 Calculating effective interest

End of period	Compound interest quarterly	Effective rate	Compound interest monthly	Effective rate
1	1050.945	5.095	1051.162	5.116
2	1104.486	5.095	1104.941	5.116
3	1160.755	5.095	1161.472	5.116
4	1219.890	5.095	1220.895	5.116
5	1282.037	5.095	1283.359	5.116
6	1347.351	5.095	1349.018	5.116
7	1415.992	5.095	1418.036	5.116
8	1488.131	5.095	1490.585	5.116
9	1563.944	5.095	1566.847	5.116
10	1643.619	5.095	1647.009	5.116
11	1727.354	5.095	1731.274	5.116
12	1815.355	5.095	1819.849	5.116

We see that δ is independent of time t but influenced by the periodicity of compounding m. A higher frequency of compounding results in a higher interest rate. Here δ, the effective interest rate, is a function of m.

If compounding is annual, meaning $m = 1$, then the effective interest rate δ the is equal to the constant i, the stated interest rate. The values in Column 5 in Table 12.1, which is calculated base on balance of \$1,000 invested at 5.0% over 12 years, is constant at 5% over the life of the investment.

If interest is compounded quarterly then

$$\delta = \left(1 + \frac{0.05}{4}\right)^4 1 = 0.05095 \quad \text{or} \quad 5.095\%$$

and if $m = 12$ then

$$\delta = \left(1 + \frac{0.05}{12}\right)^{12} 1 = 0.05116 \quad \text{or} \quad 5.116\%$$

These results are given in the third and the fifth column of Table 12.2.

If compounding is continuous, then the effective interest rate would be

$$\delta = \frac{Pe^{it} - Pe^{i(t-1)}}{Pe^{i(t-1)}} = e^i - 1$$

When i is 5% the effective interest rate for continuous compounding is 5.127%.

12.3 Present Value

The concept of time value of money plays a fundamental role in economics and business decisions. Time value of money pertains to the fact that *a dollar received in the future is worth less than a dollar in hand today*. Different projects, investments, or wholesale/retail activities may generate different streams of profits, incomes, or sales revenues over different time horizons. A profit-maximizing firm must be able to analyze and compare streams of earnings generated by competing investment alternatives and select the one with the highest rate of return. To make the comparison possible, the firm must adjust the value of these streams to a common point in time. Given that nobody can predict the future, of all "common points" in time the "present" is the most reliable. Although the future values of economic or business factors affecting a project are subject to forecasting errors, their current values are known. As we will see, the present value problems are duals or "inverses" of future value problems.

To illustrate the concept let's work through several examples. In our discussion of compound interest we developed in Eq. (12.11) a general expression for the "future value" of P dollars invested at i percent interest, compounded m times a year:

$$P_t = P \left(1 + \frac{i}{m} \right)^{mt} \qquad \text{for} \quad t = 1, 2, 3, \ldots, T$$

Let's inverse this problem by asking "how much do I need to invest *now* in order to have P_t dollars after t, years if the interest rate is i compounded m times a year?" Here the unknown is P. We solve for P from (12.11)

$$P = \frac{P_t}{(1 + \frac{i}{m})^{mt}} = P_t \left(1 + \frac{i}{m} \right)^{-mt} \tag{12.14}$$

Whereas P_t is the *future value* of P dollars invested today at the given rate with a given compounding periodicity, P is the *present value* of P_t dollars earned in t years time. To better convey these concepts it is common to use FV_t (future value at time t) in place of P_t and PV (present value) for P. Using the new designations, (12.14) is written as

$$PV = FV_t \left(1 + \frac{i}{m} \right)^{-mt} \tag{12.15}$$

It should be clear that the present value of a sum of money earned t years in the future is the amount that if invested today will grow to that sum in t years.

As an example, what is the present value of \$10,000 due in 5 years if the interest rate is 6% compounded quarterly? Here $FV_t = FV_5 = 10000$, $i = 0.06$, $m = 4$, and $t = 5$. Using (12.15)

$$PV = 10000 \left(1 + \frac{0.06}{4}\right)^{-4*5} = 10000(1.015)^{-20} = 7424.7$$

You need to invest \$7,425 at 6% compounded quarterly to have \$10,000 in 5 years.

What would the present value of \$10,000 be if the interest rate is 6.5%, ceteris paribus?

$$PV = 10000 \left(1 + \frac{0.065}{4}\right)^{-4*5} = 10000(1.01625)^{-20} = 7244.2$$

What if the interest rate is compounded monthly, ceteris paribus?

$$PV = 10000 \left(1 + \frac{0.06}{12}\right)^{-12*5} = 10000(1.005)^{-60} = 7413.7$$

What if the amount is due in 6 years, ceteris paribus?

$$PV = 10000 \left(1 + \frac{0.06}{4}\right)^{-4*6} = 10000(1.015)^{-24} = 6995.4$$

It is clear from (12.15) and the above numerical examples that the longer the investment horizon (or investment holding period), the higher the interest rate, and higher the frequency of compounding, the smaller would be the present value of a future sum. That is, *PV* is inversely related to the interest rate, the periodicity of compounding, and the length of investment.

To fully understand the significance of present value analysis in the investment decision making process in the financial market, let's consider the following scenario. You are offered to invest \$5,000 and receive \$250 every year for two years. After two years you are repaid your \$5,000. If the interest rate in the economy is 6.0% compounded monthly, are you willing to make this investment? The answer to this question is rather straightforward. This investment is offering you $(250/5000) * 100 = 5\%$ simple interest rate, while you have the option of investing your \$5,000 elsewhere at 6% interest compounded. You are not, of course, going to accept this offer.

Then comes a slight change in the offer. You get your \$250 for two years and \$5,000 at the end of the second year. How much are you willing to invest? The key to this decision is the present value of the stream of payments you will receive compared to the amount of the investment or alternative investments. The *PV* of this investment is the sum of the present values of future payments, \$250 for two years, and the present value of the \$5,000 you'll receive in two years, using the available 6% interest rate. With the help of (12.15) and assuming monthly compounding, we calculate the investment's *PV* as

$$PV = 250(1 + \frac{0.06}{12})^{-12*1} + 250(1 + \frac{0.06}{12})^{-12*2} + 5000(1 + \frac{0.06}{12})^{-12*2}$$
$$= 250(1.005)^{-12} + 250 * (1.005)^{-24} + 5000(1.005)^{-24}$$
$$= 235.48 + 221.80 + 4435.93 = 4893.20$$

You are willing to invest \$4,893.20! In the context of this problem, the 6% interest rate used to calculate the present value of each payment in the future is called the *discount rate* and the sum of the present values of the payments is referred to as the *discounted value* of the investment. As long as the interest rate or rate of return generated by the investment is less than alternatively available interest rates in the market, the investment would not be undertaken.

12.3.1 Financial Debt Instruments and Their Yields

Based on the above analysis, if the interest rate is 6.0% compounded monthly the value of an investment that pays the fixed amount of \$250 a year for two years plus \$5,000 at the end of the second year is \$4,893.20. This is the fundamental approach for determining the market price of interest-paying or fixed income-generating financial instruments like government and corporate bonds. The discounting process clearly shows why the market prices for these types of instruments are so sensitive to changes in the interest rates in the economy.

Financial debt instruments are classified into *short, intermediate*, and *long-term* based on the length of time it takes them to reach their terminal date, called the *maturity*. Short-term instruments mature in less than one year, and the intermediate-term instruments have a maturity between one and ten years. Long-term instruments have a maturity of more than ten years.

Assume an intermediate- or long-term bond, which reaches maturity after T years, has a stated interest rate, called the *coupon rate*,[3] i_c. What is the price or market value of this bond if the interest rate is i compounded m times?   The market price is the present value of the stream of interests this bond pays plus the present value of the *face* or *par value*[4] of the bond when it reaches maturity and is redeemed. If F denotes the face value of a bond, then this bond pays the amount of $C = i_c * F$ interest every year[5] for T years. C is called the *coupon payment*. The price of this bond, denoted by P_B, is the present value of the stream of coupon payments over T years plus the present value of bond's par value paid at maturity.

---

[3] The interest paying bonds are called *coupon bonds*.

[4] The par or face value of a bond is its denomination. Bonds are issued with various denominations, generally with increments of \$1,000, like \$1,000, \$2,000, and so on.

[5] Interests are commonly paid semiannually. We ignore this twist here.

$$P_B = PV = \sum_{t=1}^{T} \frac{C}{\left(1+\frac{i}{m}\right)^{mt}} + \frac{F}{\left(1+\frac{i}{m}\right)^{mT}}$$

$$P_B = C \sum_{t=1}^{T} \left(1+\frac{i}{m}\right)^{-mt} + F\left(1+\frac{i}{m}\right)^{-mT}$$

$\displaystyle\sum_{t=1}^{T}\left(1+\frac{i}{m}\right)^{-mT}$ is the sum of T terms of a geometric sequence with $\dfrac{1}{\left(1+\frac{i}{m}\right)^m}$ as both the first term and the common ratio. Therefore,

$$P_B = \frac{C\left[(1+\frac{i}{m})^{mT} - 1\right]}{(1+\frac{i}{m})^{mT}\left[(1+\frac{i}{m})^m - 1\right]} + \frac{F}{(1+\frac{i}{m})^{mT}} \tag{12.16}$$

We can use this expression to determine the market price of a $5,000 face value 10-year bond with 5 % coupon rate and the interest rate of 6 % compounded monthly. The bond's coupon payment is $C = 0.05 * 5000 = \$250$, so we have

$$P_B = \frac{250\left[\left(1+\frac{0.06}{12}\right)^{12*10} - 1\right]}{\left(1+\frac{0.06}{12}\right)^{12*10}\left[\left(1+\frac{0.06}{12}\right)^{12} - 1\right]} + \frac{5000}{\left(1+\frac{0.06}{12}\right)^{12*10}}$$

$$P_B = 1825.48 + 2748.16 = \$4{,}573.65$$

We can simplify (12.16) by assuming annual compounding, that is $m = 1$

$$P_B = \frac{C[(1+i)^T - 1]}{i(1+i)^T} + \frac{F}{(1+i)^T} \tag{12.17}$$

Considering the annual compounding for the above bond price problem, we get

$$P_B = \frac{197.71}{0.10745} + \frac{5000}{1.791} = 1840.02 + 2791.97 = \$4{,}632.00$$

Whether interest is compounded annually or monthly, it is clear that the price of this bond in the market is below its face value. When a bond is traded below its face value it is said that the bond is *traded at a discount*. In our example the amount of discount, using annual compounding case, is $F - P_B = \$5{,}000 - \$4{,}632 = \$368$. It is a tradition in the bond market that a bond price is quoted as the amount paid per $100 face value. In this case the bond price would be quoted as $92.64, or at a $7.36 discount per $100 par value.

In the above example the bond coupon rate is 5% but interest rate offered by instruments similar in risk and term[6] has moved up to 6%. It is natural that investors in the bond market prefer a 6% instrument to a similar instrument that only offers 5%. Consequently demand for the 5% bond declines, putting downward pressure on its price. The price of the bond is then determined by using the prevailing interest rate of 6% as the "discount rate" for computing the present value of the stream of coupon payments and its face value at maturity. Selling a bond at a discount is a form of "compensation" for its lower coupon rate.

When a bond matures, the bondholders receive the bond's face value. An investor who buys this bond at $4,632 and holds it to its maturity can realize an additional earning of $368, i.e. a $368 *capital gain.* This capital gain compensates the lower coupon rate paid by the bond and pushes the return on this bond up to 6%. The rate of return on a bond is called its *yield.* The yield of a bond *traded at discount* is higher than its stated interest or coupon rate.

If we use the bond's coupon rate as the discount rate, the price of the bond, irrespective of periodicity of compounding, will be exactly equal to its face value. This means that if the bond's interest rate is equal to the interest rate offered by similar instruments in the market, it would be traded at par. In this case the bond's yield is equal to its coupon rate.

What if the bond's interest rate is more than other similar instruments in the market? In that case the price of the bond would be more than its face value. When a bond's price is more than its par value, it is said that the bond is *selling or traded at a premium.* If the alternative interest rate available to investors in the market is 4.5%, then our bond paying 5% would command a premium. Using 4.5% as the discount rate in (12.17), we have

$$P_B = \frac{250[(1+0.045)^{10} - 1]}{0.045(1+0.045)^{10}} + \frac{5000}{(1+0.045)^{10}} = 1978.18 + 3219.64 = \$5,197.82$$

This bond would be traded at $103.96 or a $3.96 premium per $100 face value. In this case the bond yield is 4.5%. The reason for the yield dropping to 4.5% is the fact that the bondholder must absorb a $197.82 *capital loss* at maturity.

To capture the essence of the above discussion let's consider a simple example. You invest $10,000 in a bond that pays 5% interest forever. That is, this bond has no

---

[6] We are treating the subject of *fixed income (bond) valuation* rather casually here. In the real world *bond valuation* is much more complex and involves many different variables and factors. One very important factor is the financial health and creditworthiness of the bond issuing entity. Perception of investors of potential default risk by issuers of bond (or any other financial debt instrument, for that matter) plays a major role in shaping demand for these papers and their prices in the financial market. (Footnote 6 continued)

The financial crisis of 2008 proved, once again, that the investors' *trust and confidence* in the financial health of a company trumps all other factors. This manifested itself vividly when bonds issued by faltering corporations like GM and AIG were offered at a fraction of their face values and nobody was buying.

Our exposition of the subject here is elementary, providing a *first approximation* method for investment evaluation.

maturity and generates $500 interest year after year. These types of bonds are rare. An exception is the British government bond, issued for the first time in the late 18th century, and are still traded today. It is called *British Consol* or Consol for short. Your 5% forever bond generates $C = 0.05 * 10000 = \$500$ annual income year after year. What is the price of a bond with yearly coupon payment of $C forever? Let's rewrite (12.17) alternatively as

$$P_B = \frac{C}{i} - \frac{C}{i(1+i)^T} + \frac{F}{(1+i)^T}$$

As T grows to infinity the second and third terms of the above expression approach zero and the price of the forever bond approaches

$$P_B = \frac{C}{i} \qquad (12.18)$$

Then the price of your 5% bond paying $500 forever is $P_B = \frac{500}{0.05} = \$10,000$.

Now let's assume that the interest rate jumps to 6%. What would be the price of your bond? The answer is the amount you need to invest at 6% to generate the same amount of $500. That is

$$P_B = \frac{500}{0.06} = \$8,333.33$$

You don't, therefore, need to invest $10,000 in order to earn $500 each year; you only need $8,333. So the price of your bond drops to $8,333. On the other hand, if the interest rate declines to 4.5% then you need to invest more than $10,000 to earn $500 per year, and the price of your bond jumps to

$$P_B = \frac{500}{0.045} = \$11,111.11$$

A nice feature of the price of forever bond in (12.18) is that it shows how a rise in interest rates leads to a decline in the price of a bond, in other words, it shows that there is an inverse relationship between bond price, interest rate, and yield (of course, *ceteris paribus*).

12.3.2 Yield to Maturity and Current Yield

The full title of what we have been referring to as "yield" is *yield to maturity*, abbreviated to YTM. An investor receives this yield if he or she holds the bond till maturity. YTM is the interest rate that equates the present value of the stream of payments made by a debt instrument—like a bond—with its market price. In our bond price Eq. (12.16), YTM is the interest rate i that makes the RHS of the equation equal to its LHS. As bonds are traded in the market their prices fluctuate. So YTM changes as bond price changes. Given the amount of coupon C, the bond's face value F, the number of years to maturity T, and the price of the bond established in the bond market P_B, we must solve for i in (12.16) to determine the YTM. This cannot be done in closed form or analytically. Alternatives are computer software or financial calculators (both use some variation of the Newton-Raphson numerical approximation method).

There is a simple approximation method that works like this: We first calculate the Bond's Annual Earning (BAE) as

$$BAE = C + \text{Prorated Capital Gain} \quad \text{or}$$
$$BAE = C - \text{Prorated Capital Loss}$$

If the bond was purchased at a discount then $F - P_B$ is capital gain. If the bond was purchased at premium then $F - P_B$ is capital loss. Capital gain or loss will be realized after T years at the maturity date. We prorate the sum by $\dfrac{(F - PB)}{T}$. That is, we divide the future capital gain or loss by the number of years left to maturity and pretend that besides coupon the bond earns this amount (in case of capital gain) or loses this amount (in case of capital loss) annually. Next we calculate the Average Price (AP) of the bond as $AP = \dfrac{(F + PB)}{2}$. The YTM is then calculated as

$$YTM = \frac{BEA}{AP} * 100$$

A numerical example should help to understand this approximation method.

Example 6

Assume the price of a \$5,000 bond with a coupon rate of 5 % and 4 years to maturity is \$4,940. What is its approximate YTM?

This bond pays $5000 * 0.05 = \$250$ a year. It is purchased at a discount. At maturity the bondholder will realize a capital gain of $5000 - 4940 = \$60$. There are 4 years left to maturity. The prorated capital gain is then calculated as $60/4 = \$15$ a year. We calculate the Bond's Annual Earning, its Average Price, and the YTM as

$$BAE = 250 + 15 = 265$$

$$AP = \frac{(5000 + 4940)}{2} = 4970$$

$$YTM = \frac{265}{4970} * 100 = 5.33\,\%$$

This bond's YTM is 0.33 percent, or 33 basis points, more than its coupon rate.[7] This difference is due to the fact that the bond holder besides interest earns additional $60–$15 on annual basis—as capital gain.

What if the bond is traded at $101.2 per $100 par value? This means that the price of the $5,000 bond is $5,060. In this case the bondholder will have a $60 capital loss when the bond is redeemed at maturity. The prorated capital loss is $\frac{60}{4} = \$15$. Then we have

$$BEA = 250 - 15 = 235$$

$$AP = \frac{5000 + 5060}{2} = 5030$$

$$YTM = \frac{235}{5030} * 100 = 4.67\%$$

In this case the bond's YTM is 33 basis points less than it's coupon rate. This difference is due to $60 capital loss.

Using a financial calculator we get YTM $= 5.34\%$ for the first part of the example, and 4.66% for the second part. This approximation method is not that bad after all!

Another widely used measure of return or yield is the *current yield* (CY). It is defined as the ratio of the bond's coupon payments to the price of the bond.

$$CY = \frac{C}{P_B} * 100$$

The current yields for the bonds in the first and second part of the above numerical example are

$$CY_1 = \frac{250}{4940} * 100 = 5.06\% \quad \text{and} \quad CY_2 = \frac{250}{5060} * 100 = 4.94\%$$

A decline in the price of a bond leads to a higher CY for the bond. An increase in the price of a bond lowers its CY. This shows again that there is an inverse relation between yield and bond price.

In summary, when a bond is traded at par its YTM is equal to both its CY and coupon rate. If it is traded at a discount, its YTM is greater than its CY, which in turn is greater that its coupon rate. If a bond is traded at a premium, its YTM is less than its CY, which in turn is less than its coupon rate.

$P_B < F$ Bond traded at discount $YTM > CY > i_c$
$P_B = F$ Bond traded at par $YTM = CY = i_c$
$P_B > F$ Bond traded at premium $YTM < CY < i_c$

---

[7] A *basis point* is one hundredth of a percentage point, that is 0.01 percentage point is equal to 1 basis point. It is used in the financial industry to denote a rate change or the difference (*spread*) between two rates.

12.3.3 Money Market Instruments

Short-term financial instruments are called *money market instruments* and traded in the *money market*. In the money market only debt instruments with maturities less than one year are traded. The money market is for short-term borrowing and lending. Instruments used by borrowers to raise funds are commonly called "paper." Examples of money market instruments are Commercial Paper, Certificate of Deposit, Banker's Acceptance, Repurchase Agreement, Treasury Bills (T-bill), and Municipal (Tax Anticipation) Notes.

Money market instruments don't pay interest. They are traded at discount. They are offered at a price below their face values and the face values are repaid at maturity. For this reason T-bill is called a *discount bond*. Discounting is the mechanism for pricing short-term instruments. The discount rate used for discounting these instruments is generally a benchmark rate like the yield on short-term government bond (T-bills). The general formula for calculating the price of these instruments can be expressed as

$$P = \text{Price} = \frac{\text{Face value of instrument}}{\left(1 + \dfrac{\text{discount rate}}{m}\right)^{m * \left(\frac{\text{day count}}{\text{annual basis}}\right)}}$$

where the "discount rate" is the rate used for discounting the instrument, m is the number of annual compounding (which in the money market is generally 12, for monthly compounding), the "day count" is the number of days left to maturity of the instrument, and the "annual basis" is the number of days in a year. By convention the number of days in a year for Money Market instruments is 360. In the Bond market it is 365. The following example will clarify this process.

Example 7

Commercial paper is an unsecured (meaning that it is not backed by collateral) short-term debt instrument issued by corporations with high credit ratings. These papers are issued with maturities ranging from 2 to 270 days and marketed on the discount basis.

What is the market price of a $5 million 60-day commercial paper, if the yield of 60-day T-bill is 3.5%?

Assuming interest is compounded monthly and using the 60-day T-bill yield as the discount rate, we have

$$P = \frac{5000000}{(1 + \frac{0.035}{12})^{12*(\frac{60}{360})}} = \$4,970,960.44$$

At maturity this commercial paper generates $29,039.56 profit.

There are long-term bonds without coupon payments. They are called *zero-coupon bonds*. These bonds are traded at a deep discount, generating profit when they are redeemed at face value upon maturity. As the following example illustrates, calculating the zero-coupon bond yield is a bit challenging.

Example 8

What is the return or yield of a 20-year $60,000 zero-coupon bond offered at $30,000?

We must find the interest rate that helps $30,000 grow to $60,000 in 20 years. Assuming monthly compounding and using

$$P_t = P(1 + \frac{i}{m})^{mt}$$

we have

$$60000 = 30000(1 + \frac{i}{12})^{12*20} = 30000(1 + \frac{i}{12})^{240}$$

We now have to solve for i. By rewriting the above equation as

$$(1 + \frac{i}{12})^{240} = 2$$

and taking the natural logarithm of both sides, we have

$$240 \ln(1 + \frac{i}{12}) = \ln(2) = 0.69315 \quad \longrightarrow \quad \ln(1 + \frac{i}{12}) = 0.00289$$

Taking the anti-log of both sides of the last expression,

$$1 + \frac{i}{12} = e^{0.00289} = 1.00289 \quad \longrightarrow \quad \frac{i}{12} = 0.00289 \text{ and } i = 12*0.00289 = 0.0347$$

This zero coupon bond's yield is 3.47 %.

When calculating a coupon bond's YTM or a zero-coupon bond's yield, we have to make the assumption that the investor holds the bond till maturity. What if the investor decides to sell the security before it reaches its maturity? That is, what if the investor's "holding period" is shorter than the bond's remaining years of its term? How well an investor does by holding an asset over a certain period of time is measured by the *rate of return* or the *average annual rate of return*. The concept of rate of return could be equally applied to both financial and physical assets.

Assume an investor buys an asset at time t for the price of P_t. She sells this asset at time $t + k$ for the price of P_{t+k}, where k is the "holding period" of the asset. For debt securities k cannot be longer than the time left to maturity of the security T, so $k < T$. For assets with no maturity date $k > 0$. If this asset is an interest-bearing or an income-generating asset, we denote its annual earnings by C_{t+n} $(n = 1, 2, \ldots, k)$. If income generated by the asset does not vary from year to year, that is it is a fixed income asset (examples are coupon bonds paying a fixed annual coupon or a rental property paying rent), all C_{t+n} will be the same constant. We then define the *holding period rate of return* (HPRR), expressed as a percentage, as

$$HPRR = \frac{\sum_{n=1}^{k} C_{t+n} + (P_{t+k} - P_t)}{P_t} * 100 \qquad (12.19)$$

The *rate of return* (RR), or more accurately the *average annual rate of return*, of this asset is then is

$$RR = \frac{HPRR}{k}$$

Let's quickly demystify the above equations through a few examples.

Example 9

In 2009 an investor purchases a \$5,000 face value, 10-year, 5% coupon bond for \$4,850. After 3 years she sells the bond for \$4,900. What is the investor's rate of return?

In this example $t = 2009$; $P_t = P_{2009} = \$4{,}850$; $k = 3$; $P_{t+k} = P_{2012} = \$4{,}900$; $C_{2009+n} = 0.05 * 5000 = \$250$ (the bond's coupon payment) for $n = 1, 2, 3$. Using (12.18), we have

$$HPRR = \frac{\sum_{n=1}^{3} C_{2009+n} + (P_{2012} - P_{2009})}{P_{2009}} * 100$$

$$HPRR = \frac{250 + 250 + 250 + (4900 - 4850)}{4850} * 100 = 16.49\%$$

$$RR = \frac{16.49}{3} = 5.498\%$$

Example 10

An investor purchases a property in 1995 for \$300,000. She sells the property in 2005 for \$500,000. During the 10 years that she owned the property, it generated a monthly income (rent − monthly mortgage payment − maintenance cost) of \$300. What is her rate of return?

In this example $t = 1995$; $P_t = P_{1995} = \$300{,}000$; $k = 10$; $P_{t+k} = P_{2005} = \$500{,}000$; $C_{1995+n} = \$300$ for $n = 1, 2, \ldots, 10$. So using (12.18), we have

$$HPRR = \frac{10 * 12 * 300 + (500000 - 300000)}{300000} * 100 = 78.67\%$$

$$RR = \frac{78.67}{10} = 7.867\%$$

Example 11

A $10,000 face value, 15-year corporate bond with a coupon rate of 4.5 % is purchased 5 years before its maturity. If the investor holds this bond to maturity what is the bond's rate of return?

Note that it is not necessary to know the date of purchase of the asset, as long as we know the holding period. In this example we have $P_t = \$10,000$; $k = 5$; $P_{t+k} = P_{t+5} = \$10,000$; $C_{t+n} = 0.045 * 10000 = 450$ for $n = 1, 2, \ldots, 5$. Using (12.18), we have

$$HPRR = \frac{5 * 450 + (10000 - 10000)}{10000} * 100 = 22.5\%$$

$$RR = \frac{22.5}{5} = 4.5\%$$

It is clear from this example that the rate of return RR of a security purchased at par and held to maturity is exactly the same as its yield to maturity YTM. You will be asked in the exercise section to verify that RR > YTM if a security is traded at a discount and RR > YTM if it is purchased at a premium.

Example 12

In Example 7 an investor paid $4,970,960.44 for a 60-day $5,000,000 commercial paper. Now assume that after 20 days this investor decides to sell the commercial paper in the market. What would be the price if the discount rate remains the same at 3.5 %?

Using the formula for pricing money market instruments, we have

$$P = \frac{5000000}{\left(1 + \frac{0.035}{12}\right)^{12*\left(\frac{40}{360}\right)}} = \$4,980,621.51$$

leading to a $9,661 profit for the investor. In this example the holding period rate of return is

$$HPRR = \frac{9661}{4970960.44} * 100 = 0.194\%$$

0.194 % over 20 days translates to $0.194 * \frac{360}{20} = 3.492$ percent annually, that is, the annualized rate of return on this investment is 3.492 %.

In the exercise section of this chapter you will be asked to show that if the discount rate remains the same, the holding period rate of return can be expressed simply as

$$HPRR = \left(1 + \frac{i}{12}\right)^{\frac{12}{360}*n} - 1$$

where n is the number of days that the investor holds the instrument. In our example n is 20 days, so the *HPRR* is

$$HPRR = \left(1 + \frac{0.035}{12}\right)^{\frac{12*20}{360}} - 1 = 0.00194 \quad \text{or} \quad 0.194\%$$

12.3.4 Present Value and Project Selection

Application of present value is not limited to evaluation of financial instruments. It is also applied to project selection or decision making regarding investment in plant, equipment, and capital budgeting. A typical example would be the following.

A corporation is considering the purchase of a piece of equipment. Equipment price, or the original capital outlay, is P_E. Over its useful life of T years, this piece of equipment would generate C dollars of monthly revenue (net of maintenance and other costs) for the firm. After T years this equipment is sold as scarp for S dollar. Should the firm buy this equipment? What criteria should be used in making this decision? This problem is strikingly similar to investing in a coupon bond. The monthly revenue of C dollars generated by the equipment is similar to coupon payments. P_E is similar to the bond price P_B, possibly several magnitudes larger, and S is similar to the bond's face value F, possibly much smaller. Then the answer to our question is centered around the analysis of the combined present value of the generated monthly revenues, the cash flow, and the scrap, using a proper discount rate. A slightly different version of Eq. (12.16) is reproduced below. The only changes are the replacement of P_B by P_E and F by S.

$$P_E = \frac{C\left[(1+\frac{i}{m})^{mT} - 1\right]}{(1+\frac{i}{m})^{mT}\left[(1+\frac{i}{m})^{m} - 1\right]} + \frac{S}{(1+\frac{i}{m})^{mT}}$$

In the context of investment decision making, the discount rate i that equates the LHS of the equation with the RHS (the price or cost of equipment) is called the *internal rate of return*. A project is a good investment if its internal rate of return is greater than the rate of return of alternative investments, including investing in other projects, buying long-term financial instruments, or even leaving the amount in an interest bearing bank account.

Example 13

Gamma Corporation will not undertake any investment unless it returns at least 20% per year compounded monthly. Gamma is considering the development of three new products, X, Y, Z.

If Gamma intends to invest in only one product, which one should it pick?

Product	Development cost	Monthly profits
X	$450,000	$17,500 per month for 3 years
Y	$530,000	$12,000 per month for 6 years
Z	$750,000	$20,000 per month for 5 years

We must first calculate the present value of the stream of profits for each project. This problem requires a slightly different version of Eq. (12.16). Here payments (profits) are made monthly, whereas in (12.16) they are made annually. Let's denote the monthly payments by C and years of payments by T. Then there are $12 * T$ payments. The present value of the stream of these payments is

$$PV = \frac{C}{(1+\frac{i}{12})} + \frac{C}{(1+\frac{i}{12})^2} + \frac{C}{(1+\frac{i}{12})^3} + \cdots + \frac{C}{(1+\frac{i}{12})^{12T}} = C\sum_{t=1}^{12T} \frac{1}{(1+\frac{i}{12})^t}$$

$\sum_{t=1}^{12T} \frac{1}{(1+\frac{i}{12})^t}$ is the sum of $12 * T$ terms of a geometric sequence with $\frac{1}{(1+\frac{i}{12})}$ as its first term and common ratio. Then PV is

$$PV = \frac{C\left[(1+\frac{i}{12})^{12T} - 1\right]}{(\frac{i}{12})(1+\frac{i}{12})^{12T}} \tag{12.20}$$

Using (12.20) we can compute PV of Gamma Corporation's projects. Since Gamma is not willing to consider any project with a rate of return less than 20%, then 0.2 must be used as the discount rate. 20% is Gamma's *targeted internal rate of return*.

$$PV_X = \frac{17500\left[(1+\frac{0.2}{12})^{12*3} - 1\right]}{(\frac{0.2}{12})(1+\frac{0.2}{12})^{12*3}} = \$470,891.08$$

Similarly $PV_Y = \$500,984.73$ and $PV_Z = \$754,891.21$

Table 12.3 below summarizes the information Gamma needs to make a decision. The last column, **Net Present Value**, is the difference between the present value of each project and its cost. Obviously Project Y is immediately eliminated. The present value of the future stream of its earnings cannot even cover the cost of the project.

Table 12.3 Net present values of 3 projects

Product	Cost	Present value	Net present value
X	$450,000	$470,891	$20,891
Y	$530,000	$500,985	–$29,015
Z	$750,000	$754,891	$4,981

Both X and Z have positive Net PVs. However, the rational choice is clearly X, which has much larger net present value.

It should be noted that a different targeted return or discount rate may lead to a different investment decision. If Gamma, for example, were willing to accept a 15 % rate of return, then Project Z would be the best choice.

12.4 Fixed Payment Loans

Loans that their payoff require periodic (monthly, quarterly, annual, etc.) payments of a fixed amount over the term of the loan are called *fixed payment loans*. For fixed payment loans, the interest rate charged by the lending institution (bank, mortgage company, saving and loan association, credit union, finance company, ...) is fixed over the life or term of the loan. When the amount of a loan, the term or length of a loan, and the interest rate applied to the loan (often determined based on the credit history and *risk profile* of a loan applicant) is determined, then the lending institution calculates the amount of periodic payment by the borrower.

Let's assume that payments are monthly. Denote the monthly payment by C. Let L, i, m, and T denote the loan amount, annual interest rate, number of compounding, and term (life, time horizon, or number of years to maturity) of the loan, respectively. Then C is determined by equating the sum of the present values of future monthly payments to the loan amount L:

$$L = \frac{C}{(1 + \frac{i}{m})^{\frac{m}{12}}} + \frac{C}{(1 + \frac{i}{m})^{\frac{2m}{12}}} + \cdots + \frac{C}{(1 + \frac{i}{m})^{mT}}$$

$$L = \sum_{t=1}^{12*T} \frac{C}{(1 + \frac{i}{m})^{\frac{mt}{12}}}$$

This is the sum of $12T$ terms of a geometric series with the first term and common ratio

$$a = C(1 + \frac{i}{m})^{-\frac{m}{12}} \qquad\qquad r = (1 + \frac{i}{m})^{-\frac{m}{12}}$$

Remember, the sum of n terms of a geometric series with the first term a and common ratio r is

$$S_n = \frac{a(1 - r^n)}{(1 - r)}$$

So we have

$$L = \frac{C(1 + \frac{i}{m})^{-\frac{m}{12}} \left\{ 1 - \left[(1 + \frac{i}{m})^{-\frac{m}{12}} \right]^{12T} \right\}}{1 - (1 + \frac{i}{m})^{-\frac{m}{12}}}$$

Solving for C (monthly or coupon payment), we have

$$C = \frac{L\left[(1 + \frac{i}{m})^{\frac{m}{12}} - 1\right]}{1 - (1 + \frac{i}{m})^{-mT}} \tag{12.21}$$

Notice if m is 12, that is compounding is monthly (which is very common in the credit market and we assume that in the rest of this chapter) then (12.21) is reduced to

$$C = \frac{L(\frac{i}{12})}{1 - (1 + \frac{i}{12})^{-12T}} \tag{12.22}$$

or alternatively,

$$C = \frac{L(\frac{i}{12})(1 + \frac{i}{12})^{12T}}{(1 + \frac{i}{12})^{12T} - 1}$$

Example 14

What is the amount of your monthly payment if you get a 3-year $15,000 car loan at 6 percent interest compounded monthly? Here $L = 15000$, $i = 0.06$, $m = 12$, and $T = 3$. From (12.22) we have

$$C = \frac{15000\left(\frac{0.06}{12}\right)}{1 - \left(1 + \frac{0.06}{12}\right)^{-12*3}} = \frac{75}{0.16436} = \$456.33$$

You have to pay $456.33 a months for 36 month—a total of $16,426.85—to pay off your $15,000 car loan.

12.4.1 Fixed Mortgage Loan

Contrary to car loans, home mortgage loans are generally very long-term. 15, 25, and 30-year mortgage loans are very common. Mortgage rates, the interest rates charged to borrowers, are always higher for longer-term mortgages. Given the amount of a loan, the mortgage rate, and the length or term of a loan, the monthly payment that a borrower must make over the life of a loan can be determined by using (12.22). For example, a 30-year $250,000 loan with fixed mortgage rate of 5 % requires 360 monthly payments of $1,342.05. The same loan amount for a 15-year mortgage with rate of 4.5 % requires 180 monthly payments of $1,912.48.

In the 30-year loan example, the borrower pays a total of $360 * 1342.05 = \$483,138$ to pay off a $250,000 loan, meaning the borrower pays $233,138 interest over 30 years. The total payment for the 15-year loan example is $344,246 with a $94,246 interest payment.

Table 12.4 Amortization schedule of a 30-year fixed mortgage loan

Payment number	Fixed monthly payment	Amount going to interest	Amount applied to principal	Equity accumulated	Loan outstanding
1	$1342.05	$1041.67	$300.38	$300.38	$249699.62
2	1342.05	1040.42	301.63	602.01	249397.99
3	1342.05	1039.16	302.89	904.91	249095.09
4	1342.05	1037.90	304.15	1209.06	248790.94
.	.	.	.	.	.
.	.	.	.	.	.
.	.	.	.	.	.
155	1342.05	772.19	569.86	65244.57	184755.43
156	1342.05	769.81	572.24	65816.81	184183.19
.	.	.	.	.	.
.	.	.	.	.	.
.	.	.	.	.	.
250	1342.05	496.15	845.90	131770.47	118229.53
251	1342.05	492.62	849.43	132619.90	117380.10
.	.	.	.	.	.
.	.	.	.	.	.
.	.	.	.	.	.
358	1342.05	16.65	1325.40	247329.23	2670.77
359	1342.05	11.13	1330.92	248660.15	1339.85
360	1345.43	5.58	1339.85	250000.00	0.00

At the early phases of mortgage payments, most of monthly payments are applied to the interest, and only small amounts are applied to the principal. For the 30-year loan, the fist month's interest is $25000(0.05/12) = \$1041.67$. Therefore, from the $1,342.05 monthly payment only $300.38 is applied to the principal. But as the number of payments increases and the amount of outstanding loan declines, a bigger portion of the monthly payments are applied to the principal.

In Table 12.4 the amount applied to interest, the amount applied to the principal, total equity accumulated, and the amount of loan outstanding for selective months of our 30-year 5% mortgage loan are given. A chart or table similar to Table 12.4 is called an *amortization schedule or chart*. The word "amortize" means "to reduce gradually." Amortization is the process of paying off a debt in regular installments over a specified period of time.

We now develop a mathematical expression for determining the balance or amount of loan outstanding at any time period t. Notice that the balance at time t is equal to the balance at time $t - 1$ minus the amount of the last monthly payment applied to the principal. Denoting the balance of the loan outstanding by B, we have

$$B_t = B_{t-1} - (C - \frac{i}{12}B_{t-1}) = (1 + \frac{i}{12})B_{t-1} - C$$

A reader who studied Chap. 11 "*Economics Dynamics*" should be able to recognize this equation as a first-order difference equation of the form

$$y_t = ay_{t-1} + b$$

with the solution

$$y_t = a^t y_0 + b\,\frac{1 - a^t}{1 - a}$$

Here $a = (1 + \frac{i}{12})$, $b = -C$ and $y_0 = L$, which is the original loan amount or the balance outstanding at $t = 0$. Therefore

$$B_t = (1 + \frac{i}{12})^t\, L - \left[\frac{1 - (1 + \frac{i}{12})^t}{1 - (1 + \frac{i}{12})}\right] C$$

or after simplification,

$$B_t = (1 + \frac{i}{12})^t\, L - \left[\frac{(1 + \frac{i}{12})^t - 1}{\frac{i}{12}}\right] C \qquad (12.23)$$

Let's check the validity of (12.23). For our 30-year 5%, mortgage loan example, the outstanding loan after the 250th payment is

$$B_{250} = \left(1 + \frac{0.05}{12}\right)^{250} 250000 - \left[\frac{\left(1 + \frac{0.05}{12}\right)^{250} - 1}{\frac{0.05}{12}}\right] 1342.05$$

$$= \$118{,}229.53$$

exactly the number in Table 12.4 above.

We can similarly derive an expression for the amount applied to the principal for the tth payment. Denoting this amount by $Prin_t$ we have

$$Prin_t = C - \frac{i}{12}\, B_{t-1}$$

After substituting for B_{t-1} by lagging (12.23) one period, we have

$$Prin_t = C - \frac{i}{12}\left\{(1 + \frac{i}{12})^{t-1} L - \left[\frac{(1 + \frac{i}{12})^{t-1} - 1}{\frac{i}{12}}\right] C\right\}$$

which can be simplified to

$$Prin_t = (1 + \frac{i}{12})^{t-1} (C - \frac{i}{12} L) \tag{12.24}$$

In our 30-year loan example the amount applied to the principal out of 250th monthly payment is

$$Prin_{250} = (1.004167)^{249} (1342.05 - 0.004167 * 250000) = \$845.90$$

Finally, the equity accumulated after making t payments, EQ_t, can be expressed as

$$EQ_t = \sum_{k=1}^{t} Prin_k = \sum_{k=1}^{t} (1 + \frac{i}{12})^{k-1} (C - \frac{i}{12} L)$$

$$EQ_t = (C - \frac{i}{12} L) \sum_{k=1}^{t} (1 + \frac{i}{12})^{k-1}$$

$\sum_{k=1}^{t} (1 + \frac{i}{12})^{k-1}$ is the sum of t terms of a geometric series with first term 1 and the common ratio $(1 + \frac{i}{12})$. So we have

$$EQ_t = (C - \frac{i}{12} L) \left[\frac{(1 + \frac{i}{12})^t - 1}{\frac{i}{12}} \right] \tag{12.25}$$

For our example, equity accumulated after making 250 payments is

$$EQ_{250} = \left[1342.05 - \left(\frac{0.05}{12} \right) * 250000 \right] (438.6744) = \$131,770.47$$

This is exactly the value found in Table 12.4.

12.5 Refinancing

An important issue related to long-term fixed payment loans in general, and to mortgage loans in particular, is "refinancing." Lured by lower mortgage rates and lower monthly payments, each year a number of homeowners refinance their loans. But in many cases the refinancing is not to the long-term benefit of homeowners. We use our 30-year, 5 % interest rate loan example to illustrate the point.

Let's assume that after 10 years and 120 payments, we are offered to refinance the remaining outstanding loan with a new 30-year loan with a rate of 4.0 % (for now I am ignoring the closing costs associated with refinancing, more on this later). Using (12.23) after 120 payments the balance of our loan is $203,356. Refinancing this balance with the new 30-year, 4.0 % rate reduces the amount of our monthly payments, using (12.22), to $970.85. This is an apparent savings of $1342.05 - 970.85 = \$371.20$ per month. This savings is always an attractive reason for refinancing. But this is an illusory savings!

In the original 5 % loan the total payments is $360 * 1342.05 = \$483,139$. The new 4.0 % loan total payment is $360 * 970.85 = \$349,506$. This added to $120 * 1342.05 = \$161,046$ we already paid for the first loan, amounts to $510,552, an additional $27,414 payment over the original 5 % loan!

But how can we reconcile this loss with the $371.20 monthly savings? Here lies the core of the problem. With the new loan the *payment clock* or *payment counter* is *reset*. After 10 years we are already 120 payments into our original loan. But with the new loan we restart from the first payment. The resetting of the payment clock means that big portion of monthly payments of the new loan goes to interest and the small remaining part applies to the principal. As long as the difference in accumulation of equity between the two loans is smaller than the savings from the lower monthly payments the homeowner benefits. At some point in time, however, this benefit vanishes, unless the homeowner sells the property before the time threshold is crossed.

To formulate this problem assume that k periods into the original loan the homeowner is offered to refinance the outstanding amount of loan L' at a lower rate i'. Assume the new monthly payment is C'. With the original loan and monthly payment L and C, based on (12.25) the equity accumulated after $k + t$ payments is

$$EQ_{t+k} = (C - \frac{i}{12} L) \left[\frac{(1 + \frac{i}{12})^{t+k} - 1}{\frac{i}{12}} \right]$$

The amount of equity for the new loan, denoted $E\hat{Q}_t$ is

$$E\hat{Q}_t = EQ_{t+k} + (C' - \frac{i'}{12} L') \left[\frac{(1 + \frac{i'}{12})^t - 1}{\frac{i'}{12}} \right]$$

For a certain period of time the inequality

$$EQ_{t+k} - E\hat{Q}_t < (t + k)C - (kC + tC') \quad \longrightarrow \quad EQ_{t+k} - E\hat{Q}_t < t(C - C')$$

holds, but at some point in time $EQ_{t+k} - E\hat{Q}_t$ will become equal to $t(C - C')$, and beyond that the inequality will be reversed. If the homeowner's strategy is to sell the property before that time period is reached, he or she will benefit from the refinancing. Otherwise when the time threshold is crossed, no gain from refinancing will be realized. How far in time is this threshold?

To determine the time threshold we must solve for t in

$$EQ_{t+k} - E\hat{Q}_t = t(C - C')$$

In our example $L = 250000$, $i = 0.05$, $C = 1342.05$, $L' = 203356$, $i' = 0.04$, $C' = 970.85$, and $k = 120$. Therefore,

$$EQ_{t+120} = 72092.974\left[(1.004167)^{t+120} - 1\right]$$

$$= 118737.81(1.004167)^t - 72092.97$$

$$E\hat{Q}_t = EQ_{120} + 90116.22\left[(1.0033)^t - 1\right]$$

$$= 46644.84 + 90116.22(10033)^t - 90116.22$$

$$E\hat{Q}_t = 90116.22(10033)^t - 43471.38$$

$$EQ_{t+k} - E\hat{Q}_t = 118737.81(1.004167)^t - 90116.22(1.0033)^t - 28620.96$$

Given that $t(C - C') = 371.21t$, we have

$$118737.81(1.004167)^t - 90116.22(1.0033)^t - 28620.96 = 371.21t$$

leading to the equation

$$118737.81(1.004167)^t - 90116.22(1.0033)^t - 371.21t - 28620.96 = 0$$

This is a mixed exponential equation with no analytical solution in sight. But by using numerical method of linear approximation based on Taylor series expansions we determine an estimate for t approximately about 219.[8] This means that after 18 years gain of refinancing will vanish and the homeowner will start incurring losses.

Our discussion so far leads to the clear conclusion that a homeowner's decision to refinance should not be entirely a function of *rate spread*, that is, the difference between the original mortgage rate and the current rate available for refinancing, or monthly payments. It should be, to a large extent, a function of homeowner's future plan for remaining in the same house or moving to a new location, and how soon this is expected to happen.

Furthermore, the above analysis is based on the assumption that a homeowner replaces the current loan with a same loan type—e.g. a 30-year fixed with a 30-year fixed—through refinancing. Otherwise a lower rate for a loan of shorter

---

[8] To avoid disrupting the flow of our discussion, the technical detail of estimating t is deferred to the Appendix.

term or duration could be more to the advantage of the homeowner. For instance, if in the above example the 4 % rate were for a 20-year loan, the refinancing would be definitely beneficial along the entire time-line. In this case the monthly payment would be $1,232.30, a savings of $109.75 per month. The total payment over the life of 20-year loan is $240 * 1232.30 = \$295,752$. This added to the $161,046 already paid amounts to $456,798, a savings of $23,340 over the entire term of the original 5 % loan.

The predatory practice of some lending institutions, like banks and mortgage brokers/dealers, is that they lure homeowners into refinancing with the incentive of a lower rate and, most often, lower monthly payments. But as the above simple analysis indicates, in many cases refinancing with lower monthly payments is not, in the long run, to the advantage of the borrower (and we are even ignoring substantial refinancing or *closing costs*). The available market data shows that a large number of homeowners do not benefit from refinancing at certain lower mortgage rate. In the following example, we formulate a simple model for determining the rate below which refinancing is to the benefit of a home owner.

12.5.1 Bound of Rates for Advantageous Refinancing

Example 15

Assume a family takes out a *jumbo mortgage loan*[9] of $500,000 with the rate of 6.25 %. After 10 years of monthly payments of $3078.59 they decide to refinance the balance of the loan. What is the mortgage rate necessary to refinance the balance of $421,189 outstanding over a new 30-year loan which is to the benefit of this household? The new rate, i', must be such that the sum of 360 new monthly payments C' is less than the sum of the remaining 240 payments C of the original 6.25 % loan. Using (12.22),

$$360 * \left[\frac{421189 \, (\frac{i'}{12})}{1 - (1 + \frac{i'}{12})^{-12*30}} \right] \leq 240 * 3078.59$$

which can be simplified to

$$\frac{(\frac{i'}{12})}{1 - (1 + \frac{i'}{12})^{-360}} \leq 0.004873 \tag{12.26}$$

We cannot solve for i' analytically or in closed form, but by using numerical method of linear approximation we can determine a relatively accurate estimate of i'.

---

[9] The definition of "jumbo mortgage" changes over time. It is generally a loan amount above the industry standard conventional loan. Currently in some real estate markets loans larger than $417,000 are considered jumbo loans.

The value of i' that equates the left and right hand side of (12.26) is 4.18 %.[10] This means any interest rate below 4.18 % make the refinancing beneficial for the home-owner, irrespective of the family decision to stay in the house or move to a new location. The range of rates for advantageous refinancing is $i' \leq 4.18\%$, where 4.18 % is the upper bound of the advantageous rates. At 4.18 % the monthly payment is $2,054.77. In the Exercise section of this chapter you will be asked to show that if this family decides to refinance the balance of the loan after 5 years (60 payments) any rate below 5.2 % is to their benefit. The different results for refinancing after 5 years compared to 10 years leads us to the question of determining the optimal time horizon for refinancing.

12.5.2 Optimal Time Horizon for Refinancing

Example 16

Assume a home buyer takes out a T-year $L mortgage at i percent interest. The mortgage rate declines and changes to $i' < i$. What is the time frame within which refinancing the loan is to the homeowner's benefit? Let's assume that if the home owner replaces the original loan with a T-year mortgage with i' percent rate within k period she benefits from the refinancing. Our task is to determine k.

The total payment of the original loan is $12\,TC$ where C is the monthly payment. The total payment after refinancing is $kC + 12\,TC'$ where C' is the monthly payment of the refinanced loan. Given i', we should refinance within k periods if

$$kC + 12\,TC' \leq 12\,TC \tag{12.27}$$

Using (12.23) the balance of the loan after k payments is

$$B_k = \left(1 + \frac{i}{12}\right)^k L - \left[\frac{\left(1 + \frac{i}{12}\right) - 1}{\frac{i}{12}}\right] C$$

where C is given by (12.22). The monthly payment of the refinanced loan at i' percent is

$$C' = \frac{B_k\left(\frac{i'}{12}\right)\left(1 + \frac{i'}{12}\right)^{12\,T}}{\left(1 + \frac{i'}{12}\right)^{12\,T} - 1} \tag{12.28}$$

Note that C' via B_k is a function of k. We rewrite Eq. (12.27) as

$$kC + 12\,T\,C' = 12\,T\,C \quad \longrightarrow \quad kC + 12\,T\,C' - 12\,T\,C = 0 \tag{12.29}$$

---

[10] Again, to avoid disrupting our discussion, the technical details of estimating i' are presented in the Appendix.

By solving (12.29) for k, the time frame for beneficial refinancing is determined. Let's try our model with the numerical example we used before, that is $L = 250000$, $i = 0.05$, and $i' = 0.04$. What is k?

We compute the necessary components of the model as $C = 1342.05$ and

$$B_k = (1.0041667)^k \, 250000 - \left(\frac{1.004167^k - 1}{0.004167} \right) 1342.05$$

$$= 322066.2347 - 72066.2347(1.004167)^k$$

Using (12.28), we have

$$C' = 004775B_k = 1537.5442 - 344.0442(1.004167)^k$$

Equation (12.29) can now be expressed as

$$1342.05k + 360 \left[1537.5442 - 344.0442(1.004167)^k \right] - 360 * 1342.05 = 0$$

and simplified to

$$3.72792k - 344.0442(1.004167)^k + 195.4942 = 0$$

Using Microsoft Mathematic the approximate solution to this mixed exponential function is $k \approx 72$. This means that the home owner benefits if she has a chance to refinance her 5% loan within 6 years at 4%. Let's say the home owner refinances after 5 years or 60 payments. The balance of the original 5% loan after 60 payments is $B_{60} = 229571.83$ and the new monthly payments is $C' = 1096.01$. The LHS of (12.29) is $60 * 1342.05 + 12 * 30 * 1096.01 = \$475,086.60$, which is less than its RHS $12 * 30 * 1342.05 = \$483,138$. In this case the home owner saves $8,051.40.

The question of saving a certain amount of money by refinancing a mortgage at a lower rate leads to another type of mortgage problems similar to the following example.

Example 17

A family purchased a house in 2008 with a 15-year, $350,000, 5.15% mortgage loan. After the financial crisis and the Federal Reserve implementing "Quantitative Easing", the mortgage rates, along other interest rates in the economy, declined and reached a historic low in 2012. In that year, after 76 payments, a mortgage dealer offered the family a 10-year loan with the rate that the company claimed saved them $10,000 over the original loan. What was that rate?

The monthly payments of the original loan, using (12.22), is $2,795.20 and the original mortgage costs the family $180 * 2795.20 = 503,136.55$ dollars. The outstanding balance of the loan after 76 payments is, using (12.23), $234,092.41 and the remaining payment is $104 * 2795.20 = 290700.8$. The new 10-year loan with the rate

Fig. 12.1 Graph of $120 \ln(i' + 12) + \ln(0.11991 - i') - 296.0678$

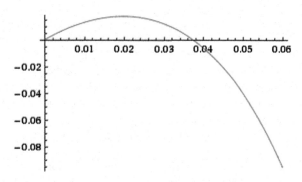

i' and monthly payments C' is structured such that $120 * C' = 290700.85 - 10000 = \$280700.8$, leading to $C' = 2339.17$. The problem now is similar to Example 15; we want to determine i' such that

$$C' = \frac{234092.41(\frac{i'}{12})}{1 - (1 + \frac{i'}{12})^{-12*10}} = 2339.17$$

$$i' = 0.11991\left[1 - (1 + \frac{i'}{12})^{-120}\right]$$

$$0.11991 - i' = 0.11991(1 + \frac{i'}{12})^{-120}$$

Taking the logarithm of both sides, we have

$$\ln(0.11991 - i') = \ln(0.11991) - 120 \ln\left[(1 + \frac{i'}{12})\right] = -2.121 - 120 \ln\left(\frac{i' + 12}{12}\right)$$

$$\ln(0.11991 - i') = -2.121 - 120 \ln(i' + 12) + 298.1888$$

Finally,

$$120 \ln(i' + 12) + \ln(0.11991 - i') - 296.0678 = 0$$

WoframAlpha provides two solutions to this logarithmic equation. The first solution is $i' = 0.000009$, which is too small and we discard it. The second solution is $i' = 0.0372$. The rate that saves the family \$10,000 is 3.72 % (Fig. 12.1).

12.5.3 Closing Costs and APR

So far in our narrative we ignored the costs associated with obtaining a loan and/or its refinancing. In the process of application and approval of a loan, the borrower is charged various fees. For a car loan (which is generally a small amount with a short duration) it may be limited to the application fee and credit checking fee. But the list of charges and fees for a mortgage loan typically is much longer and may include some or all of the following fees (*and this is not an exhaustive list*): points,[11] application, title insurance, survey, appraisals, credit checks, loan origination and documentation, commitment and processing, hazard and mortgage insurance, and interest prepayments.

Different lenders charge borrowers different set of fees. They normally present the borrowers with the option of paying these fees upfront out of pocket or agree that the amount be added to the loan. A simple example helps to clarify the idea. Let's assume a borrower applies for a $200,000 mortgage loan with a 6.5 % rate. He agrees to pay a point, $2,000, to get this rate and an additional $2,200 in other costs. At the closing, this borrower has the option of paying the sum of $4,200 out of pocket or agrees to a total loan of $204,200, by adding the $4,200 "closing costs" to the original loan amount. In either case the cost of credit extended to borrower by the lender is more than 6.5 % stated mortgage rate.

Lenders often mislead borrowers by charging hidden fees. To ensure that borrowers can make an informed decision in the loan market, the Truth in Lending Act of 1968 requires that lenders calculate and disclose the Annual Percentage Rate (APR) of a loan. The APR represents the total cost of credit to consumers. The calculation of APR is slightly complex, and different lenders use different methods for calculating APR.

APR is similar to the effective rate, with costs incurred by the borrower incorporated into the calculation. Let's consider a simple version of APR calculation by limiting the length of the loan to only one year. It helps to illustrate the point. Let's assume you borrow $10,000 for one year with the nominal interest rate of 6.9 % per year compounded monthly. You are required to pay the lender the principal and interest after one year. What is the APR of this loan if the lender charges a $350 application and processing fee and you agree that the fee be added to your loan?

Your loan is actually $10,350. The amount of interest you pay for a $10,000 loan is

$$\text{Interest} = 10350 \left(1 + \frac{0.069}{12}\right)^{12} - 10000 = 1087.17$$

The actual interest rate for the loan or APR is

$$APR = \frac{1087.17}{10000} = 0.1087 \quad \text{or} \quad 10.87\%$$

almost 4 percentage points more than the nominal rate. Compare this to a loan charging 7.1 % interest but with $250 application and processing fee added to your loan. Here

---

[11] Points (a point equals one percent of the amount of the mortgage loan) are fees that the mortgage lender charges for making the loan. Points are prepaid interest, or interest that is due when the loan is taken out. Lenders offer lower rates with the payment of more points.

$$\text{Interest} = 10250 \left(1 + \frac{0.071}{12}\right)^{12} - 10000 = 1001.91$$

In this case the APR is $\dfrac{1001.9}{10000} = 0.001$ or 10%, less than in the previous case.

What is the APR if you pay the charge up-front out of pocket? Here the total payment on a \$10,000 loan is interest plus the fees. So

$$\text{Interest} = 10000 \left(1 + \frac{0.069}{12}\right)^{12} - 10000 = 712.24$$

$$\text{Interest} + \text{fees} = 712.24 + 350 = 1062.24$$

$$APR = \frac{1062.24}{10000} = 0.1062$$

or 10.62%, slightly better than 10.87%.

12.6 Savings Plan, Annuity

Assume that P dollars is deposited with regular periodicity into an account that pays i percent interest rate compounded m times a year. How much money will be accumulated in the account after T years? Payments of a fixed amount for a specified number of periods is an example of a *savings plan, savings fund,* or *Annuity.* Contribution to college funds for children, contribution to pension funds for retirement, contribution to sinking funds for redeeming bonds or preferred stocks, and contribution to an account later used for replacement of depreciated capital equipments are all examples of savings fund.

At the end of T years the first P dollars deposited at the beginning of the first year[12] grows to $P(1 + \frac{i}{m})^{mT}$, the second deposit grows to $P(1 + \frac{i}{m})^{mT-1}$, and so on. Consequently the amount accumulated in the fund at the end of T years is

$$S = P(1 + \frac{i}{m})^{mT} + P(1 + \frac{i}{m})^{mT-1} + P(1 + \frac{i}{m})^{mT-2} + \cdots + P(1 + \frac{i}{m})$$

$$= \sum_{t=1}^{mT} P(1 + \frac{i}{m})^{t}$$

This is the sum of mT terms of a geometric series with first term $P(1 + \frac{i}{m})$ and the common ratio of $(1 + \frac{i}{m})$, that is

---

[12] I assume that deposits are made at the beginning of each period. This type of annuity is called *Annuity-due.*

$$a = P(1 + \frac{i}{m}) \qquad\qquad r = (1 + \frac{i}{m})$$

Recall that the sum of n terms of a geometric series with the first term a and common ratio r is

$$S = \frac{a(1 - r^n)}{1 - r}$$

so we have

$$S = \frac{P(1 + \frac{i}{m})\left[1 - (1 + \frac{i}{m})^{mT}\right]}{1 - (1 + \frac{i}{m})} = \frac{mP(1 + \frac{i}{m})\left[(1 + \frac{i}{m})^{mT} - 1\right]}{i} \qquad (12.30)$$

Example 18

A savings plan is created by depositing \$100 a month into an account that pays 6% per year compounded monthly. What is the value of this plan at the end of 10 years?
   In this example $P = 100$, $i = 0.06$, $m = 12$, and $T = 10$. Using (12.30), we have

$$S = \frac{12 * 100(1 + \frac{0.06}{12})\left[(1 + \frac{0.06}{12})^{12*10} - 1\right]}{0.06} = \$16,470$$

Notice that there are 5 variables in (12.30): S, P, i, m, and T. If the values for any 4 variables is gives, the 5th variable could be determined. For example, an important variant of an annuity problem is that an entity (individual, family, firm) sets a financial goal of having S dollars after T years, given i and m, and needs to determine the amount of monthly payments required to achieve this goal. Here we must find P. This requires that in (12.30) we solve for P in terms of the other variables

$$P = \frac{i S}{m(1 + \frac{i}{m})\left[(1 + \frac{i}{m})^{mT} - 1\right]}$$

Example 19

How much do we need to deposit monthly in order to have \$10,000 after 10 years, if the interest rate is 6% compounded monthly?

$$P = \frac{0.06 \, (10000)}{12(1 + \frac{0.06}{12})\left[(1 + \frac{0.06}{12})^{12*10} - 1\right]} = \$60.72$$

Given P, S, i, and m we can determine T. From (12.30) we have

$$iS = mP(1 + \frac{i}{m})[(1 + \frac{i}{m})^{mT} - 1] = mP(1 + \frac{i}{m})(1 + \frac{i}{m})^{mT} - mP(1 + \frac{i}{m})$$

$$iS + mP(1 + \frac{i}{m}) = mP(1 + \frac{i}{m})(1 + \frac{i}{m})^{mT}$$

and by taking the logarithm of both sides, we have

$$\ln(iS + mP(1 + \frac{i}{m})) = \ln(mP) + \ln(1 + \frac{i}{m}) + mT \ln(1 + \frac{i}{m})$$

Solving for T,

$$T = \frac{\ln(iS + mP(1 + \frac{i}{m})) - \ln(mP) - \ln(1 + \frac{i}{m})}{m \ln(1 + \frac{i}{m})}$$

Example 20

How long does it take for a savings fund with $100 monthly payments and an interest rate of 5 % compounded monthly to grow to $10,000?

$$T = \frac{\ln(0.05 * 10000 + 12 * 100 * 1.004167) - \ln(12 * 100) - \ln(1.004167)}{12 \ln(1.004167)}$$

$$= 6.96 \ \text{years}$$

12.7 College Fund

A practical variation of a savings fund is a college fund for children.[13] Let's assume a family wants to establish a college fund for one of their children with the plan that during the last 4 years of T years ($T \geq 4$) of monthly contributions of P dollars, they pay a constant amount of C dollars for tuition and board each year, with the last payment exhausting the fund. Given C, T, i, and m, it could be shown that the amount of monthly contribution P is

$$P = \frac{C (\frac{i}{m}) \left[(1 + \frac{i}{m})^{4m} - 1 \right]}{\left[(1 + \frac{i}{m})^{m(T-1)} - 1 \right] \left[(1 + \frac{i}{m})^{m} - 1 \right]} \tag{12.31}$$

---

[13] College funds are known as 529 Plans. A 529 plan is a tax-advantaged savings plan designed to encourage saving for future college costs. 529 plans, legally known as "qualified tuition plans," are sponsored by states, state agencies, or educational institutions and are authorized by Sect. 529 of the Internal Revenue Code (*See U.S. Securities and Exchange Commission at* www.sec.gov/investor/pubs/intro529.htm).

If the interest rate is 6% compounded monthly and $T = 22$ years (this family starts saving as soon as the child is born!) with C equal to \$30,000 for tuition and board in the last 4 years, then from (12.31) P would be \$261.63 a month.

To verify this answer (which provides a clue to how to set up and solve the problem) we should note that using (12.30) this family accumulates \$101,850 in the fund by the end of 18th year. At this point the first \$30,000 cost of tuition and board must be paid, so the fund is reduced to \$71,850. This money earns interest, and the fund also receives an additional \$3,243 from the monthly contribution and interest during the following year. By the end of the 19th year the fund has \$79,500, out of which another \$30,000 must be paid, reducing it to \$49,500. This amount grows to \$52,560, and another \$3,243 from the monthly contribution and interest puts the value of the fund at \$55,800. By making another \$30,000 payment, the fund is reduced to \$25,800. This amount grows to \$27,400 and the addition of \$3,243 for the last year's contribution puts the fund value at the end of 21st year (beginning of 22nd) equal to \$30,600, out of which the last \$30,000 is paid (the \$600 discrepancy is due to rounding errors accumulated over various stages of computations).

12.8 Exercises

1. How much will you have after 6 years if you invest \$15,000 at 6.5% per year compounded annually? Quarterly?

2. How much will you have at the end of 10 years if you invest \$20,000 at 3.5 % per year compounded monthly?

3. How much do you need to invest now at an 8% interest rate compounded semi-annually in order to have \$10,000 in 5 years?

4. Find the present value of \$12,000 due four years from now if the rate of interest is 6.0% per year and is compounded quarterly.

5. How long does it take for an amount of money invested at 4.5% interest compounded monthly to triple?

6. What rate of interest compounded quarterly doubles an investment over 10 years?

7. What rate of interest compounded continuously doubles an investment over 10 years?

Effective Interest Rate

The effective interest rate is your true interest rate cost of borrowing, stated as an annual simple rate. Frequent compounding results in a higher effective interest rate.

8. What is the effective interest rate of a nominal interest rate of 6.9% per year compounded monthly?

9. Assume a $10,000 bond with a coupon rate of 4.5 % and 5 years to maturity is priced at $9,800.

  (a) What is its approximate YTM?

  (b) What is the investor's rate of return?

10. Assume a bond with a coupon rate of 5.5 % and 7 years to maturity is traded at $102.50 per $100 face value.

  (a) What is its approximate YTM?

  (b) What is the investor's rate of return?

11. A $20,000 bond with coupon rate of 6 %, maturing in 3 years, is traded at par.

  (a) What is its approximate YTM?

  (b) What is the investor's rate of return?

12. Using results from Exercises 9, 10, and 11, express the relationship between a bond's YTM and its rate of return when it is traded at par, premium, or discount.

13. What is the market value of a 30-day $10,000,000 commercial paper if the yield of 30-day Treasury bill is 2.5 %?

14. A 60-day $10,000,000 commercial paper is offered at $9,964,000. What is the discount rate?

15. What is the day count of a $5,000,000 commercial paper sold for $4,900,000 if the discount rate is 3.5 %?

16. An investor purchases a property in 2000 for $450,000. She sells the property in 2006 for $550,000. During the 6 years that she owns the property, it generates a monthly rental income (net of monthly mortgage payments and maintenance costs) of $400 a month. What is the investor's rate of return?

17. You get a 3-year $20,000 car loan at 6.5 % interest compounded monthly. What is the amount of your monthly payment?

18. What is the monthly mortgage payment of a 25-year $250,000 fixed loan if the mortgage rate is 4.75 %?

19. What proportion of the 50th monthly payment in Exercise 18 is applied to the principal?

20. What is the accumulated equity in Exercise 18 after the 120th monthly payment? What is the amount of the loan outstanding?

21. You take out a 30-year fixed $500,000 mortgage with the rate of 6.25 %. After 5 years you decide to refinance the balance of the loan with another 30-year fixed mortgage. What is the upper bond of the advantageous rates?

22. A family borrows $350,000 from a mortgage company. The term of the mortgage is 25-year fixed with 5.8 % interest. What is the optimal time horizon for this family to refinance the balance of this loan with a new 4.3 %, 25-year fixed mortgage.

23. You purchase your home with a $200,000, 15-year fixed, 4.5 % mortgage loan. After seven years, 84 payments, you are offered to refinance the balance of your loan with a 10-year 3.75 % fix-rate mortgage. Is refinancing to your financial advantage?

24. In Exercise 23 assume that the closing cost of refinancing is $4,000, which you agree to be added to the balance of your loan. What range of mortgage rates make the refinancing financially to your advantage?

25. A family purchased a house in January 15, 2007 with a 25-year, $350,000, 6.35 % mortgage. In March 15, 2012, when the mortgage rates dropped to the historic low, they refinance the balance of the loan with a 15-year, 4.25 % mortgage. How much does this family save?

26. A family purchased a house in 2008 with a 15-year, $250,000, 5.15 % mortgage. In 2013 after 80 payments they are offered a 10-year loan with the rate that saves the family about $8,000 over the original loan. What is that rate?

27. You borrow $10,000 for one year with the nominal interest rate of 6.9 % per year compounded monthly. You are required to pay the lender the principal and interest after one year. What is the APR of this loan if the lender charges a $350 application and processing fee, and you agree that the $350 should be added to your loan? What is the APR if you instead withdraw $350 from your saving account, which pays 3.5 %, and pay the charge up front?

28. A savings plan is created by depositing $250 a month into an account that pays 4.5 % interest compounded quarterly. What is the value of this fund after 15 years?

29. Parents create a college fund for their newly born child. They deposit $800 every quarter in the fund, which pays 5.5 % interest compounded quarterly. What is the balance in the fund after 18 years?

30. A newlywed couple sets up a fund to save $80,000, to be used as a down payment for buying a house. How much do they have to contribute monthly to the fund over 12 years if the fund pays 4.5 % interest compounded monthly?

31. You contribute $200 a month to an account that pays 3.76 % interest compounded monthly. How long does it take for you to accumulate $32,000 in the account?

32. What rate of interest, compounded monthly, grow the value of a savings fund to $32,000 over 10 years, if the monthly contribution were $200?

33. You buy a $5,000 face value bond with a coupon rate of 4.5 % for $4,750. Three years left to the bond's maturity, the interest rate on similar instruments with the same risk jumps to 5.5 %. What is the value of your bond now?

34. What would be the value of your bond in Exercise 31 if the interest rate drops to 4.0 %?

35. In our example in Sect. 12.7, the annual cost of a four-year college was assumed to be $30,000, or $120,000 for four years. This figure is in the ball park of the *national average* for private universities. According to Fidelity Investments, a major financial services company who manages 529 plans, the current estimate

of the national *average* amount of money needed to cover four years of expenses (tuition, board, books, ...) at a private university is $115,000. The average amount is $51,000 for public colleges. Fidelity estimates that the cost of college is rising at an average of 5.4 % a year, twice the overall rate of inflation.

Use Eq. (12.31) and determine the monthly amount a family needs to contribute to a college fund for 18 years in order to cover the $45,000 annual cost of a 4-year college.

Appendix A: Estimating Time *t* and New Interest Rate *i′* for Sect. 12.5

In the text we found ourselves faced with solving the following scary exponential equation

$$118737.81(1.00417)^t - 90116.22(1.00333)^t - 371.21t - 28620.96 = 0$$

Solving this equation is equivalent to finding the zero(s) of the function

$$f(t) = 118737.81(1.00417)^t - 90116.22(1.00333)^t - 371.21t - 28620.96$$

The graph of the above function indicates that an acceptable solution for t is between 210 and 230 (Fig. 12.2).

The Table 12.5 further narrows down the solution for t between 215 and 220. Using Taylor linear approximation, we expand the function around $t = 220$.

$$f(220) = 96.20$$

$$f'(t) = 118737.81(1.00417)^t \ln(1.00417) - 90116.22(1.00333)^t \ln(1.00333)$$
$$- 371.21$$

Fig. 12.2 Graph of $118737.81(1.00417)^t - 90116.22(1.00333)^t - 371.21t - 28620.96$

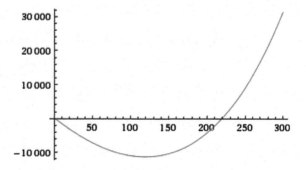

Table 12.5 Values of $f(t)$ for selected values of t

$t = Month$	$f(t)$
200	−4271.44
210	−2235.80
215	−1107.72
220	96.20
225	1377.78
230	2738.88

$$f'(220) = 240.542$$

Now we can write the linear approximation of the function as

$$f(t) = 96.20 + 240.542(t - 220) = -52822.99 + 240.542t$$

The zero of this function is

$$f(t) = 0 \quad \longrightarrow \quad -52822.99 + 240.542t = 0 \quad \text{and} \quad t = \frac{52822.99}{240.542} \approx 219.$$

Next we encountered problem of determining i' such that the it satisfies

$$\frac{(\frac{i'}{12})}{1 - (1 + \frac{i'}{12})^{-360}} \leq 0.004873$$

Assuming equality between the left and right hand sides, we simplify the expression to

$$\frac{i'}{12} = 0.004873 - 0.004873(1 + \frac{i'}{12})^{-360}$$

$$0.004873(1 + \frac{i'}{12})^{-360} = 0.004873 - \frac{i'}{12}$$

Taking the logarithm of both sides, we have

$$\ln(0.004873) - 360 \ \ln(1 + \frac{i'}{12}) = \ln(0.004873 - \frac{i'}{12})$$

$$-5.32407 - 360 \ \ln(\frac{12 + i'}{12}) = \ln(\frac{0.058475 - i'}{12})$$

$$-5.32407 - 360 \ \ln(12 + i') + 360 \ \ln(12) = \ln(0.058475 - i') - \ln(12)$$

and finally,

Table 12.6 Values of $f(i')$ and $f'(i')$ for selected values of i'

i	$f(i')$	$f'(i')$
0.0321	0.1656	−7.9947
0.0341	0.1465	−11.1106
0.0361	0.1208	−14.7827
0.0381	0.0869	−19.1747
0.0400	**0.0458**	**−24.2239**
0.0410	0.0201	−27.3268
0.0420	−0.0089	−30.8027
0.0440	−0.0786	−39.1942

Fig. 12.3 Graph of $360 \ \ln(12 + i') + \ln(0.058475 - i') - 891.7272$

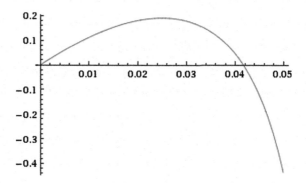

$$360 \ \ln(12 + i') + \ln(0.058475 - i') - 891.7272 = 0 \qquad (12.32)$$

Solution to this logarithmic function is equivalent of finding zero(s) of the function

$$f(i') = 360 \ \ln(12 + i') + \ln(0.058475 - i') - 891.7272$$

$$f(i') = 0 \quad \longrightarrow \quad 360 \ \ln(12 + i') + \ln(0.058475 - i') - 891.7272 = 0$$

Table 12.6 indicates that the solution must be in the neighborhood of $i' = 0.04$ (Fig. 12.3).

We expand the function around this :

$$f(0.04) = 0.45816$$

$$f'(i') = \frac{360}{12 + i'} - \frac{1}{0.058475 - i'}$$

$$f'(0.04) = -24.2239$$

The linear function is

$$f(i') = 0.045816 - 24.2239\,(i' - 0.04) = 1.01489 - 24.2239\,i'$$

The zero of this function is a linear approximation to i':

$$f'(i') = 0 \quad \longrightarrow \quad 1.01489 - 24.2239\,i' = 0 \quad \text{leading to} \quad i' = 0.04190 \approx 0.042$$

Note that we can use WolframAlpha and solve the for i' in Eq. (12.32). We enter (by using x instead of i')

$$solve[360\,log(12 + x) + log(0.058475 - x) - 891.7272 = 0]$$

There are two possible answers. The first one is a negative number that must be discarded. The second answer (to four decimal places) is $x = 0.0417$. WolframAlpha interpret log as the natural logarithm. The symbol for log at base 10 is $log10$.

Chapter 13
Matrices and Their Applications

13.1 Matrix

A *matrix* is a rectangular array of numbers. These numbers are arranged in *rows* and *columns* enclosed by parentheses, square brackets, single vertical lines, or double vertical lines (I will use the square brackets). Traditionally, an uppercase letter of the English alphabet is used to denote a matrix. The following is an example of a matrix:

$$B = \begin{bmatrix} 10 & -2 & 12.5 \\ 32 & -10 & 50 \\ -1 & 12 & -2 \\ 55 & 54 & 21 \end{bmatrix} \tag{13.1}$$

Matrix B is a rectangular array of 12 numbers. These numbers are referred to as the *elements* of the matrix.[1] Matrix B's elements are arranged in 4 rows and 3 columns.

As a more concrete example, consider a supper market chain with 5 stores in a small town. Assume that the management of the chain collects and organizes the weekly sales of the stores in 3 different categories of food products (meat, dairy, vegetable) and 1 category of non-food product. The average weekly sales (in $1,000) are shown in the following table:

Average weekly sales of various products at different stores (in $1,000)

Products	Store 1	Store 2	Store 3	Store 4	Store 5
Meat	3.5	2.2	5.1	6.7	1.2
Dairy	5.1	3.2	4.1	4.2	2.7
Vegetable	4.1	2.1	3.5	3.2	1.9
Non-food	8.5	6.3	10.2	12.3	5.8

---

[1] Elements of a matrix can be any object.

S. Vali, *Principles of Mathematical Economics*,
Mathematics Textbooks for Science and Engineering 3,
DOI: 10.2991/978-94-6239-036-2_13, © Atlantis Press and the authors 2014

The 'numbers' in this table constitute a matrix of 4 rows (products) and 5 columns (stores).

$$D = \begin{bmatrix} 3.5 & 2.2 & 5.1 & 6.7 & 1.2 \\ 5.1 & 3.2 & 4.1 & 4.2 & 2.7 \\ 4.1 & 2.1 & 3.5 & 3.2 & 1.9 \\ 8.5 & 6.3 & 10.2 & 12.3 & 5.8 \end{bmatrix}$$

Generally, a matrix can be written as

$$A = \begin{bmatrix} a_{11} & a_{12} & a_{13} & \cdots & a_{1n} \\ a_{21} & a_{22} & a_{23} & \cdots & a_{2n} \\ a_{31} & a_{32} & a_{33} & \cdots & a_{3n} \\ \vdots & \vdots & \vdots & \vdots & \vdots \\ a_{m1} & a_{m2} & a_{m3} & \cdots & a_{mn} \end{bmatrix} = [a_{ij}]$$

Elements of a matrix have two subscripts or indexes to indicate their "location" or "address". A typical element of A is denoted by a_{ij}, where the first subscript or index i stands for the row position and the second subscript or index j stands for the column position. Thus a_{32} indicates that the element is located in the third row and the second column of matrix A. The element located in the first row and second column of matrix B is b_{12}. This element is -2. For ease of exposition and simplicity, a matrix A is sometimes written as $[a_{ij}]$. Readers should note that without the square brackets, a_{ij} is only a number or element, not a matrix.

The number of rows and columns of a matrix determine its *size*, *dimension*, or *order*. A matrix with m rows and n columns is an $m \times n$ (read m by n) matrix and has mn elements. Matrix B in (13.1) is a 4 by 3 matrix and has $4 \times 3 = 12$ elements. A matrix with the same number of rows and columns ($m = n$) is called a *square matrix*. Matrix C below is a 3 by 3 square matrix, also called a square matrix of size or dimension 3.

$$C = \begin{bmatrix} 11 & 2 & 2.5 \\ 3 & -10 & 30 \\ 1 & 22 & -21 \end{bmatrix}$$

When $m = 1$ we have a $1 \times n$ matrix. This is a matrix with 1 row and n columns. Similarly, when $n = 1$ we have a $m \times 1$ matrix, a matrix with m rows and 1 column. Matrices with 1 row (*row-matrices*) or 1 column (*column-matrices*) are called *vectors*. A $1 \times n$ vector is called a *row vector* and an $m \times 1$ vector is referred to as *column vector*. Below, V is a row vector of size n and W is a column vector of size m.

$$V = [v_1 \quad v_2 \quad v_3 \quad \cdots \quad v_n]$$

$$W = \begin{bmatrix} w_1 \\ w_2 \\ w_3 \\ \vdots \\ w_m \end{bmatrix}$$

If $m = n = 1$ then we have a matrix with only 1 element. $A = [a_{11}] = a$. A 1×1 matrix is called a *scalar*. A scalar is a single number.

A matrix of order m by n can be considered an assembly of n column vectors of size m arranged side by side, or m row vectors of size n stacked on top of each other.

Two matrices are said to be *equal* if they are of the same size and have the same elements in the same position. Matrices $C = [c_{ij}]$ and $D = [d_{ij}]$ are equal ($C = D$) if and only if we have $c_{ij} = d_{ij}$ for all i and j. Thus the equality of two $m \times n$ matrices required a system of mn equalities, one for each corresponding pair of elements. If, for example, C is

$$C = \begin{bmatrix} -4 & 2 \\ -10 & 30 \\ 11 & -21 \end{bmatrix}$$

and D is defined as

$$D = \begin{bmatrix} d_{11} & d_{12} \\ d_{21} & d_{22} \\ d_{31} & d_{32} \end{bmatrix}$$

then equality of C and D requires the following $3 * 2 = 6$ equalities

$$d_{11} = -4 \quad d_{12} = 2 \quad d_{21} = -10 \quad d_{22} = 30 \quad d_{31} = 11 \quad d_{32} = -21$$

As another example, if the column vector of unknowns X is equal to a column vector of values K

$$X = \begin{bmatrix} x_1 \\ x_2 \\ x_3 \end{bmatrix} \quad K = \begin{bmatrix} 123 \\ -17 \\ 0.056 \end{bmatrix}$$

then $X = K$ leads to

$$\begin{bmatrix} x_1 \\ x_2 \\ x_3 \end{bmatrix} = \begin{bmatrix} 123 \\ -17 \\ 0.056 \end{bmatrix}$$

and the following solution: $x_1 = 123$; $x_2 = -17$; and $x_3 = 0.056$.

13.2 Matrix Operation

13.2.1 Matrix Addition, Scalar Multiplication and Subtraction

We can add two matrices, or subtract one from another, if they are of exactly the same dimension. If this requirement is satisfied, then the matrices are *conformable for* addition and subtraction. Consider two $m \times n$ matrices $A = [a_{ij}]$ and $B = [b_{ij}]$. A and B are of the same dimension, so they are conformable for addition and subtraction. Lets denote the sum of these matrices by C. The matrix C is obtained by adding the corresponding elements of A and B, that is

$$c_{ij} = a_{ij} + b_{ij} \qquad i = 1, 2, 3, \ldots, m; \quad j = 1, 2, 3, \ldots, n$$

$$C = A + B = \begin{bmatrix} a_{11} + b_{11} & a_{12} + b_{12} & a_{13} + b_{13} & \cdots & a_{1n} + b_{1n} \\ a_{21} + b_{21} & a_{22} + b_{22} & a_{23} + b_{23} & \cdots & a_{2n} + b_{2n} \\ \vdots & \vdots & \vdots & \vdots & \vdots \\ a_{m1} + b_{m1} & a_{m2} + b_{m2} & a_{m3} + b_{m3} & \cdots & a_{mn} + b_{mn} \end{bmatrix}$$

Example 1

Assume

$$A = \begin{bmatrix} -4 & 2 \\ -10 & 30 \\ 11 & -21 \end{bmatrix} \quad \text{and} \quad B = \begin{bmatrix} 8 & 3 \\ 15 & 5 \\ -3 & -5 \end{bmatrix} \quad \text{then}$$

$$C = A + B$$

$$C = \begin{bmatrix} -4 & 2 \\ -10 & 30 \\ 11 & -21 \end{bmatrix} + \begin{bmatrix} 8 & 3 \\ 15 & 5 \\ -3 & -5 \end{bmatrix} = \begin{bmatrix} -4 + 8 & 2 + 3 \\ -10 + 15 & 30 + 5 \\ 11 + (-3) & -21 + (-5) \end{bmatrix} = \begin{bmatrix} 4 & 5 \\ 5 & 35 \\ 8 & -26 \end{bmatrix}$$

The *scalar multiplication* of a matrix A is the product of A by a scalar k. We obtain the product by multiplying each element of A by k. Assume $k = 5$, then we have

$$kA = 5A = 5 \begin{bmatrix} -4 & 2 \\ -10 & 30 \\ 11 & -21 \end{bmatrix} = \begin{bmatrix} -20 & 10 \\ -50 & 150 \\ 55 & -105 \end{bmatrix}$$

If $k = -1$ then $-A = (-1)A$ is the negative of the matrix A. Now we can express the difference between matrix A and B as

$$D = A - B = A + (-B)$$

$$D = \begin{bmatrix} -4 & 2 \\ -10 & 30 \\ 11 & -21 \end{bmatrix} - \begin{bmatrix} 8 & 3 \\ 15 & 5 \\ -3 & -5 \end{bmatrix} = \begin{bmatrix} -4 - 8 & 2 - 3 \\ -10 - 15 & 30 - 5 \\ 11 - (-3) & -21 - (-5) \end{bmatrix}$$

$$= \begin{bmatrix} -12 & -1 \\ -25 & 25 \\ 14 & -16 \end{bmatrix}$$

13.2.2 Matrix Multiplication

The product of two matrices A and B is defined if they are *conformable for multiplication*. This condition requires that the number of columns of the first matrix be the same as the number of rows of the second matrix. Consider matrices A and B with dimensions $m \times n$ and $n \times p$, respectively. The number of columns of A is exactly the same as the number of rows of B, therefore they are conformable for multiplication, provided we *premultiply* B by A. This means we form the product $C = AB$. The matrix C resulting from AB is of order $m \times p$. The product of B and A, when B is *postmultiplied* by A to form BA, is not defined. In this case, the number of columns of B (p) is not the same as the number of rows of A (m), therefore they are not conformable for BA multiplication. Only square matrices of the same size are conformable for both pre- and post-multiplications.

Lets begin with a simple case: the product of a row vector V with n elements and a column vector W of the same size. This product, which is sometimes referred to as the *dot product* or the *inner product*, is obtained by multiplying the elements of the two vectors with the same index value and adding them up, that is

$$VW = \begin{bmatrix} v_1 & v_2 & v_3 & \cdots & v_n \end{bmatrix} \begin{bmatrix} w_1 \\ w_2 \\ w_3 \\ \vdots \\ w_n \end{bmatrix} = v_1 w_1 + v_2 w_2 + v_3 w_3 + \cdots + v_n w_n$$

$$= \sum_{k=1}^{n} v_k w_k$$

It should be immediately noted that the result of VW is a scalar, a number. For this treason the product of a row and a column vector is also called the *scalar product*. It should be also clear why we need the number of columns of V to be equal to the number of rows of W to satisfy comfortability.

Example 2

If

$$V = \begin{bmatrix} 1 & -10 & 3 & 25 \end{bmatrix} \quad \text{and} \quad W = \begin{bmatrix} -2 \\ 7 \\ 10 \\ 5 \end{bmatrix} \quad \text{then we have,}$$

$$VW = \begin{bmatrix} 1 & -10 & 3 & 25 \end{bmatrix} \begin{bmatrix} -2 \\ 7 \\ 10 \\ 5 \end{bmatrix} = 1 * (-2) + (-10) * 7 + 3 * 10 + 25 * 5 = 83$$

Recall from Chap. 1 that a household expenditure E on a bundle or basket of n goods and services can be expressed as

$$E = p_1 q_1 + p_2 q_2 + \cdots + p_n q_n = \sum_{i=1}^{n} p_i q_i$$

where q_i $i = 1, 2, ..., n$, denote the quantity of the ith item purchased by this household and p_i is the corresponding price of each unit of q_i. If we use the column vector Q to denote the quantities of goods and services that the household buys and row vector P to denote the associated unit prices of these goods and services, then, using matrix notation, the household expenditure can be expressed as

$$E = PQ = \begin{bmatrix} p_1 & p_2 & p_3 & \cdots & p_n \end{bmatrix} \begin{bmatrix} q_1 \\ q_2 \\ q_3 \\ \vdots \\ q_n \end{bmatrix} = \sum_{i=1}^{n} p_i q_i$$

Now consider a slightly more complicated case of matrix multiplication. Assume V is a row vector of size 1 by m and A a matrix of order $m \times n$. Since the number of columns of V is equal to the number of rows of A the product VA is defined. The product of V and A is another vector W of size 1 by n.

$$W_{1 \times n} = V_{1 \times m} \times A_{m \times n}$$

$$W = \begin{bmatrix} v_1 & v_2 & v_3 \cdots & v_m \end{bmatrix} \begin{bmatrix} a_{11} & a_{12} & \cdots & a_{1n} \\ a_{21} & a_{22} & \cdots & a_{2n} \\ \vdots & \vdots & \vdots & \vdots \\ a_{m1} & a_{m2} & \cdots & a_{mn} \end{bmatrix}$$

Elements of W are obtained by multiplying vector V with each column of A, treating it as a column vector. For example, w_1 is obtained by multiplying V with the first column of A,

$$w_1 = \begin{bmatrix} v_1 & v_2 & \cdots & v_m \end{bmatrix} \begin{bmatrix} a_{11} \\ a_{21} \\ \vdots \\ a_{m1} \end{bmatrix} = v_1 a_{11} + v_2 a_{21} + \cdots + v_m a_{m1} = \sum_{k=1}^{m} v_k a_{k1}$$

Similarly w_2 is determined by multiplying V with the second column of A, and so on.

Example 3

As a numerical example, assume

$$V = \begin{bmatrix} 1 & 5 & 9 \end{bmatrix} \quad \text{and} \quad A = \begin{bmatrix} 1 & 12 \\ 7 & 8 \\ 13 & 11 \end{bmatrix}$$

$$W = VA = \begin{bmatrix} 1 & 5 & 9 \end{bmatrix} \begin{bmatrix} 1 & 12 \\ 7 & 8 \\ 13 & 11 \end{bmatrix}$$

$$w_1 = \begin{bmatrix} 1 & 5 & 9 \end{bmatrix} \begin{bmatrix} 1 \\ 7 \\ 13 \end{bmatrix} = 1*1 + 5*7 + 9*13 = 153$$

$$w_2 = \begin{bmatrix} 1 & 5 & 9 \end{bmatrix} \begin{bmatrix} 12 \\ 8 \\ 11 \end{bmatrix} = 1*12 + 5*8 + 9*11 = 151$$

And $W = [153\ \ 151]$.

Now consider the general case of matrix multiplication. Assume matrix A is m by n and matrix B is n by p. The matrix C, the product of A and B, is $m \times p$ and obtained by a series of vector multiplications: rows of A (row vectors) multiplied by columns of B (column vectors).

$$C = AB$$

$$\begin{bmatrix} c_{11} & c_{12} & \cdots & c_{1p} \\ c_{21} & c_{22} & \cdots & c_{2p} \\ \vdots & \vdots & \vdots & \vdots \\ c_{m1} & c_{m2} & \cdots & c_{mp} \end{bmatrix} = \begin{bmatrix} a_{11} & a_{12} & \cdots & a_{1n} \\ a_{21} & a_{22} & \cdots & a_{2n} \\ \vdots & \vdots & \vdots & \vdots \\ a_{m1} & a_{m2} & \cdots & a_{mn} \end{bmatrix} \begin{bmatrix} b_{11} & b_{12} & \cdots & b_{1p} \\ b_{21} & b_{22} & \cdots & b_{2p} \\ \vdots & \vdots & \vdots & \vdots \\ b_{n1} & b_{n2} & \cdots & b_{np} \end{bmatrix}$$

where a typical element c_{ij} of matrix C is the product of the ith row of A and the jth column of B,

$$c_{ij} = \begin{bmatrix} a_{i1} & a_{i2} & a_{i3} & \cdots & a_{in} \end{bmatrix} \begin{bmatrix} b_{1j} \\ b_{2j} \\ b_{3j} \\ \vdots \\ b_{nj} \end{bmatrix} = \sum_{k=1}^{n} a_{ik}b_{k,j}$$

For example c_{11}, the first element of C, is computed as

$$c_{11} = \begin{bmatrix} a_{11} & a_{12} & a_{13} & \cdots & a_{1n} \end{bmatrix} \begin{bmatrix} b_{11} \\ b_{21} \\ b_{31} \\ \vdots \\ b_{n1} \end{bmatrix} \quad \text{first row of A} \times \text{first column of B}$$

leading to

$$c_{11} = a_{11}b_{11} + a_{12}b_{21} + a_{13}b_{31} + \cdots + a_{1n}b_{n1} = \sum_{k=1}^{n} a_{1k}b_{k1}$$

Or similarly c_{23} is computed as

$$c_{23} = \begin{bmatrix} a_{21} & a_{22} & a_{23} & \cdots & a_{2n} \end{bmatrix} \begin{bmatrix} b_{13} \\ b_{23} \\ b_{33} \\ \vdots \\ b_{n3} \end{bmatrix} \quad \text{second row of A} \times \text{third column of B}$$

resulting in

$$c_{23} = a_{21}b_{13} + a_{22}b_{23} + a_{23}b_{33} + \cdots + a_{2n}b_{n3} = \sum_{k=1}^{n} a_{2k}b_{k3}$$

The following example illustrates the matrix multiplication process.

Example 4

Assume

$$A_{2\times2} = \begin{bmatrix} 2 & 4 \\ 8 & -2 \end{bmatrix} \quad \text{and} \quad B_{2\times3} = \begin{bmatrix} 10 & 5 & 3 \\ 2 & 6 & -1 \end{bmatrix} \quad \text{then}$$

$$C_{2\times3} = \begin{bmatrix} 2 & 4 \\ 8 & -2 \end{bmatrix} \begin{bmatrix} 10 & 5 & 3 \\ 2 & 6 & -1 \end{bmatrix}$$

$$c_{11} = \begin{bmatrix} 2 & 4 \end{bmatrix} \begin{bmatrix} 10 \\ 2 \end{bmatrix} = 2 * 10 + 4 * 2 = 28$$

$$c_{12} = \begin{bmatrix} 2 & 4 \end{bmatrix} \begin{bmatrix} 5 \\ 6 \end{bmatrix} = 2 * 5 + 4 * 6 = 34$$

$$c_{13} = \begin{bmatrix} 2 & 4 \end{bmatrix} \begin{bmatrix} 3 \\ -1 \end{bmatrix} = 2 * 3 + 4 * (-1) = 2$$

$$c_{21} = \begin{bmatrix} 8 & -2 \end{bmatrix} \begin{bmatrix} 10 \\ 2 \end{bmatrix} = 8 * 10 + (-2) * 2 = 76$$

$$c_{22} = \begin{bmatrix} 8 & -2 \end{bmatrix} \begin{bmatrix} 5 \\ 6 \end{bmatrix} = 8 * 5 + (-2) * 6 = 28$$

$$c_{23} = \begin{bmatrix} 8 & -2 \end{bmatrix} \begin{bmatrix} 3 \\ -1 \end{bmatrix} = 8 * 3 + (-2) * (-1) = 26$$

Subsequently the matrix C is

$$C = \begin{bmatrix} 28 & 34 & 2 \\ 76 & 28 & 26 \end{bmatrix}$$

Note that BA is not defined; the number of columns of B, 3, is more than the number of rows of A, 2.

If A and B are square matrices of the same order, then both AB and BA are defined, but generally they are not equal. In scalar algebra $ab = ba$ (for example $5 * 6 = 6 * 5 = 30$). This is the *commutative law* of multiplication. In general, in matrix algebra, $AB \neq BA$. In matrix algebra multiplication is not commutative. For example, consider the following 2 by 2 matrices

$$A = \begin{bmatrix} 5 & 10 \\ -1 & 2 \end{bmatrix} B = \begin{bmatrix} -2 & 12 \\ 3 & 5 \end{bmatrix}$$

$$AB = \begin{bmatrix} 5 & 10 \\ -1 & 2 \end{bmatrix} \begin{bmatrix} -2 & 12 \\ 3 & 5 \end{bmatrix} = \begin{bmatrix} 5 * (-2) + 10 * 3 & 5 * 12 + 10 * 5 \\ -1 * (-2) + 2 * 3 & -1 * 12 + 2 * 5 \end{bmatrix} = \begin{bmatrix} 20 & 110 \\ 8 & -2 \end{bmatrix}$$

$$BA = \begin{bmatrix} -2 & 12 \\ 3 & 5 \end{bmatrix} \begin{bmatrix} 5 & 10 \\ -1 & 2 \end{bmatrix} = \begin{bmatrix} -2 * 5 + 12(-1) & -2 * 10 + 12 * 2 \\ 3 * 5 + 5(-1) & 3 * 10 + 5 * 2 \end{bmatrix} = \begin{bmatrix} -22 & 4 \\ 10 & 40 \end{bmatrix}$$

Obviously

$$\begin{bmatrix} 20 & 110 \\ 8 & -2 \end{bmatrix} \neq \begin{bmatrix} -22 & 4 \\ 10 & 40 \end{bmatrix}$$

13.2.3 Transpose of a Matrix

The *Transpose* of an $m \times n$ matrix A is another matrix A' obtained by interchanging rows and columns of A. Several notations are commonly used in mathematics for denoting the transpose of a matrix, such as A^T, $\tilde{A}$, A^{tr}, A^t, and A'. We use A' to denote the transpose of A. Lets start with simple case of row or column matrices, that is row or column vectors. If V is a row vector, its transpose is a column vector. If V is a column vector, its transpose is a row vector. If

$$V = \begin{bmatrix} v_1 & v_2 & v_3 & \cdots & v_n \end{bmatrix} \quad \text{then} \quad V' = \begin{bmatrix} v_1 \\ v_2 \\ v_3 \\ \vdots \\ v_n \end{bmatrix}$$

Similarly, if

$$V = \begin{bmatrix} v_1 \\ v_2 \\ v_3 \\ \vdots \\ v_n \end{bmatrix} \quad \text{then} \quad V' = \begin{bmatrix} v_1 & v_2 & v_3 & \cdots & v_n \end{bmatrix}$$

Consider the row vector V

$$V = \begin{bmatrix} 10 & -2 & 3 & 6 \end{bmatrix} \quad \text{its transpose is} \quad V' = \begin{bmatrix} 10 \\ -2 \\ 3 \\ 6 \end{bmatrix}$$

The transpose of an $m \times n$ matrix A is an $n \times m$ matrix A' obtained by writing the rows of A as columns of A', or equivalently, by writing the columns of A as rows of A'. The transpose of $A = [a_{ij}]$ is $A' = [a_{ji}]$, $\forall i = 1, 2, \ldots, m$; $j = 1, 2, \ldots, n$. If

$$A = \begin{bmatrix} 12 & 3 & -10 & -2 \\ 5 & 7 & 24 & 0 \\ 0 & -7 & 3 & 15 \end{bmatrix} \quad \text{then} \quad A' = \begin{bmatrix} 12 & 5 & 0 \\ 3 & 7 & -7 \\ -10 & 24 & 3 \\ -2 & 0 & 15 \end{bmatrix}$$

It should be self evident that the transpose of a transposed matrix is the original matrix, that is $(A')' = A$. Although not self evident, it is also true that

$$(AB)' = B'A' \tag{13.2}$$

13.2.4 Some Special Vectors and Matrices

A vector of 1, that is a vector with all its elements 1, is a *summation* vector. If as a row vector it is premultiplied by a column vector of the same size[2] or as column vector it is postmultiplied by a row vector, the result would be a scalar which is the sum of elements of that vector. Assume S is a 1 by n row vector with all entries 1 and V is column vector of size n

$$S = \begin{bmatrix} 1 & 1 & 1 & \cdots & 1 \end{bmatrix} \qquad V = \begin{bmatrix} v_1 \\ v_2 \\ v_3 \\ \vdots \\ v_n \end{bmatrix}$$

then the dot or inner product of S and V is

$$SV = \begin{bmatrix} 1 & 1 & 1 & \cdots & 1 \end{bmatrix} \begin{bmatrix} v_1 \\ v_2 \\ v_3 \\ \vdots \\ v_n \end{bmatrix} = \sum_{i=1}^{n} v_i$$

We get the same result if we premultiply S', the transpose of S, by V', the transpose of V (a simple validation of (13.2))

$$V'S' = \begin{bmatrix} v_1 & v_2 & v_3 & \cdots & v_n \end{bmatrix} \begin{bmatrix} 1 \\ 1 \\ 1 \\ \vdots \\ 1 \end{bmatrix} = \sum_{i=1}^{n} v_i = SV$$

Premultiplying an $m \times n$ matrix A by a $1 \times m$ row vector S leads to a $1 \times n$ row vector with elements that are the sums of matrix A's columns. Postmutiplying A by an $n \times 1$ column vector S generates a column vector of size m by 1 with elements formed by summing each row of A.

$$SA = \begin{bmatrix} 1 & 1 & 1 & \cdots & 1 \end{bmatrix} \begin{bmatrix} a_{11} & a_{12} & a_{13} & \cdots & a_{1n} \\ a_{21} & a_{22} & a_{23} & \cdots & a_{2n} \\ a_{31} & a_{32} & a_{33} & \cdots & a_{3n} \\ \vdots & \vdots & \vdots & \cdots & \vdots \\ a_{m1} & a_{m2} & a_{m3} & \cdots & a_{mn} \end{bmatrix}$$

---

[2] To avoid repeating 'the same size' hereafter I assume that vectors and matrices are conformable for multiplication.

$$= \left[\sum_{k=1}^{n} a_{k1} \quad \sum_{k=1}^{n} a_{k2} \quad \sum_{k=1}^{n} a_{k3} \cdots \sum_{k=1}^{n} a_{kn}\right]$$

$$AS' = \begin{bmatrix} a_{11} & a_{12} & a_{13} & \cdots & a_{1n} \\ a_{21} & a_{22} & a_{23} & \cdots & a_{2n} \\ a_{31} & a_{32} & a_{33} & \cdots & a_{3n} \\ \vdots & \vdots & \vdots & \cdots & \vdots \\ a_{m1} & a_{m2} & a_{m3} & \cdots & a_{mn} \end{bmatrix} \begin{bmatrix} 1 \\ 1 \\ 1 \\ \vdots \\ 1 \end{bmatrix} = \begin{bmatrix} \sum_{k=1}^{n} a_{1k} \\ \sum_{k=1}^{n} a_{2k} \\ \sum_{k=1}^{n} a_{3k} \\ \cdots \\ \sum_{k=1}^{n} a_{nk} \end{bmatrix}$$

For example, if we postmultiply the matrix D of Sect. 13.1, which shows the average weekly sales of food and non-food products at 5 different stores of a supper market chain, by a 5×1 sum vector S'_1, we obtain the total sales of each products,

$$DS'_1 = \begin{bmatrix} 3.5 & 2.2 & 5.1 & 6.7 & 1.2 \\ 5.1 & 3.2 & 4.1 & 4.2 & 2.7 \\ 4.1 & 2.1 & 3.5 & 3.2 & 1.9 \\ 8.5 & 6.3 & 10.2 & 12.3 & 5.8 \end{bmatrix} \begin{bmatrix} 1 \\ 1 \\ 1 \\ 1 \\ 1 \end{bmatrix} = \begin{bmatrix} 18.7 \\ 19.3 \\ 14.9 \\ 43.1 \end{bmatrix}$$

Similarly, if we premultiply D by a 4×1 sum vector S_2, we obtain the total average weekly sales of each store.

$$S_2 D = \begin{bmatrix} 1 & 1 & 1 & 1 \end{bmatrix} \begin{bmatrix} 3.5 & 2.2 & 5.1 & 6.7 & 1.2 \\ 5.1 & 3.2 & 4.1 & 4.2 & 2.7 \\ 4.1 & 2.1 & 3.5 & 3.2 & 1.9 \\ 8.5 & 6.3 & 10.2 & 12.3 & 5.8 \end{bmatrix}$$

$$= \begin{bmatrix} 21.2 & 13.8 & 22.9 & 26.4 & 11.6 \end{bmatrix}$$

A matrix with only zero entries is called a zero or *null* matrix. Assuming conformability, adding a null matrix to or subtracting it from a matrix does not change the elements of that matrix. The result of multiplying a matrix by a null matrix is another null matrix. In short, the null matrix plays the role of 0 in the regular (scalar) algebra.

In a matrix $A = [a_{ij}]$ the entries where i is equal j, that is entries $a_{11}, a_{22}, \ldots, a_{nn}$, are called *diagonal* entries. These entries start with the element on the top left corner of the matrix, a_{11}, down to the element on the bottom right corner, a_{nn}. These entries constitute the *main, leading,* or *principle diagonal* of a matrix. In the following 3 by 3 matrix, the diagonal elements, which are shown in bold face, are 3, 6, and -1.

$$A = \begin{bmatrix} \mathbf{3} & 4 & -5 \\ 12 & \mathbf{6} & 27 \\ 0 & 2 & \mathbf{-1} \end{bmatrix}$$

If all the non-diagonal or off the main diagonal elements of a square matrix are zero, the matrix is called a *diagonal matrix*. The matrix A below is an example of an n by n diagonal matrix

$$A = \begin{bmatrix} a_{11} & 0 & 0 & \cdots & 0 \\ 0 & a_{22} & 0 & \cdots & 0 \\ 0 & 0 & a_{33} & \cdots & 0 \\ \vdots & \vdots & \vdots & \cdots & \vdots \\ 0 & 0 & 0 & \cdots & a_{nn} \end{bmatrix}$$

The following 3 by 3 matrix B is a numerical example of a diagonal matrix

$$B = \begin{bmatrix} 2 & 0 & 0 \\ 0 & -3 & 0 \\ 0 & 0 & -1 \end{bmatrix}$$

A diagonal matrix is sometimes denoted by

$$A = \text{diag}\,(a_{11}, a_{22}, \ldots, a_{nn})$$

For example B may be represented by

$$B = \text{diag}\,(2, -3, -1)$$

A diagonal matrix with all diagonal entires equal to 1 is a *unit* or *identity matrix*. The notation for identity matrix is I, or I_n if we need to specify the size of the matrix.

$$I_2 = \begin{bmatrix} 1 & 0 \\ 0 & 1 \end{bmatrix} \qquad I_3 = \begin{bmatrix} 1 & 0 & 0 \\ 0 & 1 & 0 \\ 0 & 0 & 1 \end{bmatrix} \qquad \cdots \qquad I_n = \begin{bmatrix} 1 & 0 & 0 & \cdots & 0 \\ 0 & 1 & 0 & \cdots & 0 \\ 0 & 0 & 1 & \cdots & 0 \\ \vdots & \vdots & \vdots & \cdots & \vdots \\ 0 & 0 & 0 & \cdots & 1 \end{bmatrix}$$

If A is any square matrix and I the unit matrix of the same size, then

$$IA = AI = A$$

The identity matrix is the counterpart of 1 in matrix algebra. As 1 is the multiplicative identity in scalar algebra ($1 \times a = a \times 1 = a$), I is the multiplicative identity in matrix algebra.

A square matrix A is *symmetric* if $A' = A$. In a symmetric matrix the mirror image of elements with respect to the main diagonal are the same, $a_{ij} = a_{ji}$. The following are examples of 2×2 and 3×3 symmetric matrices

$$A = \begin{bmatrix} 3 & 5 \\ 5 & -1 \end{bmatrix} \quad B = \begin{bmatrix} -3 & 4 & -3 \\ 4 & 2 & 7 \\ -3 & 7 & 0 \end{bmatrix}$$

Assume B is an $m \times n$ matrix. It can be shown that the product $C = B'B$ is an $n \times n$ symmetric matrix and the product $D = BB'$ is an $m \times m$ symmetric matrix.

If all entries above the main diagonal of a square matrix are zero it is called a *lower triangular* matrix. When all entries below the main diagonal are zero, the matrix is called *upper triangular*. The following matrices A and B are lower and upper triangular, respectively:

$$A = \begin{bmatrix} -1 & 0 & 0 & 0 \\ 3 & 2 & 0 & 0 \\ 7 & -1 & 2 & 0 \\ 12 & 9 & 10 & 4 \end{bmatrix} \quad B = \begin{bmatrix} 2 & 6 & 1 \\ 0 & 8 & 3 \\ 0 & 0 & 4 \end{bmatrix}$$

13.3 Matrix Representation of a Linear System

Consider the following system of n equations with n unknowns

$$\begin{aligned} a_{11}x_1 + a_{12}x_2 + \cdots + a_{1n}x_n &= b_1 \\ a_{12}x_1 + a_{22}x_2 + \cdots + a_{2n}x_n &= b_2 \\ a_{31}x_1 + a_{32}x_2 + \cdots + a_{3n}x_n &= b_3 \\ &\vdots \\ a_{n1}x_1 + a_{n2}x_2 + \cdots + a_{nn}x_n &= b_n \end{aligned} \tag{13.3}$$

where $x_1, x_2, \cdots, x_n$ are the unknowns, $a_{11}, a_{12}, \cdots, a_{nn}$ are the coefficients of the unknowns, and $b_1, b_2, \cdots, b_n$ are the constants.

We form the matrix of coefficients, A, the column vector of unknowns, X, and the column vector of constants, $\mathbf{b}$

$$A = \begin{bmatrix} a_{11} & a_{12} & a_{13} & \cdots & a_{1n} \\ a_{21} & a_{22} & a_{23} & \cdots & a_{2n} \\ \vdots & \vdots & \vdots & \vdots & \vdots \\ a_{n1} & a_{n2} & a_{n3} & \cdots & a_{nn} \end{bmatrix} \quad X = \begin{bmatrix} x_1 \\ x_2 \\ \vdots \\ x_n \end{bmatrix} \quad \mathbf{b} = \begin{bmatrix} b_1 \\ b_2 \\ \vdots \\ b_n \end{bmatrix}$$

Using these components, the system in (13.3) can be equivalently expressed as

$$A X = \mathbf{b}.$$

Example 5

In the following system of 3 equations with 3 unknowns

$$3x_1 - 2x_2 + 3.5x_3 = 10$$
$$-2x_1 + 5x_3 = 5 \qquad\qquad (13.4)$$
$$3.6x_1 + 3x_2 - 2x_3 = 2$$

matrix A, vectors X and $\mathbf{b}$ are

$$A = \begin{bmatrix} 3 & -2 & 3.5 \\ -2 & 0 & 5 \\ 3.6 & 3 & -2 \end{bmatrix} \quad X = \begin{bmatrix} x_1 \\ x_2 \\ x_3 \end{bmatrix} \quad \mathbf{b} = \begin{bmatrix} 10 \\ 5 \\ 2 \end{bmatrix}$$

The system in matrix form is

$$AX = \mathbf{b} = \begin{bmatrix} 3 & -2 & 3.5 \\ -2 & 0 & 5 \\ 3.6 & 3 & -2 \end{bmatrix} \begin{bmatrix} x_1 \\ x_2 \\ x_3 \end{bmatrix} = \begin{bmatrix} 10 \\ 5 \\ 2 \end{bmatrix}$$

Note that x_2 does not appear in the second equation, so its coefficient is entered as zero in matrix A.

13.4 Matrix Inversion and Solution to a Linear System

There are several familiar methods for solving a system of linear equations similar to (13.4). In the *substitution* method, we express one of the variables in terms of other variables in one of the equations and substitute it in the remaining equations. This reduces the number of unknowns and equations by one. We do the same for the second, third,..., $n - 1$th variable and we end up with one equation in terms of one of the unknowns. We solve for this variable and in reverse order (recursively) determine the solution for the other variables.

But solving a system of equations by substitution is a tedious and time consuming process, especially if the system consists of a large number of unknowns and equations. A more efficient methods is the *Gaussian elimination* method. In this method, through a series of steps known as *row operations* or *elementary row operations*, we reduce the original system to a simple system where each equation contains only one of the variables (more about Gaussian elimination method in Sect. 13.4.4). Although superior to the substitution method, solving systems of equations using the elimination method is still computationally time consume and tedious. The most efficient method for solving systems of equations is the matrix method. The matrix method is easily adaptable to computer applications, to the extent that a large system of equations that would be almost impossible to solve by hand calculations can be solved in a fraction of a second.

Consider a number a. The reciprocal of a is $\dfrac{1}{a}$. We also express the reciprocal of a by a^{-1} and call it *inverse* of a. Note that

$$a\,\frac{1}{a} = \frac{1}{a}\,a = a\,a^{-1} = a^{-1}a = 1 \tag{13.5}$$

Now consider a simple equation with one unknown $ax = b$. We solve for x by writing

$$x = \frac{b}{a} = \frac{1}{a}\,b = a^{-1}\,b \qquad a \neq 0 \tag{13.6}$$

Now imagine that instead of a number a, one unknown x, and a constant b, we have an $n \times n$ matrix, A, a vector of n unknowns, X, and a vector of n constants, **b**. So instead of

$$ax = b$$

we have

$$AX = \mathbf{b}$$

which is a system of n equations with n unknowns like (13.3) represented in the matrix form. Can we solve for X using $X = A^{-1}\mathbf{b}$? The answer is a 'conditional' yes. It is possible if the matrix A^{-1}, the *inverse* of A, exists. If A^{-1} exists then the following expression, which is the analog of (13.5) in matrix algebra, must be true

$$AA^{-1} = A^{-1}A = I$$

where I is an $n \times n$ unit matrix. If A^{-1} exists, the matrix A is said to be *invertible* or it is *non-singular*.

Whether a square matrix A is invertible or singular depends on the value of its *determinant*. Associated with any square matrix is a number (scalar) called its determinant, denoted by $det(A)$ or $|A|$. If $|A| \neq 0$ then matrix A is invertible or nonsingular. Otherwise if $|A| = 0$ then A is singular and has no inverse. Therefore, the required condition for a system on n equations with n unknowns

$$AX = \mathbf{b}$$

to a have solution is $|A| \neq 0$. If this condition is satisfied, then the solution to the system will be

$$X = A^{-1}\,\mathbf{b}. \tag{13.7}$$

13.4.1 Determinants

Calculating the determinants of *order* 1 and 2 is easy and straightforward. Order here refers to the size of the matrix for which we want to calculate the determinant. A 1×1 matrix $A = [a_{11}]$ has only one element (it is a scalar) and its determinant is

$$|A| = |a_{11}| = a_{11}$$

The determinant of a 2×2 matrix $A = \begin{bmatrix} a_{11} & a_{12} \\ a_{21} & a_{22} \end{bmatrix}$ is defined as

$$|A| = \begin{vmatrix} a_{11} & a_{12} \\ a_{21} & a_{22} \end{vmatrix} = a_{11}a_{22} - a_{12}a_{21}$$

That is, the order 2 determinant is computed as the product of the two elements on the main diagonal minus the product of the two off diagonal elements. For example, if $A = \begin{bmatrix} 8 & 3 \\ 2 & 1 \end{bmatrix}$, then

$$|A| = \begin{vmatrix} 8 & 3 \\ 2 & 1 \end{vmatrix} = 8 \times 1 - 3 \times 2 = 2$$

To find the determinants of order 3 or higher we use the *Laplace expansion*, named after the French mathematician and astronomer Pierre Simon Laplace. The Laplace expansion process reduces the order of the determinants until we get to order 2, which we know how to calculate. We work through an order 3 determinant and then highlight the extension to higher orders.

Assume we want to determine

$$|A| = \begin{vmatrix} a_{11} & a_{12} & a_{13} \\ a_{21} & a_{22} & a_{23} \\ a_{31} & a_{32} & a_{33} \end{vmatrix}$$

We pick a row or a column (any row or column) and we expand by that row or column. Assume we want to expand by the first row. We pick the first element of the first row, a_{11}, and eliminate the first row and the first column of the matrix, that is the row and column that this element belongs to. We end up with a sub-determinant of $|A|$. This sub-determinant is called the *minor* of element a_{11}, denoted by $|M_{11}|$, which is an order 2 determinant. We find $|M_{12}|$, the minor of element a_{12} by eliminating the first row and the second column, and similarly $|M_{13}|$ by eliminating the first row and the third column. The minors of the elements of the first row (the row we choose for expansion) are

$$|M_{11}| = \begin{vmatrix} a_{22} & a_{23} \\ a_{32} & a_{33} \end{vmatrix} \qquad |M_{12}| = \begin{vmatrix} a_{21} & a_{23} \\ a_{31} & a_{33} \end{vmatrix} \qquad |M_{11}| = \begin{vmatrix} a_{21} & a_{22} \\ a_{31} & a_{32} \end{vmatrix}$$

Note that the minors of the first row elements are determinants of order 2 and their values are

$$|M_{11}| = a_{22}a_{33} - a_{23}a_{32}$$
$$|M_{12}| = a_{21}a_{33} - a_{23}a_{31}$$
$$|M_{13}| = a_{21}a_{32} - a_{22}a_{31}$$

In general, if we expand an order n determinant (the determinant of an $n \times n$ matrix) by its ith row, the minor associated with element a_{ij} is a sub-determinant $|M_{ij}|$ obtained by eliminating the ith row and the jth column.

Next we write the *cofactors* of the elements of the first row. A cofactor, denoted $|C_{ij}|$, is simply a "signed" minor. They are related to minors by the following expression:

$$|C_{ij}| = (-1)^{i+j} |M_{ij}|$$

For example, the cofactor associated with element a_{11}, where $i = 1$ and $j = 1$, is

$$|C_{11}| = (-1)^{1+1} |M_{11}| = (-1)^2 |M_{11}| = |M_{11}|$$

But the cofactor associated with the element a_{12} is

$$|C_{12}| = (-1)^{1+2} |M_{12}| = (-1)^3 |M_{11}| = -|M_{12}|$$

The following matrix of signs gives the sign for minors of a 3 order determinant

$$\begin{bmatrix} + & - & + \\ - & + & - \\ + & - & + \end{bmatrix}$$

We can now express an order 3 (or third order) determinant as

$$|A| = a_{11}|C_{11}| + a_{12}|C_{12}| + a_{13}|C_{13}| = \sum_{j=1}^{3} a_{1j} |C_{1j}|$$

$$= a_{11}|M_{11}| - a_{12}|M_{12}| + a_{13}|M_{13}| = \sum_{j=1}^{3} a_{1j} \left[(-1)^{1+j} |M_{1j}| \right]$$

$$= a_{11}(a_{22}a_{33} - a_{23}a_{32}) - a_{12}(a_{21}a_{33} - a_{23}a_{31}) + a_{13}(a_{21}a_{32} - a_{22}a_{31})$$

Note that if we decide to find the determinant of A by expanding the second row, we must evaluate

$$|A| = a_{21}|C_{21}| + a_{22}|C_{22}| + a_{23}|C_{23}| = \sum_{j=1}^{3} a_{2j} |C_{2j}|$$

Similarly, calculating the determinant by expanding along the third column

$$|A| = a_{13}|C_{13}| + a_{23}|C_{23}| + a_{33}|C_{33}| = \sum_{j=1}^{3} a_{2j}\,|C_{2j}|$$

The general expression for an order n determinant expanded by the ith row is

$$|A| = \sum_{j=1}^{n} a_{ij}\,|C_{ij}| \tag{13.8}$$

The same determinant is obtained through expansion by the jth column, using

$$|A| = \sum_{i=1}^{n} a_{ij}\,|C_{ij}| \tag{13.9}$$

The following numerical examples should help to demonstrate the process of computing the determinants.

Example 6

Assume $|A| = \begin{vmatrix} 5 & 2 & 0 \\ 3 & 1 & 6 \\ -9 & 2 & -1 \end{vmatrix}$

Let us expand by the first column. The minors and cofactors associated with elements $a_{11} = 5$, $a_{21} = 3$, and $a_{31} = -9$ are

$$|M_{11}| = \begin{vmatrix} 1 & 6 \\ 2 & -1 \end{vmatrix} = 1(-1) - 6(2) = -13$$

$$|C_{11}| = (-1)^{1+1}|M_{11}| = (-1)^2(-13) = -13$$

$$|M_{21}| = \begin{vmatrix} 2 & 0 \\ 2 & -1 \end{vmatrix} = 2(-1) - 0(2) = -2$$

$$|C_{21}| = (-1)^{2+1}|M_{11}| = (-1)^3(-2) = 2$$

$$|M_{31}| = \begin{vmatrix} 2 & 0 \\ 1 & 6 \end{vmatrix} = 2(6) - 0(1) = 12$$

$$|C_{31}| = (-1)^{3+1}(12) = 12$$

Thus $|A|$ is

$$|A| = a_{11}|C_{11}| + a_{21}|C_{21}| + a_{31}|C_{31}| = 5(-13) + 3(2) + (-9)(12) = -167$$

Note that we would have achieved the same result by expanding along the first row. Since a_{13} is 0 the third term in

$$|A| = a_{11}|C_{11}| + a_{12}|C_{12}| + a_{13}|C_{13}|$$

is 0 and we have to only calculate the first two terms. For example, the quickest way to find the determinant

$$|A| = \begin{vmatrix} 0 & 1 & 1 \\ 2 & -1 & 1 \\ 0 & 1 & -1 \end{vmatrix}$$

is by expanding it by the first column

$$|A| = \begin{vmatrix} 0 & 1 & 1 \\ 2 & -1 & 1 \\ 0 & 1 & -1 \end{vmatrix} = (-1)^{1+2} 2 \begin{vmatrix} 1 & 1 \\ 1 & -1 \end{vmatrix} = (-2)(-2) = 4$$

13.4.2 Matrix Inversion, Method of Adjoint

The significance of the condition $|A| \neq 0$ for the existence of A^{-1} and by extension the solution to a system of equations like (13.3) will be fully realized when evaluation of A^{-1} is discussed. One inversion method which is adequate for small matrices is the *method of adjoint*. The adjoint matrix of a square matrix A, denoted $Adj\,(A)$, is the transpose of the matrix of cofactors of A. The matrix of cofactors of A, denoted by C, is

$$C = \begin{bmatrix} |c_{11}| & |c_{12}| & \cdots & |c_{1n}| \\ |c_{21}| & |c_{22}| & \cdots & |c_{n2}| \\ \vdots & \vdots & \cdots & \vdots \\ |c_{n1}| & |c_{2n}| & \cdots & |c_{nn}| \end{bmatrix}$$

and

$$Adj(A) = C' = \begin{bmatrix} |c_{11}| & |c_{21}| & \cdots & |c_{n1}| \\ |c_{12}| & |c_{22}| & \cdots & |c_{2n}| \\ \vdots & \vdots & \cdots & \vdots \\ |c_{1n}| & |c_{n2}| & \cdots & |c_{nn}| \end{bmatrix}$$

The inverse of the matrix A is then expressed as

$$A^{-1} = \frac{1}{|A|} Adj(A) \tag{13.10}$$

It is immediately clear that if $|A| = 0$ in (13.10), the fraction $\dfrac{1}{|A|}$ is not defined and subsequently A^{-1} does not exist, meaning that A is a singular matrix. If A is the matrix of coefficients of a system of equations like (13.7), its singularity means that the system has no solution.

Example 7

Assume the following system of 3 equations with 3 unknowns

$$x_1 - x_2 + 2x_3 = 100$$
$$3x_1 + 2x_2 - 2x_3 = 20$$
$$x_1 + 2x_3 = 10$$

This system in matrix form is

$$AX = b = \begin{bmatrix} 1 & -1 & 2 \\ 3 & 2 & -2 \\ 1 & 0 & 2 \end{bmatrix} \begin{bmatrix} x_1 \\ x_2 \\ x_3 \end{bmatrix} = \begin{bmatrix} 100 \\ 20 \\ 10 \end{bmatrix}$$

We find the determinant of the coefficient matrix by expanding the third row

$$|A| = \begin{vmatrix} 1 & -1 & 2 \\ 3 & 2 & -2 \\ 1 & 0 & 2 \end{vmatrix} = 1\,(-1)^{3+1} \begin{vmatrix} -1 & 2 \\ 2 & -2 \end{vmatrix} + 2\,(-1)^{3+3} \begin{vmatrix} 1 & -1 \\ 3 & 2 \end{vmatrix}$$

$$|A| = (2 - 4) + 2(2 + 3) = 8$$

Since $|A| = 8 \neq 0$, the inverse of A exists and the solution to the system is

$$\begin{bmatrix} x_1 \\ x_2 \\ x_3 \end{bmatrix} = \begin{bmatrix} 1 & -1 & 2 \\ 3 & 2 & -2 \\ 1 & 0 & 2 \end{bmatrix}^{-1} \begin{bmatrix} 100 \\ 20 \\ 10 \end{bmatrix}$$

To find the inverse of A we form the matrix of cofactors of A

$$|c_{11}| = (-1)^{1+1} \begin{vmatrix} 2 & -2 \\ 0 & 2 \end{vmatrix} = 4 \qquad |c_{12}| = (-1)^{1+2} \begin{vmatrix} 3 & -2 \\ 1 & 2 \end{vmatrix} = -8$$

$$|c_{13}| = (-1)^{1+3} \begin{vmatrix} 3 & 2 \\ 1 & 0 \end{vmatrix} = -2 \qquad |c_{21}| = (-1)^{2+1} \begin{vmatrix} -1 & 2 \\ 0 & 2 \end{vmatrix} = 2$$

$$|c_{22}| = (-1)^{2+2} \begin{vmatrix} 1 & 2 \\ 1 & 2 \end{vmatrix} = 0 \qquad |c_{23}| = (-1)^{2+3} \begin{vmatrix} 1 & -1 \\ 1 & 0 \end{vmatrix} = -1$$

$$|c_{31}| = (-1)^{3+1} \begin{vmatrix} -1 & 2 \\ 2 & -2 \end{vmatrix} = -2 \qquad |c_{32}| = (-1)^{3+2} \begin{vmatrix} 1 & 2 \\ 3 & -2 \end{vmatrix} = 8$$

$$|c_{33}| = (-1)^{3+3} \begin{vmatrix} 1 & -1 \\ 3 & 2 \end{vmatrix} = 5$$

$$C = \begin{bmatrix} 4 & -8 & -2 \\ 2 & 0 & -1 \\ -2 & 8 & 5 \end{bmatrix} \longrightarrow Adj(A) = C' = \begin{bmatrix} 4 & 2 & -2 \\ -8 & 0 & 8 \\ -2 & -1 & 5 \end{bmatrix}$$

Finally

$$A^{-1} = \frac{1}{|A|} Adj(A) = \frac{1}{8} \begin{bmatrix} 4 & 2 & -2 \\ -8 & 0 & 8 \\ -2 & -1 & 5 \end{bmatrix} = \begin{bmatrix} \frac{1}{2} & \frac{1}{4} & -\frac{1}{4} \\ -1 & 0 & 1 \\ -\frac{1}{4} & -\frac{1}{8} & \frac{5}{8} \end{bmatrix}$$

And solution to the system $X = A^{-1}\mathbf{b}$ is

$$\begin{bmatrix} x_1 \\ x_2 \\ x_3 \end{bmatrix} = \begin{bmatrix} \frac{1}{2} & \frac{1}{4} & -\frac{1}{4} \\ -1 & 0 & 1 \\ -\frac{1}{4} & -\frac{1}{8} & \frac{5}{8} \end{bmatrix} \begin{bmatrix} 100 \\ 20 \\ 10 \end{bmatrix} = \begin{bmatrix} 52.50 \\ -90 \\ -21.25 \end{bmatrix}$$

The above problem clearly shows the potential computational complexity of matrix inversion. With increases in the size of system of equations, and in turn matrices to be inverted, the number of operations dramatically increase to the extent that for large matrices the sheer number of required operations exceed the computational ability of individuals. There are a number of computer programs that are readily available for handling all of the important matrix operations. We introduce some of these tools after discussing some of the properties of determinant and the Gauss-Jordan inversion method.

13.4.3 Properties of Determinants

The following properties are helpful in computing the value of the determinant of a matrix. These properties are based on the definition and evaluation process of the determinant obtained through expansion by row or column as expressed in (13.8) and (13.9).

1. If all elements of a row or column of a matrix are zero, then the determinant of the matrix is zero.

$$\begin{vmatrix} 3 & 1 & 1 \\ 0 & 0 & 0 \\ -10 & 1 & -1 \end{vmatrix} = \begin{vmatrix} 5 & 1 & 0 \\ 1 & -2 & 0 \\ -10 & 1 & 0 \end{vmatrix} = 0$$

2. If the corresponding elements of any two rows (or two columns) of matrix A are equal, or one is the multiple of another, then the determinant of A is zero.

$$A = \begin{bmatrix} 3 & -1 & 2 \\ 3 & -1 & 2 \\ 5 & 0 & 1 \end{bmatrix} \longrightarrow |A| = 0 \quad \text{first and second row are the same}$$

$$A = \begin{bmatrix} 6 & -2 & 4 \\ 3 & -1 & 2 \\ 5 & 0 & 1 \end{bmatrix} \quad \longrightarrow \quad |A| = 0 \quad \text{first row is a multiple of second row}$$

3. Determinants of diagonal, upper and lower triangular matrices are the product of the elements of their main diagonal. If A is an $n \times n$ diagonal, upper or lower triangular matrix, then

$$|A| = a_{11} \times a_{22} \times a_{33} \times \cdots \times a_{nn} = \prod_{i=1}^{n} a_{ii}$$

It should be immediately clear that if any of the main diagonal elements of a diagonal, upper or lower triangular matrix is zero, the determinant of the matrix must be zero, leading to the conclusion that the matrix is singular and not invertible.

4. If all elements of *a row* or *a column* of a square matrix A are multiplied by a constant k, the value of the determinant of A will change to $k\,|A|$. In the following example, the elements of the second row of $|A|$ are multiplied by 2. This changes $|A|$ from 5 to 10.

$$|A| = \begin{vmatrix} 3 & 1 & 1 \\ 2 & 5 & 1 \\ 0 & 8 & 1 \end{vmatrix} = 5 \quad \longrightarrow \quad \begin{vmatrix} 3 & 1 & 1 \\ 2 \times 2 & 2 \times 5 & 2 \times 1 \\ 0 & 8 & 1 \end{vmatrix} = 2(5) = 10$$

5. If *all elements of an $n \times n$ matrix A* are multiplied by a constant k, the value of the determinant of A will change to $k^n\,|A|$. That is, $|kA| = k^n\,|A|$. In the following example, the elements of A are multiplied by 2. This changes $|A|$ from 5 to $2^3 \times 5 = 40$.

$$2A = \begin{bmatrix} 6 & 2 & 2 \\ 4 & 10 & 2 \\ 0 & 16 & 2 \end{bmatrix} \quad \longrightarrow \quad |2A| = 2^3 |A| = 8 * 5 = 40$$

6. The determinant of a matrix A remains unchanged if k times the elements of any row (column) are added to the corresponding elements of another row (column).

$$A = \begin{bmatrix} 2 & 10 & -2 \\ -1 & -5 & 3 \\ 0 & 16 & 2 \end{bmatrix} \quad \longrightarrow \quad |A| = -64$$

If we multiply the second row of A by $k = 2$ and add it to the first row, we get

$$A = \begin{bmatrix} 0 & 0 & 4 \\ -1 & -5 & 3 \\ 0 & 16 & 2 \end{bmatrix}$$

Now we can expand the determinant of A by the first row

$$|A| = 0 \, |C_{11}| + 0 \, |C_{12}| + 4 \, |C_{13}| = 4 \, (-1)^{1+3} \, |M_{13}|$$

where $|M_{13}| = \begin{vmatrix} -1 & -5 \\ 0 & 16 \end{vmatrix} = -16$. Thus

$$|A| = 4(-16) = -64$$

7. The determinant of the product of two matrices A and B is the product of their determinants, that is

$$|AB| = |A| \, |B|$$

$$\text{Let } A = \begin{bmatrix} 1 & 3 \\ 2 & 8 \end{bmatrix} \text{ and } B = \begin{bmatrix} 5 & 7 \\ 3 & 2 \end{bmatrix}$$

Then

$$AB = \begin{bmatrix} 1 & 3 \\ 2 & 8 \end{bmatrix} \begin{bmatrix} 5 & 7 \\ 3 & 2 \end{bmatrix} = \begin{bmatrix} 14 & 13 \\ 34 & 30 \end{bmatrix}$$

The determinant of A is $1 * 8 - 3 * 2 = 2$, the determinant of B is $5 * 2 - 7 * 3 = -11$, and the determinant of AB is $14 * 30 - 13 * 34 = -22$, which is equal to the product of $|A|$ and $|B|$, $2(-11)$.

8. The determinant of the inverse of a matrix A is the inverse of the determinant of A, that is,

$$|A^{-1}| = \frac{1}{|A|}$$

For example, the determinant of matrix

$$A = \begin{bmatrix} 3 & 1 & 1 \\ 2 & 5 & 1 \\ 0 & 8 & 1 \end{bmatrix} \quad \text{is } |A| = 5$$

The inverse of A is

$$A = \begin{bmatrix} -0.6 & 1.4 & -0.8 \\ -0.4 & 0.6 & -0.2 \\ 3.2 & -4.8 & 2.6 \end{bmatrix} \quad \text{and the determinant of this matrix is } 0.2$$

which is $\frac{1}{5}$. We can use Property 7 and show that Property 8 is true for any nonsingular square matrix. From Property 7,

$$|AB| = |A| \, |B|$$

Now replace B by A^{-1}, which gives

$$|A\,A^{-1}| = |A|\,|A^{-1}|, \quad \text{but} \quad A\,A^{-1} = I \quad \text{and the determinat of a unit}$$

matrix is 1, therefore

$$1 = |A|\,|A^{-1}| \quad \longrightarrow \quad |A^{-1}| = \frac{1}{|A|}$$

13.4.4 Matrix Inversion, the Gauss-Jordan Elimination Method

The Gauss-Jordan inversion method is an extended application of the Gaussian elimination method which was originally introduced by Gauss, a German mathematician, for solving a system of equations. This method requires fewer operations compared to the method of adjoint, so it is considered more efficient. We use a simple example to demonstrate the mechanics of the method. We wish to solve the following system of two equations with two unknowns:

$$2x + 3y = 28$$
$$x + 2y = 17$$

We conduct a series of operations that are generally known as *elementary row operations*, with the objective of reducing the number of variables in each equation to only one with coefficient of 1. We can eliminate variable x from the second equation through the following steps:

(1) Divide both side of the first equation by 2, the coefficient of x.

$$x + \frac{3}{2}y = 14$$
$$x + 2y = 17$$

(2) Subtract the first equation from the second equation.

$$x + \frac{3}{2}y = 14$$
$$0 + \frac{1}{2}y = 3$$

We can eliminate variable y from the first equation by the following row operation:

(3) Multiply both sides of the second equation by 3 and subtract it from the first equation

$$x + 0 = 5$$
$$0 + \frac{1}{2}y = 3$$

(4) Multiply both sides of the second equation by 2

$$x + 0 = 5 \qquad\qquad (13.11)$$
$$0 + y = 6$$

The solution to the system is $x = 5$ and $y = 6$.
If we write the original system of equations in the matrix form

$$\begin{bmatrix} 2 & 3 \\ 1 & 2 \end{bmatrix} \begin{bmatrix} x \\ y \end{bmatrix} = \begin{bmatrix} 28 \\ 17 \end{bmatrix}$$

and the solution to the system in (13.11) in the matrix form

$$\begin{bmatrix} 1 & 0 \\ 0 & 1 \end{bmatrix} \begin{bmatrix} x \\ y \end{bmatrix} = \begin{bmatrix} 5 \\ 6 \end{bmatrix}$$

it becomes clear that the Gaussian elimination method solves the system by converting the matrix of coefficients to an identity matrix through a series of elementary row operation. The elementary row operations that are most frequently used are (1) multiplication of all elements in a row by a nonzero scalar(a number); (2) interchanging any two rows; (3) constant multiple of elements in a row are added to or subtracted from another row. These mechanics are implemented in the Gauss-Jordan inversion method.

We place matrix A that we want to invert side by side with an identity matrix I forming what is known as an *augmented matrix* $[A|I]$. We apply the elementary row operations for converting A to an identity matrix to both matrices. At the end of the process when A is converted to I then I in the augmented matrix $[A|I]$ is converted to A^{-1}, that is $[A|I] \longrightarrow [I|A^{-1}]$. Let's apply the method and find the inverse of the coefficient matrix in the above system,

$$[A|I] = \begin{bmatrix} 2 & 3 & | & 1 & 0 \\ 1 & 2 & | & 0 & 1 \end{bmatrix}$$

(1) Interchange the first and the second rows

$$\begin{bmatrix} 1 & 2 & | & 0 & 1 \\ 2 & 3 & | & 1 & 0 \end{bmatrix}$$

(2) Multiply the first row by -2 and add it to the second row

$$\begin{bmatrix} 1 & 2 & | & 0 & 1 \\ 0 & -1 & | & 1 & -2 \end{bmatrix} \quad \text{New row 2} = \text{Old row 2} - 2* \text{row 1}$$

(3) Multiply the second row by 2 and add it to the first row

$$\begin{bmatrix} 1 & 0 & | & 2 & -3 \\ 0 & -1 & | & 1 & -2 \end{bmatrix} \quad \text{New row 1} = \text{Old row 1} + 2* \text{row 2}$$

(4) Multiply the second row by -1

$$\begin{bmatrix} 1 & 0 & | & 2 & -3 \\ 0 & 1 & | & -1 & 2 \end{bmatrix} \quad \text{New row 2} (= -1)* \text{Old row 2}$$

Matrix $\begin{bmatrix} 2 & -3 \\ -1 & 2 \end{bmatrix}$ is the inverse of $\begin{bmatrix} 2 & 3 \\ 1 & 2 \end{bmatrix}$.

As another example let us consider a slightly larger problem that requires inversion of a 3 by 3 matrix.

Example 6

Consider the following system of 3 equations with 3 unknowns

$$2x_1 + x_3 = 4$$
$$4x_1 + x_2 + 2x_3 = 10$$
$$3x_1 + x_2 + 2x_3 = 5$$

This system in the matrix form is

$$Ax = b \quad \longrightarrow \quad \begin{bmatrix} 2 & 0 & 1 \\ 4 & 1 & 2 \\ 3 & 1 & 2 \end{bmatrix} \begin{bmatrix} x_1 \\ x_2 \\ x_3 \end{bmatrix} = \begin{bmatrix} 4 \\ 10 \\ 5 \end{bmatrix}$$

If the matrix of coefficients is invertible then the solution to the system is

$$x = A^{-1}b \quad \longrightarrow \quad \begin{bmatrix} x_1 \\ x_2 \\ x_3 \end{bmatrix} = \begin{bmatrix} 2 & 0 & 1 \\ 4 & 1 & 2 \\ 3 & 1 & 2 \end{bmatrix}^{-1} \begin{bmatrix} 4 \\ 10 \\ 5 \end{bmatrix}$$

To invert the matrix of coefficients A, we form the augmented matrix $[A|I]$

$$\begin{bmatrix} 2 & 0 & 1 & | & 1 & 0 & 0 \\ 4 & 1 & 2 & | & 0 & 1 & 0 \\ 3 & 1 & 2 & | & 0 & 0 & 1 \end{bmatrix}$$

and apply the following set of row operations:

(1) Divide the first row elements by 2

$$\left[\begin{array}{ccc|ccc} 1 & 0 & \frac{1}{2} & \frac{1}{2} & 0 & 0 \\ 4 & 1 & 2 & 0 & 1 & 0 \\ 3 & 1 & 2 & 0 & 0 & 1 \end{array}\right]$$

(2) Multiply the first row by (-4) and add it to the second row

$$\left[\begin{array}{ccc|ccc} 1 & 0 & \frac{1}{2} & \frac{1}{2} & 0 & 0 \\ 0 & 1 & 0 & -2 & 1 & 0 \\ 3 & 1 & 2 & 0 & 0 & 1 \end{array}\right]$$

(3) Multiply the first row by (-3) and add it to the third row

$$\left[\begin{array}{ccc|ccc} 1 & 0 & \frac{1}{2} & \frac{1}{2} & 0 & 0 \\ 0 & 1 & 0 & -2 & 1 & 0 \\ 0 & 1 & \frac{1}{2} & -\frac{3}{2} & 0 & 1 \end{array}\right]$$

(4) Multiply the second row by (-1) and add it to the third row

$$\left[\begin{array}{ccc|ccc} 1 & 0 & \frac{1}{2} & \frac{1}{2} & 0 & 0 \\ 0 & 1 & 0 & -2 & 1 & 0 \\ 0 & 0 & \frac{1}{2} & \frac{1}{2} & -1 & 1 \end{array}\right]$$

(5) Multiply the third row by 2

$$\left[\begin{array}{ccc|ccc} 1 & 0 & \frac{1}{2} & \frac{1}{2} & 1 & -1 \\ 0 & 1 & 0 & -2 & 1 & 0 \\ 0 & 0 & 1 & 1 & -2 & 2 \end{array}\right]$$

(6) Multiply the third row by $-\frac{1}{2}$ and add it to the first row

$$\left[\begin{array}{ccc|ccc} 1 & 0 & 0 & 0 & 1 & -1 \\ 0 & 1 & 0 & -2 & 1 & 0 \\ 0 & 0 & 1 & 1 & -2 & 2 \end{array}\right]$$

We can now find the solution to the system as

$$\begin{bmatrix} x_1 \\ x_2 \\ x_3 \end{bmatrix} = \begin{bmatrix} 0 & 1 & -1 \\ -2 & 1 & 0 \\ 1 & -2 & 2 \end{bmatrix} \begin{bmatrix} 4 \\ 10 \\ 5 \end{bmatrix} = \begin{bmatrix} 5 \\ 2 \\ -6 \end{bmatrix}$$

Inverting a 2 by 2, 3 by 3, or even a 4 by 4 matrix by hand is relatively easy, but as the size of the matrix grows the number of operations needed to invert the matrix rapidly increases. Inverting large matrices requires an extremely large number of mathematical operations that are most conveniently carried out by a computer, making hand calculation unnecessary. In the following section matrix operations using EXCEL, R, WolframAlpha, and Microsoft Mathematics are presented. Readers are urged to read this section carefully before proceeding with the applications of matrices in economics.

13.5   Matrix Operations Using EXCEL, R, WolframAlpha, and Microsoft Mathematics

13.5.1   Matrix Operations in EXCEL©, Microsoft Mathematics, and WolframAlpha

Assume two 5×5 matrices are in cells $A1$ to $E5$ and $G1$ to $K5$. To find the sum of these matrices we highlight the cells where the sum should be placed (these are called the *destination cells*). Assume we highlight cells $A8$ to $E12$. Next press $=$ to let the software know that a function is going to be used on the destination cells. Next we type **a1 : e5 + g1 : k5** (here and in the rest of this section I bold face formulas or expression that we must type) and then press Ctrl + Shift + Enter (hold down Ctrl, Shift with one hand and press Enter with the other hand.)[3] The destination cells will be filled with the sums of elements of matrices. If we want to add more than two matrices, we simply add the matrices range to the list. If, for example, the third matrix is in the range of $M1 : Q5$ then the sum is **a1 : e5 + g1 : k5 + m1 : q5**.

Alternatively, after selecting the destination cells and pressing $=$, we can highlight one of the matrices to be added, press $+$, highlight the other matrix, and then press Ctrl + Shift + Enter.

The third alternative is to use *named ranges* in the formulas and functions. The safest way to assign a name to a range of cells is to first highlight it, then on the 'Formula' tab click 'Define Name'. The New Name dialog box opens. In the 'Name' box type a name for the range. It is safe to start a name by underscore. For example, we can name the matrix we have entered in the cell range $A1 : G5$ as _mat1, and the second matrix in $G1 : K5$ as _mat2. We can then add these two matrices at the destination cells by typing $=$ **_mat1 + _mat2** followed by Ctrl+Shift+Enter.

To multiply two matrices in EXCEL we highlight the destination cells, cells where the product matrix should be placed. Next we press $=$ to let the software know that a function is going to be used on the destination cells, and then call the function to multiply the matrices by typing **mmult(range of first matrix, range of second matrix)** which will appear in function bar above the spreadsheet grid. Next

---

[3] If you can press all three keys with one hand, call the *Guinness Book of World Records*.

we press Ctrl + Shift + Enter. The destination cells will be filled with the appropriate product. For our 5 × 5 matrices we used above for addition, assume we pick cells $A8 : E12$ for product destination. We highlight these cells, press =, and type **mmult(a1 : e5, g1 : k5)** then press Ctrl + Shift + Enter. Alternatively, we can use the name ranges that we specified for two matrices, _mat1 and _mat2, and type **mmult(_mat1, _mat2)** followed by Ctrl + Shift + Enter.

To invert a matrix enter the matrix elements on any convenient space on the spreadsheet and then highlight the destination cells, the cells where the inverse should be placed. Next press = followed by the function to invert the matrix by typing **minverse(range of matrix)**, which will appear in the function bar above the spreadsheet grid. Next we press Ctrl + Shift + Enter. The destination cells will be filled with the appropriate inverse if it exists. Recall, not all matrices have inverses. If the matrix is singular, a series of "#NUM!" appear in the destination celles. To find the inverse of the matrix we entered in cells $A1 : E5$, let choose $N1 : R5$ as destination cells. We highlight these cells and then type either **minverse(a1 : e5)** or **minverse(_mat1)** followed by Ctrl + Shift + Enter.

To transpose a matrix stored in the range $A10 : E13$ (a 4 × 5 matrix), highlight the destination cells, say, $G10 : J14$ (a 5 × 4 matrix), press =, and then type **transpose(a10:e13)** followed by Ctrl + Shift + Enter. The destination cells will be filled with the transposed element of the matrix. Next we multiply the 4 by 5 matrix in the range $A10 : E13$ by its transpose, the 5 by 4 matrix in the range $G10 : J14$. The product of these two matrices is a 4 by 4 square symmetric matrix. We obtain a 5 by 5 symmetric matrix if we multiply the 5 by 4 matrix in the range $G10 : J14$. by the 4 by 5 matrix in $A10 : E13$.

Microsoft Mathematics (MM) provides a graphing calculator that plots in 2D and 3D, equation solving, and other useful tools including matrix operations. This software was added to Windows after 2010 and can be downloaded free from the Microsoft official website.

Inside the homepage of MM click on 'Insert' and then click on 'Matrix'. A small window 'Insert Matrix' opens. Change the number of Rows and Columns to the size of the matrix you wish to invert (say 3 by 3), then click 'OK' (Fig. 13.1).

Inside the white area to the right of the calculator a matrix of size 3 by 3 by elements as empty boxes appears. Type the numbers for the matrix elements

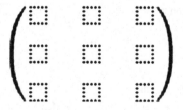

and click 'Enter' on the right bottom corner of the screen. Your matrix appears as 'input' and 'output', followed by 'Would you like to find the following value: determinant or inverse or trace or transpose or size or reduce?' Click on 'inverse'.

Fig. 13.1 Partial screen shot of MS mathematics "Insert Matrix" window

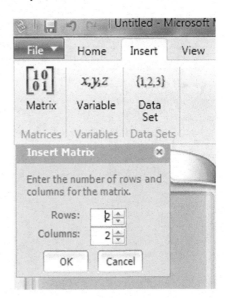

To invert an $n \times n$ matrix by WolframAlpha, in the query box we type 'inv' or 'inverse' followed by a left curly bracket, followed by elements of the matrix separated by ',' and enclosed in curly brackets, and finally the whole expression closed by a right curly bracket:

```
inv{{first row},{second row}, ... , {nth row}}
```

If, for example, we want to solve the following 4 by 4 system

$$Ax = b \longrightarrow \begin{bmatrix} 2 & -1 & 0 & 4 \\ 5 & -2 & 1 & 3 \\ 1 & -2 & 0 & 8 \\ -1 & 2 & 5 & 4 \end{bmatrix} \begin{bmatrix} x_1 \\ x_2 \\ x_3 \\ x_4 \end{bmatrix} = \begin{bmatrix} 12 \\ 35 \\ 2 \\ 10 \end{bmatrix}$$

we find the solution to the system $x = A^{-1}b$ by entering

```
x = inv{{2,-1,0,4},{5,-2,1,3},{1,-2,0,8},{-1,2,5,4}}.{12,35,2,10}
```

as shown in Fig. 13.2. Note that the matrix multiplication operator in WolframAlpha is '.', a dot.

13.5.2 Matrix Operations and Solutions to Linear Systems in R

R operates on named data structures. One of the structures in **R** is a vector of numbers, which is an entity consisting of an ordered collection of numbers. To create a row vector named v, consisting of 4 numbers 11.2, -3, 5, 4, we use the **R** command

```
> v <- matrix(c(11.2, -3, 5, 4),nrow = 1, ncol = 4)
```

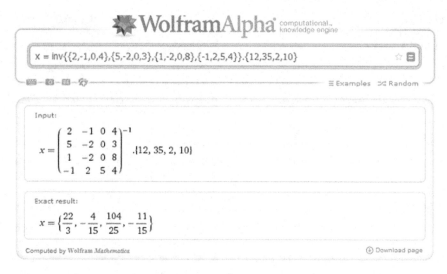

Fig. 13.2 Partial screen shot of WolframAlpha page

which is an "assignment" statement using functions 'matrix()' and 'c()'. The function c() takes the number of vector arguments (the numbers specified inside the prantheses, seperated by commas ',') and *concatenate* them end to end (concatenate means 'linking together in a series or chain'). The 'matrix' command relays to the software that v is a matrix with 'nrow', number of rows, equal to 1 and 'ncol', number of columns, equal to 4, which is a row vector of size 4. A single number (scalar) is treated as a vector of length 1.

The assignment operator ('< −'), consists of the two characters '<' (less than) and '−' (minus) side-by-side, forming an arrow that points to the object receiving the value of the expression. For all practical purpose, the equal sign '=' operator can be used as an alternative.

A more general format for 'declaring' an $m \times n$ matrix A is

```
> A = matrix(c(elements of matrix entered here), nrow=m, ncol=n)
```

By default the elements of matrix entered in c() function are assumed to be by the columns. If we wish to enter the elements by row, then we must add 'byrow = TRUE' in the 'matrix' function

```
> A=matrix(c(elements of matrix by row),nrow=m,ncol=n,byrow=TRUE)
```

There are a number of abbreviated 'matrix' commands that make entering elements of square matrices faster. If, for example, we have a 3×3 matrix A, we can enter it using

```
> A = matrix(c(9 elements by columns), 3)
```

By default the first number following 'c(),' is the number of rows of A, which in this case is 3. There are 9 numbers in the matrix with 3 rows, so the matrix must have

3 columns. The software is smart enough to recognize this. Finally, if you wish to enter the matrix by row, you can use

```
> A = matrix(c(9 elements by row), 3, b=T)
```

Here 'b = T' is the abbreviation for 'byrow = TRUE'.

A Sample R Session

When R starts you will see a window called the RConsole. This is where you type your commands and see the text results. You are prompted to type commands with the "greater than" > symbol. This is a sample of R session:

```
R version 2.15.1 (2012-06-22) -- "Roasted Marshmallows
Copyright (C) 2012 The R Foundation for Statistical Computing
ISBN 3-900051-07-0
Platform: x86_64-pc-mingw32/x64 (64-bit)
R is free software and comes with ABSOLUTELY NO WARRANTY.
You are welcome to redistribute it under certain conditions.
Type 'license()' or 'licence()' for distribution details.

 Natural language support but running in an English locale

R is a collaborative project with many contributors.
Type 'contributors()' for more information and
'citation()' on how to cite R or R packages in publications.

Type 'demo()' for some demos, 'help()' for on-line help, or
'help.start()' for an HTML browser interface to help.
Type 'q()' to quit R.

> # Any statement  started  with the pound sign '#' is treated
> # as 'comment'  by R and ignored.
> # R prompt is >, which indicates that R is ready to receive
> # new commands.
> # R works like a calculator, you type an expression and get
> # the answer
> 2+3
[1] 5
>
> # R treats a number (scalar) as an array of 1 by 1. Here [1]
> # appearing on the side of 5 indicates its position which is
> # first row and column (more below)
>
> 5*3
[1] 15
> 2/3
[1] 0.6666667
> 2^3   # 2 to power 3. Caret ^ is used for exponentiation
[1] 8
> log(3)        # log is base e or natural log
```

```
[1] 1.098612
> log10(3)      # log10 is base 10 or common log
[1] 0.4771213

> x = 10
> y=x^2-3*x+25
> # Typing a name prints its content or value
> y
[1] 95
> # Next I will enter a 4  by 1 column vector V.
>
> V = matrix(c(10,-3,6,18))
>
> # Since by default R reads the vector by column and no number
> # of row is specified, then V considered to be a column vector.
> V
     [,1]
[1,]   10
[2,]   -3
[3,]    6
[4,]    1
>
> # Numbers in the square bracket on the side indicate row
> # position while numbers in bracket on the top indicate
> # the column position.
>
> # Next I multiply V with scalar 5, forming V1 = 5V
> V1 = 5*V
> V1
     [,1]
[1,]   50
[2,]  -15
[3,]   30
[4,]   90
>
> # * is used as multiplication operator, which is used for
> # element by element multiplication
>
> # Next I transpose the column vector V to a row vector Vt,
> # using the 'transpose' function t().
>
> Vt = t(V)
> Vt
     [,1] [,2] [,3] [,4]
[1,]   10   -3    6   18
>
> # Next we calculate the inner product of Vt and V
> # In R the operator for vector and matrix multiplication
> # is '%*%', percentage sign followed by * followed by
> # percentage sign. Regular multiplication operator * is
> # for element by element multiplication
>
> Vt%*%V
```

```
         [,1]
[1,]   469
> # Which is a scalar.
>
> # Note that if we use * we get an error message
> V1 = V*Vt
Error in V * Vt : non-conformable arrays

> # This is not because V and Vt are not conformable for
> # multiplication, but because we use * which is for
> # element by element multiplication, as the following
> # examples show
>
> V*V
       [,1]
[1,]   100
[2,]     9
[3,]    36
[4,]   324
>
> Vt*Vt
       [,1] [,2] [,3] [,4]
[1,]   100    9   36  324
>
> # Next I will enter the matrix  A, matrix of coefficients in
> # a system, by rows, using the abbreviated version of matrix
> # function.
>
> A = matrix(c(1,2,1,0,0,3,2,1,1,1,3,0,0,1,0,2),4,b=T)
> A
      [,1] [,2] [,3] [,4]
[1,]    1    2    1    0
[2,]    0    3    2    1
[3,]    1    1    3    0
[4,]    0    1    0    2
>
> # Next I enter the vector of right hand side constants b
> b = matrix(c(25,18,23,15))
> b
       [,1]
[1,]    25
[2,]    18
[3,]    23
[4,]    15
>
> # To find the inverse of A we use the 'solve()' function
> Ainv = solve(A)
> # Ainv is inverse of A
> Ainv
             [,1]        [,2]        [,3]         [,4]
[1,]   0.7857143 -0.7142857   0.2142857   0.35714286
[2,]   0.2857143  0.2857143  -0.2857143  -0.14285714
[3,]  -0.3571429  0.1428571   0.3571429  -0.07142857
```

```
[4,] -0.1428571 -0.1428571  0.1428571  0.57142857
>
> # It worth repeating that the numbers in the square bracket
> # on the side indicate row position while numbers in bracket
> # on the top indicate the column position of matrix elements.
>
> # Now we can determine the values of the unknown variables
> # by multiplying Ainv with b. Don't forget in R the operator
> # for  matrix multiplication is '%*%', percentage sign
> # followed by * followed by percentage sign.
>
> X = Ainv%*%b
>
> # The solution  is stored in vector X
>
> X
            [,1]
[1,] 17.0714286
[2,]  3.5714286
[3,]  0.7857143
[4,]  5.7142857
>
> # which means x_1 = 17.071, x_2= 3.571, x_3 = 0.786, and
> # x_4 =5.714
>
> # We can achieve the same result by directly using the
> # solve function
>
> X =solve(A,b)
>
> # In general solve(A,b) solves a system of linear equations
> # with the coefficient matrix A and the vector of constant b
> # (hence the designation 'solve'). So we can obtain the vector
> # of unknown variables by directly issuing  X = solve(A,b).
> # But calculating and saving the inverse of A has advantages.
> # One such advantage is that we can calculate the determinant
> # of inverse of A that, as we will see shortly, prove to be
> # the model's autonomous multiplier in the national income
> # determination models. To find the determinant of A, we can
> # use the 'det()' function
> d =det(A)
> d
[1] 14
>
> # The determinant of matrix A is 14.
```

13.6 Application of Matrices in Economics

Example 7

Electric General, Inc. produces four different types of home appliances—refrigerators (R), dish washers (DW), washers (W), and dryers (D)—using 4 different types of materials[4] M_i $i = 1, 2, 3, 4$ and labor L. The number of units of raw materials and labor needed to produce one unit of each appliance is given in Table 13.1.

Table 13.1, which indicates the input requirements for the production of one unit output is called a *technical input-output matrix* or simply a *technical matrix*. The entries in the table are called *technical input-output coefficients* or *technical coefficients*. To use more precise and clear language, the technical matrix is often referred to as the *matrix of technical coefficients*.

Electric General produced 10,000 refrigerators, 15,000 dish washers, 8,000 washers, and 7,000 dryers during the first 6 months of 2012. The second half of the year the company produced 12,000, 16,000, 9,000, and 6,000 units of each of the appliances. Determine the total number of units of materials and labor used to produce all different types of appliances for the entire year.

Let us denote the matrix of material and labor inputs per unit of output of appliances (the matrix of technical coefficients) by A and the vectors of output of the first half and the second half of the year by Q_1 and Q_2, respectively. A matrix expression that provides the total number of units of materials and labor used in the annual production of appliances, denoted M, is

$$M = A \times Q_1 + A \times Q_2$$

or alternatively,

$$M = A \times (Q_1 + Q_2) = AQ \quad \text{where} \quad Q = Q_1 + Q_2$$

Table 13.1 Input of materials and labor per unit of output of appliances

M/L	R	DW	W	D
M_1	3	2	5	0
M_2	5	2	0	4
M_3	4	1	5	2
M_4	0	1	2	1
L	15	10	8	5

---

[4] I am using 'material' as a generic name for both primary and intermediate inputs.

$$
\begin{bmatrix} M_1 \\ M_2 \\ M_3 \\ M_4 \\ L \end{bmatrix}
=
\begin{bmatrix} 3 & 2 & 5 & 0 \\ 5 & 2 & 0 & 4 \\ 4 & 1 & 5 & 2 \\ 0 & 1 & 2 & 1 \\ 15 & 10 & 8 & 5 \end{bmatrix}
\left(\begin{bmatrix} 10000 \\ 15000 \\ 8000 \\ 7000 \end{bmatrix}
+
\begin{bmatrix} 12000 \\ 16000 \\ 9000 \\ 6000 \end{bmatrix} \right)
$$

$$
\begin{bmatrix} M_1 \\ M_2 \\ M_3 \\ M_4 \\ L \end{bmatrix}
=
\begin{bmatrix} 3 & 2 & 5 & 0 \\ 5 & 2 & 0 & 4 \\ 4 & 1 & 5 & 2 \\ 0 & 1 & 2 & 1 \\ 15 & 10 & 8 & 5 \end{bmatrix}
\begin{bmatrix} 22000 \\ 31000 \\ 17000 \\ 13000 \end{bmatrix}
=
\begin{bmatrix} 213000 \\ 224000 \\ 230000 \\ 78000 \\ 841000 \end{bmatrix}
$$

Assume the price of a unit of material inputs, mp_i $i = 1, 2, 3, 4$, and labor, w are $mp_1 = 60, mp_2 = 40, mp_3 = 30, mp_4 = 90, w = 35$. Further, assume that the market price of appliances are $p_R = 2000$, $p_{DW} = 900$, $p_W = 1000$, and $p_D = 950$. Determine the firm's profit.

We write the column vectors of unit prices of inputs (material and labor) MP and market prices of outputs P as

$$
MP = \begin{bmatrix} mp_1 \\ mp_2 \\ mp_3 \\ mp_4 \\ w \end{bmatrix} = \begin{bmatrix} 60 \\ 40 \\ 30 \\ 90 \\ 35 \end{bmatrix}
\qquad
P = \begin{bmatrix} p_R \\ p_{DW} \\ p_W \\ p_D \end{bmatrix} = \begin{bmatrix} 2000 \\ 900 \\ 1000 \\ 950 \end{bmatrix}
$$

Since profit is the difference between the total revenue TR and the total cost TC, we must find

$$
\pi = TR - TC
$$

Denoting the transpose of appliances price vector by P' and the transpose of resources price vector by MP', we have

$$
TR = P'Q = \begin{bmatrix} 2000 & 900 & 1000 & 950 \end{bmatrix} \begin{bmatrix} 22000 \\ 31000 \\ 17000 \\ 13000 \end{bmatrix} = \$101{,}250{,}000
$$

$$
TC = MP'\, M = \begin{bmatrix} 60 & 40 & 30 & 90 & 35 \end{bmatrix} \begin{bmatrix} 213000 \\ 224000 \\ 230000 \\ 78000 \\ 841000 \end{bmatrix} = 65{,}095{,}000
$$

and

$$
\pi = 101250000 - 65095000 = \$36{,}155{,}000
$$

To determine which product is most profitable, we must find the unit profit of each product. The unit profit is obtained from the difference between price of the product and the unit cost of the product. Denoting the vector of unit profits by $U\pi$, and the vector of unit costs by UC, we have

$$U\pi = P - UC = P - A' \, MP$$

where the vector of unit cost is calculated as the transpose of the matrix of technical coefficients times the unit price of resources.

$$U\pi = \begin{bmatrix} 2000 \\ 900 \\ 1000 \\ 950 \end{bmatrix} - \begin{bmatrix} 3 & 5 & 4 & 0 & 15 \\ 2 & 2 & 1 & 1 & 10 \\ 5 & 0 & 5 & 2 & 8 \\ 0 & 4 & 2 & 1 & 5 \end{bmatrix} \begin{bmatrix} 60 \\ 40 \\ 30 \\ 90 \\ 35 \end{bmatrix} = \begin{bmatrix} 975 \\ 230 \\ 90 \\ 465 \end{bmatrix}$$

It is clear that refrigerators have the highest unit profit, so the firm decides to expand production of this item. It does so by increasing the allocating to materials, specially M_2 and M_1 with large technical coefficients, to the production of refrigerators. Assume the new vector of total material NM is given as

$$NM = \begin{bmatrix} 220000 \\ 230000 \\ 240000 \\ 85000 \end{bmatrix}$$

In this vector only the allocated amounts of materials are given. The labor input required for production of various items is computed after we determine the new outputs of each appliances. A new matrix of technical coefficients NA is defined by eliminating the last row of the original matrix, which is the labor input per unit of output. Finally, we let the column vector NQ denote the new vector of outputs of various appliances that needs to be determined. We can now express our problem in matrix form as

$$NA \times NQ = NM = \begin{bmatrix} 3 & 2 & 5 & 0 \\ 5 & 2 & 0 & 4 \\ 4 & 1 & 5 & 2 \\ 0 & 1 & 2 & 1 \end{bmatrix} \begin{bmatrix} q_1 \\ q_2 \\ q_3 \\ q_4 \end{bmatrix} = \begin{bmatrix} 220000 \\ 230000 \\ 240000 \\ 85000 \end{bmatrix}$$

We solve for NQ by

$$NQ = NA^{-1} \times NM = \begin{bmatrix} 3 & 2 & 5 & 0 \\ 5 & 2 & 0 & 4 \\ 4 & 1 & 5 & 2 \\ 0 & 1 & 2 & 1 \end{bmatrix}^{-1} \begin{bmatrix} 220000 \\ 230000 \\ 240000 \\ 85000 \end{bmatrix}$$

$$NQ = \begin{bmatrix} 0.1124 & 0.0787 & 0.0674 & -0.4494 \\ 0.4157 & 0.1910 & -0.5506 & 0.3371 \\ -0.0337 & -0.1236 & 0.1798 & 0.1348 \\ -0.3483 & 0.0562 & 0.1910 & 0.3933 \end{bmatrix} \begin{bmatrix} 220000 \\ 230000 \\ 240000 \\ 85000 \end{bmatrix} \approx \begin{bmatrix} 20787 \\ 31910 \\ 18764 \\ 15562 \end{bmatrix}$$

To determine the amount of labor input for producing new volume of appliances, we form the diagonal matrix of labor input L from the last row of matrix of technical coefficients A,

$$L = \begin{bmatrix} 15 & 0 & 0 & 0 \\ 0 & 10 & 0 & 0 \\ 0 & 0 & 8 & 0 \\ 0 & 0 & 0 & 5 \end{bmatrix}$$

The required labor input for each product is then obtained by

$$LI = NQ' \times L = \begin{bmatrix} 20787 & 31910 & 18764 & 15562 \end{bmatrix} \begin{bmatrix} 15 & 0 & 0 & 0 \\ 0 & 10 & 0 & 0 \\ 0 & 0 & 8 & 0 \\ 0 & 0 & 0 & 5 \end{bmatrix}$$

$$LI = \begin{bmatrix} 311805 & 319100 & 150112 & 77810 \end{bmatrix}$$

and total labor TL employed by the firm is calculated as,

$$TL = LI \times S = \begin{bmatrix} 311805 & 319100 & 150112 & 77810 \end{bmatrix} \begin{bmatrix} 1 \\ 1 \\ 1 \\ 1 \end{bmatrix} = 858827 \quad \text{units}$$

It is easy to verify that the new production plan leads to total revenue of $103,840,900 and total cost of $67,308,945, raising the total profit from $36,155,000 to $36,531,955.

13.6.1 Matrix Representation of a 3-Commodity Market Model

In Chap. 4 we discussed an n-commodity market model. This model connects markets for two or more related goods together and studies how the equilibrium quantities and prices are simultaneously determined. Here we offer a 3-market model as another step in the direction of a more extensive general equilibrium n-commodity model and show how the model can be cast in matrix framework.

Example 8

Consider 3 goods with prices P_1, P_2, and P_3. Assume demand and supply equations for these goods are given by

$$\begin{cases} Q_{d1} = 80 - 5P_1 + 3P_2 - 3P_3 \\ Q_{s1} = -30 + 2P_1 \end{cases}$$

$$\begin{cases} Q_{d2} = 100 - 2P_2 + 3P_1 \\ Q_{s2} = -20 + 3P_2 \end{cases}$$

and

$$\begin{cases} Q_{d3} = 50 - 4P_3 + 2P_2 - 2P_1 \\ Q_{s3} = -40 + 5P_3 \end{cases}$$

where Q_{di} $i = 1, 2, 3$ and Q_{si} $i = 1, 2, 3$ denote quantity demanded and supplied of good 1, 2, and 3. Note that the prices of each good 1 and 2 are entered with positive coefficients in the demand equation for the other, while the prices of good 1 and good 3 are entered with negative coefficients in each other's demand equations. This is a clear indication that the good 1 and 2 are substitute goods and good 1 and 3 are complementary goods. Goods 2 and 3 are non-related.

To determine the equilibrium prices and quantities, we utilize the standard market equilibrium condition

$$Q_{d1} = Q_{s1}, \quad Q_{d2} = Q_{s2}, \quad \text{and} \quad Q_{d3} = Q_{s3}$$

$$80 - 5P_1 + 3P_2 - 3P_3 = -30 + 2P_1$$
$$100 + 2P_1 - 2P_2 = -20 + 3P_2$$
$$50 - 4P_3 + 2P_2 - 2P_1 = -40 + 5P_3$$

After simplification, we obtain the following system of 3 equations with 3 unknowns

$$\begin{cases} 7P_1 - 3P_2 + 3P_3 = 110 \\ -2P_1 + 5P_2 = 120 \\ 2P_1 - 2P_2 + 9P_3 = 90 \end{cases}$$

To represent this system in matrix form we construct the coefficient matrix, A, vector of unknowns, P, and vector of constant, b

$$A = \begin{bmatrix} 7 & -3 & 3 \\ -2 & 5 & 0 \\ 2 & -2 & 9 \end{bmatrix} \quad P = \begin{bmatrix} P_1 \\ P_2 \\ P_3 \end{bmatrix} \quad b = \begin{bmatrix} 110 \\ 120 \\ 90 \end{bmatrix}$$

$$AP = b = \begin{bmatrix} 7 & -3 & 3 \\ -2 & 5 & 0 \\ 2 & -2 & 9 \end{bmatrix} \begin{bmatrix} P_1 \\ P_2 \\ P_3 \end{bmatrix} = \begin{bmatrix} 110 \\ 120 \\ 90 \end{bmatrix}$$

We find the solution to this system by

$$P = A^{-1}\mathbf{b} \quad \longrightarrow \quad \begin{bmatrix} P_1 \\ P_2 \\ P_3 \end{bmatrix} = \begin{bmatrix} 7 & -3 & 3 \\ -2 & 5 & 0 \\ 2 & -2 & 9 \end{bmatrix}^{-1} \begin{bmatrix} 110 \\ 120 \\ 90 \end{bmatrix}$$

We enter the matrix A and vector $\mathbf{b}$ in **R** and use the *solve* function for determining prices:

```
> A =matrix(c(7,-3,3,-2,5,0,2,-2,9),3,b=T)
> A
      [,1] [,2] [,3]
[1,]     7   -3    3
[2,]    -2    5    0
[3,]     2   -2    9
> b=matrix(c(110,120,90))
> b
      [,1]
[1,]   110
[2,]   120
[3,]    90
> P = solve(A,b)
> P
           [,1]
[1,] 25.18519
[2,] 34.07407
[3,] 11.97531
```

With $P_1 = 25.19$, $P_2 = 34.07$, and $P_3 = 11.98$, the values of output for each good are determined to be $Q_1 = 20.38$, $Q_2 = 82.21$, and $Q_3 = 19.9$.

We get the same result by entering the matrix A and vector $\mathbf{b}$ into the inquiry box of of WolframAlpha, as show in Fig. 13.3 in the next page. Notice again the manner in which rows of matrix A and vector $\mathbf{b}$ are entered and the use of '.' as the matrix multiplication operator.

13.6.2 Matrix Representation of National Income Models

A simple national income determination model assumes a closed economy with investment and government expenditures exogenously determined,

$$Y = C + I_0 + G_0$$
$$C = C_0 + bY_D$$
$$T = T_0 + tY$$

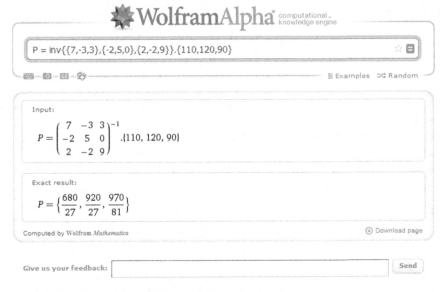

Fig. 13.3 Partial screen shot of WolframAlpha page

After substituting $Y - T$ for Y_D in the consumption function, the model becomes

$$Y = C + I_0 + G_0$$
$$C = C_0 + bY - bT$$
$$T = T_0 + tY$$

By transferring the endogenous variables to the left hand side and the exogenous variables and constants to the right hand side, we rewrite the model as

$$Y - C = I_0 + G_0$$
$$-bY + C + bT = C_0$$
$$-tY + T = T_0$$

If we denote the matrix of coefficients of the model by Coe, vector of unknown endogenous variables by EN, and vector of of constants—or given values of the exogenous variables—by EX

$$Coe = \begin{bmatrix} 1 & -1 & 0 \\ -b & 1 & b \\ -t & 0 & 1 \end{bmatrix}, \quad EN = \begin{bmatrix} Y \\ C \\ T \end{bmatrix}, \quad EN = \begin{bmatrix} I_0 + G_0 \\ C_0 \\ T_0 \end{bmatrix}$$

we can represent the model in the matrix form as

$$Coe * EN = EX$$

with the solution

$$EN = Coe^{-1} EX = \begin{bmatrix} Y \\ C \\ T \end{bmatrix} = \begin{bmatrix} 1 & -1 & 0 \\ -b & 1 & b \\ -t & 0 & 1 \end{bmatrix}^{-1} \begin{bmatrix} I_0 + G_0 \\ C_0 \\ T_0 \end{bmatrix}$$

An important fact about the matrix version of the national income model is that the determinant of the inverse of the coefficient matrix, $|Coe^{-1}|$, is the model's autonomous spending multiplier. Alternatively, using Property 8 of the determinants discussed in Sect. 13.4.3, the inverse of the determinant of Coe is the model's multiplier. Expanding by the first row, the determinant of the coefficient matrix Coe is

$$|Coe| = (-1)^2 1 \begin{vmatrix} 1 & b \\ 0 & 1 \end{vmatrix} + (-1)^3 (-1) \begin{vmatrix} -b & b \\ -t & 1 \end{vmatrix} = 1 + [-b - (-bt)] = 1 - b + bt$$

$$m = \frac{1}{|Coe|} = \frac{1}{1 - b + bt}$$

This is the same result that we obtained for the model in Chap. 4.

A more elaborate model involves investment, similar to exercise number 24 at the end of Chap. 4.

$$\begin{aligned} Y &= C + I + G_0 \\ C &= C_0 + bY_D = C_0 + bY - bT \\ I &= I_0 + iY - kr_0 \\ T &= T_0 + tY \end{aligned} \tag{13.12}$$

By rearranging the terms of equations in (13.12) system, we have

$$\begin{aligned} Y - C - I &= G_0 \\ -bY + C + bT &= C_0 \\ -iY + I &= I_0 - kr_0 \\ -tY + T &= T_0 \end{aligned} \tag{13.13}$$

Thus the matrix of coefficient in (13.13), Coe, is

$$Coe = \begin{bmatrix} 1 & -1 & -1 & 0 \\ -b & 1 & 0 & b \\ -i & 0 & 1 & 0 \\ -t & 0 & 0 & 1 \end{bmatrix}$$

and the vector of unknown endogenous variables, EN, and vector of constants—or given values of the exogenous variables—EX, are

$$EN = \begin{bmatrix} Y \\ C \\ I \\ T \end{bmatrix} \qquad EX = \begin{bmatrix} G_0 \\ C_0 \\ I_0 - kr_0 \\ T_0 \end{bmatrix}$$

The matrix representation of the national income model in (13.12) is

$$Coe * EN = EX \longrightarrow \begin{bmatrix} 1 & -1 & -1 & 0 \\ -b & 1 & 0 & b \\ -i & 0 & 1 & 0 \\ -t & 0 & 0 & 1 \end{bmatrix} \begin{bmatrix} Y \\ C \\ I \\ T \end{bmatrix} = \begin{bmatrix} G_0 \\ C_0 \\ I_0 - kr_0 \\ T_0 \end{bmatrix} \qquad (13.14)$$

The solution to (13.14) is obtained by

$$EN = Coe^{-1}EX = \begin{bmatrix} Y \\ C \\ I \\ T \end{bmatrix} = \begin{bmatrix} 1 & -1 & -1 & 0 \\ -b & 1 & 0 & b \\ -i & 0 & 1 & 0 \\ -t & 0 & 0 & 1 \end{bmatrix}^{-1} \begin{bmatrix} G_0 \\ C_0 \\ I_0 - kr_0 \\ T_0 \end{bmatrix}$$

Example 9

In a numerical version of this model given in exercise 25 of Chap. 4, the consumption, investment, and tax functions are expressed as

$$C = 60 + 0.75Y_D$$
$$I = 30 + 0.1Y - 7r_0$$
$$T = 15 + 0.1Y$$

where $C_0 = 60$, $b = 0.75$, $I_0 = 30$, $i = 0.1$, $k = 7$, $T_0 = 15$, and $t = 0.1$. If we assume that the interest rate, r_0, is 5 % and the government expenditure G_0 is 125, then the system in (13.14) is

$$\begin{bmatrix} 1 & -1 & -1 & 0 \\ -0.75 & 1 & 0 & 0.75 \\ -0.1 & 0 & 1 & 0 \\ -0.1 & 0 & 0 & 1 \end{bmatrix} \begin{bmatrix} Y \\ C \\ I \\ T \end{bmatrix} = \begin{bmatrix} 125 \\ 60 \\ -5 \\ 15 \end{bmatrix}$$

and its solution is obtained by

$$\begin{bmatrix} Y \\ C \\ I \\ T \end{bmatrix} = \begin{bmatrix} 1 & -1 & -1 & 0 \\ -0.75 & 1 & 0 & 0.75 \\ -0.1 & 0 & 1 & 0 \\ -0.1 & 0 & 0 & 1 \end{bmatrix}^{-1} \begin{bmatrix} 125 \\ 60 \\ -5 \\ 15 \end{bmatrix}$$

Using the following **R** snippet, we determine the model's equilibrium solution as $Y = 750, C = 555, I = 70$, and $T = 90$.

```
> Coe=matrix(c(1,-1,-1,0,-0.75,1,0,0.75,-0.1,0,1,0,-0.1,0,0,1),
+ 4,b=T)
>
# Note that the "+" at the start of the 2nd line is not really
# a plus sign, but R's way of denoting a command that goes
# onto multiple lines.
>
> EX = matrix(c(125,60,-5,15))
> Coein = solve(Coe)    # Coein is inverse of Coe
> EN = Coein%*%EX
> EN
        [,1]
[1,]   750
[2,]   555
[3,]    70
[4,]    90
> m = abs(det(Coein))
> m
[1] 4.44444
```

Note that we also calculated the absolute value of the determinant of Coe inverse, using $abs()$ function and $m = abs(det(Coein))$, which is equal to 4.444. m is the model's multiplier, or to be more accurate, the model's autonomous spending multiplier. Also note that the tax and interest rate multipliers are $-bm = -3.333$ and $-km = -31.11$, respectively. Finally, note that this economy is running a $G_0 - T = 125 - 90 = \$35$ billion budget deficit, about 4.6 % of the country's GDP.[5]

Now assume that the budget hawks in this country's parliament and the officials of the Super State pressure the government of this country to implement an austerity plan by cutting the government expenditure by 30 % to \$87.5 billion and raising the marginal tax rate t to 14.15 % (in the Exercise section of this Chapter, you are asked to show how we determine this value for t).

```
> Coe=matrix(c(1,-1,-1,0,-0.75,1,0,0.75,-0.1,0,1,0,-0.1415,0,0,1),
+ 4,b=T)
> EX = matrix(c(87.5,60,-5,15))
> EN = solve(Coe,EX)    # solve(Coe,EX) invert Coe and multiply the
                        # inverse by EX
> EN
[1,] 512.44510
[2,] 378.70059
[3,]  46.24451
[4,]  87.51098
>
> m = abs{det(solve(Coe))}
```

---

[5] The size of this economy is roughly 1/20th of that of the US. According to the Congressional Budget Office latest report (May 2013), the size of US budget deficit is about \$642 billion. Based on the latest BEA's estimate released on June 26, 2013, the US GDP is \$15,984 billion. The US deficit is, therefore, about 4 % of the GDP.

```
> m
[1] 3.906632
```

The result, calculated by the above **R** snippet, is a balanced budget albeit a 32 % decline in the country's GDP from 750 to 512.45, and consequently a dramatic increase in the unemployment rate.[6]

After massive nationwide demonstrations against the austerity measures the government is forced to use a more balanced approach by a combination of monetary policy: reducing the interest rate to 3 %, and fiscal policy: setting the tax rate to 11 % and increasing the government expenditure to $100 billion. The result, calculated by the following **R** snippet, shows a big improvement in the GDP with a budget deficit of only 1.7 % of the GDP.

```
> Coe=matrix(c(1,-1,-1,0,-0.75,1,0,0.75,-0.1,0,1,0,-0.11,0,0,1),
+ 4,b=T)
> EX = matrix(c(100,60,7,15))
> EN = solve(Coe,EX)
> EN
          [,1]
[1,]  669.89247
[2,]  495.90323
[3,]   73.98925
[4,]   88.68817
```

13.7 The National Input–Output Model

The Input-Output (I-O) accounts present a consistent and complete picture of the interrelationships among sectors, industries, and commodities in an economy. It is a set of information about the interworkings of an economy and provides a powerful tool for analyzing these interworkings. In the U.S. the Bureau of Economic Analysis (BEA) is responsible for providing timely, relevant, and accurate economic accounts. The BEA produces economic accounts statistics that enable government and business decision-makers, researchers, and the American public to track and understand the performance of the economy.

"The input-output (I-O) accounts are an integral and essential element of the U.S. economic accounts. First, they are the building blocks for other economic accounts. Prominent among these are BEA's national income and product accounts (NIPAs), which feature the estimates of gross domestic product (GDP). Second, the I-O accounts provide detailed statistics on economic processes and relationships.

---

[6] Note that in spite of a balanced budget the model's multiplier is 3.907. The interesting case of the balanced budget multiplier of 1, first discovered by J. Gelting and later elaborated on by the Nobel laureate Trygve Haavelmo in his article "Multiplier effects of a balanced budget" in *Econometrica*, is based on the assumption that both investment and taxes are exogenous.

They incorporate a complete, balanced set of economic statistics, and they present a full accounting of industry and final-use transactions".[7]

"The fundamental aim of national economic accounting is to provide a coherent and comprehensive picture of the nation's economy. More specifically, national economic accountants want to answer two questions. First, what is the output of the economy? Its size, its composition, and its use? Second, what is the economic process or mechanism by which this output is produced and distributed?"[8]

The Input-Output (I-O) analysis, sometimes referred to as "inter-industry analysis," is an economic tool that measures the relationships between various sectors and industries in the economy. For illustration, consider a closed economy consisting of four industries (or groups of industries referred to as sectors): Manufacturing (Manu), Utilities (Util), Financial and Insurance (Fin & In), and Information (Info).[9] The following hypothetical Input-Output table (matrix) captures the essence of intersectoral or inter-industry transactions and relationships in this economy.

Figures in any row of Table 13.2 show the shipment of industry's output to various industries (intermediate use or output) and to consumers (Final demand). For example, the numbers in the first row indicate that from the total output of $800 million of Manufacturing products, $180 million is used by the industry itself in

Table 13.2 Hypothetical input-output table (Millions of Dollars) intermediate use (Output)

	Manu	Util	Fin & In	Info	Final demand	Gross output
Manu	180	40	50	30	500	800
Util	80	20	30	30	140	300
Fin & In	60	30	100	90	360	640
Info	70	30	60	30	110	300
Labor	200	50	140	70	10	470

---

[7] US Bureau of Economic Analysis, *Consepts and Methods of the US Input-Output Accounts,* Washington DC: US Government Printing Office, September 2006, Updated April 2009.

[8] US Bureau of Economic Analysis, *An Introduction to National Economic Accounting, Methodology Paper Series MP-1,* Washington DC: US Government Printing Office, March 1985.

[9] These sectors are part of a 20-sector classification known as The North American Industry Classification System (NAICS). According to the Unitaed States' Census Bureaue "NAICS is the standard used by Federal statistical agencies in classifying business establishments for the purpose of collecting, analyzing, and publishing statistical data related to the U.S. business economy. NAICS was developed under the auspices of the Office of Management and Budget (OMB), and adopted in 1997 to replace the Standard Industrial Classification (SIC) system. It was developed jointly by the U.S. Economic Classification Policy Committee (ECPC), Statistics Canada , and Mexico's Instituto Nacional de Estadistica y Geografia , to allow for a high level of comparability in business statistics among the North American countries. NAICS is a 2- through 6-digit hierarchical classification system, offering five levels of detail. Each digit in the code is part of a series of progressively narrower categories, and the more digits in the code signify greater classification detail. The first two digits designate the economic sector, the third digit designates the subsector, the fourth digit designates the industry group, the fifth digit designates the NAICS industry, and the sixth digit designates the national industry." See Table 13.4 at the end of this chapter.

the process of manufacturing these products, $40 million is purchased by the Utility industry, $50 million is sold to the Financial and Insurance industry, $30 million is used by the Information industry, and $500 million is sold to final consumers. Reading down the first column, the figures indicate that for the Manufacturing sector to produce $800 million of gross output, it must use $180 million worth of its own product, $80 million of Utility, $60 million of Financial and Insurance, $70 million of Information, and $200 million worth of labor.

By including Labor (or rather the Household sector that provides this Labor) we are presenting a version of Input-Output model that is known as *closed model*. The household sector consumes all the goods and services captured by the column of *final demand* and supplies a primary input, labor.[10] If we exclude the household sector, which exogenously determines the final demand, we arrive at the *open* version of the input-output model. In the open model, changes in the economy are driven by exogenous changes in the household's consumption which are reflected in changes in the final demand.

To formally present the open input-output model, assume that the economy consists of n industries. We denote the matrix of *inter-industry transactions or flows* by T, where t_{ij} denote the value of output of industry i supplied to industry j. Note that the gross output of industry i, Q_i, can be written as

$$Q_i = \sum_{j=1}^{n} t_{ij} + d_i \qquad \text{for } i = 1, 2, \ldots, n \qquad (13.15)$$

where d_i is the final demand for that industry. We can write (13.15) in matrix form as

$$
\begin{bmatrix} Q_1 \\ Q_2 \\ \vdots \\ Q_n \end{bmatrix} =
\begin{bmatrix} t_{11} & t_{12} & \cdots & t_{1n} \\ t_{21} & t_{22} & \cdots & t_{2n} \\ \vdots & \vdots & \cdots & \vdots \\ t_{n1} & t_{n2} & \cdots & t_{nn} \end{bmatrix}
\begin{bmatrix} 1 \\ 1 \\ \vdots \\ 1 \end{bmatrix} +
\begin{bmatrix} d_1 \\ d_2 \\ \vdots \\ d_n \end{bmatrix}
$$

or,

$$Q = TS + D \qquad (13.16)$$

where Q is the column vector of outputs, S is the sum vector (column vector of all 1s) and D is the vector of final demand for various industries.

Next we construct a matrix A with entries indicating the input requirements for the production of *one dollar of output* for each industry. This matrix is the *matrix of input coefficient*. Entries of A, a_{ij}, are called *input-output coefficients* or simply *input coefficients*. This matrix captures and displays the technology of production of various industries in this economy. If entries in the interindustry transaction matrix (Table 13.2) are given in *physical units*, then a_{ij} are referred to as *technical coefficients* and A is called the *matrix of technical coefficients*. The word "technical" more

---

[10] The closed model can include the government and foreign trade, treated as two industries.

emphatically directs attention to the fact that the technologies of production of various industries are displayed in A.

We construct the elements of A by dividing the various inputs of each sector by the gross output of that sector,

$$a_{ij} = \frac{t_{ij}}{Q_j} \tag{13.17}$$

and subsequently

$$t_{ij} = a_{ij} Q_j \tag{13.18}$$

Now, corresponding to the interindustry flow matrix T, we have the interindustry input matrix A, computed using (13.17):

$$A = [a_{ij}] = \begin{bmatrix} a_{11} & a_{12} & \cdots & a_{1n} \\ a_{21} & a_{22} & \cdots & a_{2n} \\ \vdots & \vdots & \cdots & \vdots \\ a_{n1} & a_{n2} & \cdots & a_{nn} \end{bmatrix} = \begin{bmatrix} \dfrac{t_{11}}{Q_1} & \dfrac{t_{12}}{Q_2} & \cdots & \dfrac{t_{1n}}{Q_n} \\ \dfrac{t_{21}}{Q_1} & \dfrac{t_{22}}{Q_2} & \cdots & \dfrac{t_{2n}}{Q_n} \\ \vdots & \vdots & \cdots & \vdots \\ \dfrac{t_{n1}}{Q_1} & \dfrac{t_{n2}}{Q_2} & \cdots & \dfrac{t_{nn}}{Q_n} \end{bmatrix}$$

For our hypothetical economy with the matrix of inter-sectoral transaction presented in Table 13.2, the input coefficients are calculated and given in Table 13.3.

Figures in the columns of Table 13.3 are dollar amount of inputs from various industries needed to produce one dollar worth of output of that industry. For example, figures for Utility sector in the second column are obtained by dividing the second column of Table 13.2 by the gross output of the Utility sector ($300 million). These figures indicate the dollar amounts of various inputs or intermediate goods (Manu = $0.133 or 13.3 cents, Util = $0.067 or 6.7 cents, Fin & In = $0.100 or 10.0 cents, Info = $0.100 or 10.0 cents) needed to produce $1 of Utility output. Note that the sum of values of inputs to produce $1 of Utility output is $0.133 + 0.067 + 0.100 + 0.100 = \$0.40$ or 40 cents. The remaining 60 cents must be payments to primary factors, labor ($\frac{50}{300} = 0.167$ or 16.7 cents), capital (in form of dividend, interest, or rent), and taxes to government.

As it was mentioned earlier, the matrix of input coefficients A captures the technology of production. The fundamental assumption of the input-output model is that the industries or sectors of the economy have the *fixed-proportion production*

Table 13.3 Input–output coefficients

	Manu	Util	Fin & In	Info
Manu	0.225	0.133	0.078	0.100
Util	0.100	0.067	0.047	0.100
Fin & In	0.075	0.100	0.156	0.300
Info	0.088	0.100	0.094	0.100

function (FPPF). As the name implies, a firm or an industry with the fixed-proportion production function uses a fixed input ratio or factor combination for the production of its output. This function does not allow substitution among factors even if prices of inputs change. Production of firms and industries with FPPF is subject to constant returns to scale. For example, if production of $1 worth of output in the Information industry requires the combination of ($0.10, $0.10, $0.30, $0.10) worth of (Manu, Util, Fin & In, Info) products, producing $100 worth of output requires the combination of ($10, $10, $30, $10) worth of products of the same industries.

By using (13.18), we can rewrite (13.16) as

$$Q = AQ + D$$
$$Q - AQ = D \quad \rightarrow \quad (I - A)Q = D$$

And solving for Q, we have

$$Q = (I - A)^{-1} D \tag{13.19}$$

For our hypothetical economy matrix A and matrix $(I - A)$ are

$$A = \begin{bmatrix} 0.225 & 0.133 & 0.078 & 0.100 \\ 0.100 & 0.067 & 0.047 & 0.100 \\ 0.075 & 0.100 & 0.156 & 0.300 \\ 0.088 & 0.100 & 0.094 & 0.100 \end{bmatrix}$$

$$(I - A) = \begin{bmatrix} 1 & 0 & 0 & 0 \\ 0 & 1 & 0 & 0 \\ 0 & 0 & 1 & 0 \\ 0 & 0 & 0 & 1 \end{bmatrix} - \begin{bmatrix} 0.225 & 0.133 & 0.078 & 0.100 \\ 0.10 & 0.067 & 0.047 & 0.100 \\ 0.075 & 0.100 & 0.156 & 0.300 \\ 0.088 & 0.100 & 0.094 & 0.100 \end{bmatrix}$$

$$= \begin{bmatrix} 1 - 0.225 & 0.133 & 0.078 & 0.100 \\ 0.100 & 1 - 0.067 & 0.047 & 0.100 \\ 0.075 & 0.100 & 1 - 0.156 & 0.300 \\ 0.088 & 0.100 & 0.094 & 1 - 0.100 \end{bmatrix}$$

$$= \begin{bmatrix} 0.775 & -0.133 & -0.078 & -0.100 \\ -0.100 & 0.933 & -0.047 & -0.100 \\ -0.075 & -0.100 & 0.844 & -0.300 \\ -0.088 & -0.100 & -0.094 & 0.900 \end{bmatrix}$$

Subsequently (13.19) $Q = (I - A)^{-1} D$ is

$$Q = \begin{bmatrix} 0.775 & -0.133 & -0.078 & -0.100 \\ -0.100 & 0.933 & -0.047 & -0.100 \\ -0.075 & -0.100 & 0.844 & -0.300 \\ -0.088 & -0.100 & -0.094 & 0.900 \end{bmatrix}^{-1} \begin{bmatrix} 500 \\ 140 \\ 360 \\ 110 \end{bmatrix}$$

Using **R**, we compute $(I - A)^{-1}$ as

$$(I - A)^{-1} = \begin{bmatrix} 1.3633235 & 0.2369992 & 0.16512634 & 0.2328557 \\ 0.1750393 & 1.1263623 & 0.09866839 & 0.1774896 \\ 0.2037473 & 0.2152279 & 1.26790480 & 0.4691877 \\ 0.1740318 & 0.1708040 & 0.15953445 & 1.2026044 \end{bmatrix}$$

and

$$Q = \begin{bmatrix} 1.3633235 & 0.2369992 & 0.16512634 & 0.2328557 \\ 0.1750393 & 1.1263623 & 0.09866839 & 0.1774896 \\ 0.2037473 & 0.2152279 & 1.26790480 & 0.4691877 \\ 0.1740318 & 0.1708040 & 0.15953445 & 1.2026044 \end{bmatrix} \begin{bmatrix} 500 \\ 140 \\ 360 \\ 110 \end{bmatrix} \approx \begin{bmatrix} 800 \\ 300 \\ 640 \\ 300 \end{bmatrix}$$

Equation (13.19) can now be used to determine the level of output of various sectors or industries when the there are changes in the final demand. Consider, for example, a 2 % increase in the final demand for *only* Manufacturing products, the level of outputs of various sectors are

$$Q = \begin{bmatrix} 1.3633235 & 0.2369992 & 0.16512634 & 0.2328557 \\ 0.1750393 & 1.1263623 & 0.09866839 & 0.1774896 \\ 0.2037473 & 0.2152279 & 1.26790480 & 0.4691877 \\ 0.1740318 & 0.1708040 & 0.15953445 & 1.2026044 \end{bmatrix} \begin{bmatrix} 510 \\ 140 \\ 360 \\ 110 \end{bmatrix} \approx \begin{bmatrix} 814 \\ 302 \\ 642 \\ 302 \end{bmatrix}$$

Note that increase in demand for the manufacturing products not only leads to increase in output of this sector, but also outputs of other sectors. The reason is that for manufacturing to increases its output it must purchase additional input from the other sectors. The other sectors must now increase their output in response to additional demand from the manufacturing industries which, in turn, increase their demand for inputs from the other sectors, including manufacturing. This process creates a multiplier effect throughout the system, leading to increases in the output of all sectors. Matrix $(I - A)^{-1}$ captures this multiplier effect by incorporating all the *direct* and *indirect* output requirements for the production of one unit of final demand Table 13.4.

The input-output model provides a tool for macroeconomic policy makers to determine the level of output of various sectors or industries that can support a certain desirable level of final demands. If, for example, the desired vector of final demand is $\begin{bmatrix} 520 \\ 150 \\ 370 \\ 120 \end{bmatrix}$ then the vector of gross output for different sectors is

$$Q = (I - A)^{-1} \begin{bmatrix} 520 \\ 150 \\ 370 \\ 120 \end{bmatrix} = \begin{bmatrix} 833.5 \\ 317.8 \\ 663.7 \\ 319.5 \end{bmatrix}$$

Table 13.4 NAISC 20-sector classification

2012 NAICS

The following table provides detailed information on the structure of NAICS.

Sector	Description
11	Agriculture, Forestry, Fishing and Hunting
21	Mining, Quarrying, and Oil and Gas Extraction
22	Utilities
23	Construction
31-33	Manufacturing
42	Wholesale Trade
44-45	Retail Trade
48-49	Transportation and Warehousing
51	Information
52	Finance and Insurance
53	Real Estate and Rental and Leasing
54	Professional, Scientific, and Technical Services
55	Management of Companies and Enterprises
56	Administrative and Support and Waste Management and Remediation Services
61	Educational Services
62	Health Care and Social Assistance
71	Arts, Entertainment, and Recreation
72	Accommodation and Food Services
81	Other Services (except Public Administration)
92	Public Administration

Similarly, if the policy makers' objective is to achieve a 3 % balance growth for the economy, the the required final demand vector must be

$$
D = (I - A)Q = \begin{bmatrix} 0.775 & -0.133 & -0.078 & -0.100 \\ -0.100 & 0.933 & -0.047 & -0.100 \\ -0.075 & -0.100 & 0.844 & -0.300 \\ -0.088 & -0.100 & -0.094 & 0.900 \end{bmatrix} \begin{bmatrix} 824.0 \\ 309.0 \\ 659.2 \\ 309.0 \end{bmatrix} = \begin{bmatrix} 515.18 \\ 144.02 \\ 370.96 \\ 112.72 \end{bmatrix}
$$

13.8 Exercises

1. Given

$$
A = \begin{bmatrix} 3 & 18 & -2 \\ 5 & 4 & 0 \end{bmatrix}, \quad B = \begin{bmatrix} 0 & -1 & 3 \\ 4 & 2 & 0 \\ 9 & 7 & 8 \end{bmatrix}, \quad \text{and} \quad C = \begin{bmatrix} 2 & 0 & 2 \\ -1 & 5 & 3 \\ 4 & -3 & 1 \end{bmatrix}
$$

(a) Find A', B', and C'.

(b) Find $B + C$ and $B - C$. Is $A + B$ defined?

(c) Find AB, BA', $A'A$, AA' and BC. Is BA defined?

(d) Show that $(AB)' = B'A'$

2. Evaluate the determinant of the following matrices:

$$
A = \begin{bmatrix} 0 & -5 & 2 \\ 4 & 1 & 0 \\ 8 & 5 & 9 \end{bmatrix}, \quad B = \begin{bmatrix} 3 & 0 & -1 \\ 1 & 3 & 2 \\ 4 & -2 & 5 \end{bmatrix}, \quad \text{and} \quad C = \begin{bmatrix} 2 & 0 & 2 & -2 \\ -1 & 5 & 3 & 1 \\ 4 & 0 & 1 & -4 \\ 0 & 2 & 0 & 0 \end{bmatrix}
$$

3. Which one of the matrices in Exercise 2 is not invertible?

4. Find the inverse of matrices in Exercise 2 that are nonsingular.

5. Find $a, b,$ and c such that matrix S is symmetric,

$$
S = \begin{bmatrix} 0 & -5 & a \\ b & 1 & c \\ 8 & 5 & 9 \end{bmatrix}
$$

6. Find $a, b,$ and c such that matrix T is symmetric,

$$
T = \begin{bmatrix} 4 & -a & 2b - 4 \\ a & 1 & c + 2 \\ b & 5 & 9 \end{bmatrix}
$$

7. For a square matrix A, show that $A'A$, AA', and $A + A'$ are symmetric.

8. Write the following system in matrix form and solve it

$$
\begin{aligned}
2x_1 - 2x_2 + 4.5x_3 &= 20 \\
-2x_1 + 4x_3 &= 5 \\
3x_2 - 2x_3 &= 2
\end{aligned}
$$

9. Write the following system in matrix form and solve it

$$2x_1 - 2x_2 + 4x_3 + 2x_4 = 30$$
$$-2x_1 + 4x_3 + 3x_4 = 10$$
$$3x_1 - 2x_3 = 2$$
$$x_2 + 2x_3 + 3x_4 = 15$$

10. A chain of grocery stores has 5 stores and 3 warehouses. The following table gives the distances from each warehouse to the five stores.

	St1	St2	St3	St4	St5
Warehouse 1	10	15	25	30	12
Warehouse 2	25	20	10	22	15
Warehouse 3	30	22	15	10	30

Assume cost of shipping one pound of an item per mile is $0.02 or 2 cents. If the weights of items each store must receive per week are given by the row vector $W = [10000 \ 25000 \ 15000 \ 30000 \ 20000]$, rank the warehouses in terms of cost effectiveness of shipping items to all stores.

11. In Exercise 9, write a matrix expression that finds the average distance of stores from each warehouse.

12. Find the inverse of the following matrices by method of adjoint,

$$A = \begin{bmatrix} 3 & 2 \\ 5 & 4 \end{bmatrix}, \quad B = \begin{bmatrix} 0 & -1 \\ 4 & 2 \end{bmatrix}, \quad \text{and} \quad C = \begin{bmatrix} 1 & 2 \\ 5 & 3 \end{bmatrix}$$

13. Find the inverse of the following matrices using elementary row operations

$$A = \begin{bmatrix} 0 & -1 & 3 \\ 4 & 2 & 0 \\ 9 & 7 & 8 \end{bmatrix}, \quad \text{and} \quad B = \begin{bmatrix} 2 & 0 & 2 \\ -1 & 5 & 3 \\ 4 & -3 & 1 \end{bmatrix}$$

14. Assume the following amount of raw materials and labor required for production of a unit of products A and B,

	Product A	Product B
Pound of raw materials	5	7
Hours of labor	3	4

If the available amount of raw materials and labor are 13400 pounds and 7800 hours, determine the outputs of A and B.

15. How do the outputs of A and B in Exercise 13 change if the amount of available raw materials and labor hours are increased to 13800 and 8000?

16. In Exercise 13, what amounts of raw materials and labor hours are needed for producing 1300 units of A and 1100 units of B?

17. The demand and supply functions of two related goods are given by

$$Q_{d1} = 30 - 8P1 + 4P2$$
$$Q_{s1} = -60 + 6P1$$
$$Q_{d2} = 200 + 4P1 - 4P2$$
$$Q_{s2} = -40 + 6P2$$

Write the system in the matrix form and solve for the equilibrium prices and quantities.

18. The demand and supply function of three related goods are given by

$$Q_{d1} = 60 - 5P_1 + 2P_2 - 3P_3$$
$$Q_{s1} = -10 + 4P_1$$
$$Q_{d2} = 70 + 4P_1 - 4P_2 + 2P_3$$
$$Q_{s2} = -20 + 2P_2$$
$$Q_{d3} = 50 - 2P_1 + 4P_2 - 2P_3$$
$$Q_{s3} = -70 + 5P_3$$

Write the system in the matrix form and find the equilibrium prices and quantities.

19. Consider an open economy with the following consumption, investment, and tax functions

$$C = 40 + 0.80Y_D$$
$$T = 10 + 0.10Y$$
$$I = 38 + 0.15Y$$

Write the model in matrix form. Assume the government expenditure $G_0 = 110$, export $X_0 = 40$, and import $M_0 = 90$. What are the equilibrium output, consumption, investment, and taxes in this economy?

20. In Example 9 in Sect. 13.6.2, it was shown that to achieve a balanced budget, cutting the government expenditure by 30 % from \$125 to \$87.5 billion must be accompanied by raising the marginal tax rate from 10 to 14.15 %. Show how the required tax rate of 0.1415 is computed.

21. Consider the following extended version of a national income model presented in Chap. 4:

$$Y = C + I + G_0 + NX_0 \qquad (NX_0 \text{ is the net export})$$
$$C = -228.78 + 0.832Y_D$$
$$I = -41.951 + 0.255\,Y_D - 11.511\,r$$
$$r = -0.178 + 0.010\,Y - 0.012\,M_0$$

where r is the corporate bond rate used as proxy for interest rate and M_0 is the real money supply. Assume the tax equation is expressed as

$$T = T_0 + tY = 135 + 0.15Y$$

and the values (in billion dollars) of exogenous variables G_0, NX_0, and M_0 are 820.8, -37.5, and 3323.3, respectively.

(a) Write the model in the matrix form $EN = Coe^{-1} EX$.

(b) Solve the model for the equilibrium values of the endogenous variables Y, C, I, r, and T. Make sure you obtain the following result:

$$\begin{bmatrix} Y \\ C \\ I \\ r \\ T \end{bmatrix} = \begin{bmatrix} 4325.837 \\ 2718.132 \\ 824.405 \\ 3.200 \\ 783.876 \end{bmatrix}$$

(c) Calculate the model's autonomous (exogenous) expenditure, tax, and interest rate multipliers.

22. In Exercise 21, assume the Federal Reserve Bank expands the money supply to $3,500 billion. What would be the new equilibrium values of the endogenous variables? How does the interest rate change?

23. Now assume that, in Exercise 22, simultaneously in conjunction with the expansion of the money supply, the government increases its expenditure by an additional $40 billion dollars. What are the new equilibrium levels of income and the interest rate? Is there evidence of "crowding out"?

24. Go back to the original model in Exercise 21 and change the tax function to

$$T = 100 + 0.13Y$$

That is, lower the marginal tax rate from 15 to 13 % and autonomous taxes from 130 to 100. Redo Exercises 21, 22, and 23. Compare the equilibrium solutions.

25. Below is the inter-sectoral transaction table for the United States, aggregated from data provided by BEA Use Table (after redefinition) 2011. The sectors are Agriculture (AG), Construction and Mining (including utilities) (C&M), Manufacturing (Manu), and Services (Ser) (excluding local, state, and federal government).[11]

---

[11] Imports and exports are also excluded.

4-Sector Input-Output Table of the US (Millions of Dollars)

	AG	C&M	Manu	Ser
AG	82550	1109	246867	14595
C&M	9418	143422	653526	204375
Manu	100120	322107	1834966	868271
Ser	65865	213120	861584	4863250

The vector of Final Demand (FD) for the 4 sectors is

$$FD = \begin{bmatrix} 69980 \\ 897154 \\ 1845577 \\ 850000 \end{bmatrix}$$

(a) Use the data and prepare the matrix of inter-sectoral input coefficients. (Hint: use (13.16) and find total sectoral outs Q. Form a diagonal matrix with the sectoral outputs on the main diagonal and zero everywhere else. Multiply inter-sectoral transaction matrix T by the inverse of this matrix. This gives you matrix A)

(b) Determine the sectoral gross outputs if the final demand for only Agricultural sector's products increases by 3%. (Hint: create a unit matrix I of size 4. Calculate $(I - A)$ and $(I - A)^{-1}$. Note that you must get the original vector of output if you calculate $(I - A)^{-1} FD$)

(c) What is the growth rate of the economy?

(d) Repeat parts (b) and (c) for other sectors. Rank sector based on their impact on the economy. Can this be determined by the column elements of $(I - A)^{-1}$ that provides the direct and indirect output requirements per unit of final demand?

26. In Exercise 25, assume that the final demand for all sectors of the economy increases by 3%.

(a) determine the gross outputs of all sectors.

(b) What is the growth rate of the economy?

Index

S. Vali, *Principles of Mathematical Economics,*
Mathematics Textbooks for Science and Engineering 3,
DOI: 10.2991/978-94-6239-036-2, © Atlantis Press and the authors 2014

CPSIA information can be obtained
at www.ICGtesting.com
Printed in the USA
LVOW06*1659110916

504154LV00001B/16/P